Josephson Junctions

Josephson Junctions

History, Devices, and Applications

edited by
Edward Wolf
Gerald Arnold
Michael Gurvitch
John Zasadzinski

Pan Stanford Publishing

Published by

Pan Stanford Publishing Pte. Ltd.
Penthouse Level, Suntec Tower 3
8 Temasek Boulevard
Singapore 038988

Email: editorial@panstanford.com
Web: www.panstanford.com

British Library Cataloguing-in-Publication Data
A catalogue record for this book is available from the British Library.

Josephson Junctions: History, Devices, and Applications

ISBN 978-981-4745-47-5 (Hardcover)
ISBN 978-1-315-36452-0 (eBook)

Printed in the USA

Contents

Foreword

Following the half-century interval since the discovery of superconductivity in 1911, the diverse field of superconducting electronics was created in a remarkably short time, 1961 to 1964. The first key step, in 1961, was the experimental observation that the magnetic flux in a closed superconducting loop is quantized in units of the flux quantum, Φ_0. About a year later, Brian Josephson published his celebrated paper predicting the coherent tunneling of Cooper pairs of electrons via a tunnel barrier separating two superconductors. This effect was observed experimentally in 1963. Finally, quantum interference between two Josephson junctions connected in parallel on a superconducting loop—the dc superconducting quantum interference device (SQUID)—was observed in 1964.

It is important to realize, however, that the first observation of Josephson tunneling by Anderson and Rowell was made on a junction consisting of two thin films of tin and lead separated by an oxide layer. Not only is the superconducting transition temperature of tin, 3.7 K, below the boiling point of liquid helium at atmospheric pressure, 4.2 K, a serious inconvenience in some applications, but thin films of tin—a soft superconductor—tend to be rather short-lived and in particular to cycle poorly between room and liquid helium temperatures. Thus, the Josephson devices of the first two decades of their history were based on technologies that were crude by today's standards, ranging from machined niobium (Nb) point-contacts and blobs of tin–lead solder on Nb wires to Nb silicon Nb and Nb niobium–oxide lead junctions. Although much pioneering science was performed with these early devices, the field required two decades to realize the highly reproducible, manufacturable Nb aluminum–oxide Nb junction—the trilayer junction—that drives today's technology. The transition temperature of niobium is about

9.5 K, so that its energy gap is almost fully developed at 4.2 K. The evolution of this junction and a discussion of its broad range of applications are the subjects of this book.

The first chapter, appropriately, is Brian Josephson's story of his invention of the effect that bears his name, along with a brief account of his attempts to demonstrate it experimentally. He brings out much of the little-known contemporary thinking on the subject, which I found fascinating. In the next chapter, Ed Wolf provides us with an overview of the rest of the book, and introduces us to the other important refractory tunnel junction, namely one involving NbN. He also gives us our first glimpse of applications. In chapter 3, John Zasadzinski describes experiments on quasiparticle tunneling into Nb films overlaid with Al that may or may not have been oxidized to completion, while in the next chapter Gerald Arnold elucidates the associated theory. Such studies were important in the development of the trilayer junction because the aluminum film deposited on the base electrode is generally not oxidized to completion, so that one has to account for the proximity effect—superconductivity induced in the metallic aluminum film by the underlying niobium film.

Michael Gurvitch, in chapter 5, presents a detailed historical account of the development of the Nb aluminum–oxide Nb junction and of the many people involved as well as of early days of Bell Telephone Laboratories. In chapter 6 Jean-Claude Villegier describes junctions based on NbN—most commonly—NbN–MgO–NbN. Since niobium nitride has a transition temperature of about 16.5 K, these junctions operate well at temperatures up to 10 K.

The last four chapters are concerned with devices and applications. Valery Koshelets and colleagues describe state-of-the art superconductor–insulator–superconductor (SIS) mixers and local oscillators used as terahertz and submillimeter detectors for both ground-based and balloon-based radioastronomy. In chapter 8, Dietmar Drung and Jörn Beyer describe the theory and practice of dc SQUIDs. Applications of SQUIDs include biomagnetism—for example, magnetoencephalography and ultralow-field magnetic resonance imaging—geophysics, metrology—for example, thermometry and cryogenic current converters—amplifiers to read out transition edge sensors for astronomy and cosmology, and a commercially available magnetic property measurement system.

Siyuan Han, in chapter 9, discusses one approach to quantum computing known as quantum annealing, which has been demonstrated using rf-SQUID-based flux qubits based on Nb-trilayer junctions. Finally, Johannes Kohlmann summarizes the development of today's Josephson voltage standard, in which the standard volt is defined in terms of frequency using the Josephson voltage–frequency relation. Remarkably, arrays of 100,000 or more trilayer junctions are required for some of these standards.

I cannot emphasize too strongly that the realization of these various applications has depended utterly on the extraordinarily high yield and reproducibility of niobium-based Josephson tunnel junctions, the maturing of which has involved a very large number of people over some three decades. The development of this field is admirably illuminated in the pages that follow.

John Clarke
Department of Physics
University of California, Berkeley

Preface

The Josephson junction is a key element in the broader area of "superconducting devices," an area that was reviewed in 1990 in the excellent volume of that name edited by S. T. Ruggiero and D. A. Rudman. The present edited volume is further specialized, to deal primarily with the development and applications of the class of Josephson junctions made of *refractory materials*, mostly those made of niobium and its compound niobium nitride, that are now dominant in Josephson junction technology. These materials, having transition temperatures, respectively, of 9.2 K and approximately 16 K, can be formed, as we will see, into junctions of superior durability and closely controllable device parameters. Excellence in the latter regard, especially, has made them much more applicable in technology than the originally promising high-T_c oxide materials. The surprising range of scientific applications of this class of Josephson junctions, growing over the past 25 years, now make the successful refractory devices and their applications well worth review.

As we will see, the applications of the refractory Josephson junctions, often as central components in the SQUID superconducting quantum interference devices, now range from voltage standards, to devices for detecting submarines, to detectors in infrared astronomy, to amplifiers in radio telescopes such as the South Pole Telescope, to medical devices for treating epilepsy, and to qubits and other circuits in more than one class of possible computing technology. The history of the refractory junctions, initially made only on elemental niobium, by overcoming substantial materials science problems, is an initial topic in this book. To start our review of Josephson junctions, we are pleased to have an account from Brian Josephson of his original theoretical discovery in 1962 of the coupled superconductor effect that bears his name and led to his Nobel Prize in Physics, shared with I. Giaever and L. Esaki, in 1973.

In preparation of the present volume we have benefited from useful communication with a number of people, including Gene Hilton, Sherry Cho, David Rudman, Samuel Benz and Jonas Zmuidzinas. Others who have helped in the work described or in the production of this book are D. K. Finnemore, B. N. Harmon, W. Kent Schubert, K.-W. Ng, S. Han, R. J. Noer, Hongjie Tao, L. Y. L. Shen, W. J. Gallagher, J. M. Rowell, L. M. Folan, DeShane Lyew, Aditya Kaushal, Cornell Anthony, and Pratyush Mishra. We thank Sarabjeet Garcha at Pan Stanford Publishing for editing assistance. One of us, E.L.W., wishes to thank his wife, Carol, for support and assistance over many years and during the preparation of this book.

February 2017

E. L. Wolf
G. B. Arnold
M. A. Gurvitch
J. F. Zasadzinski

Chapter 1

The Theoretical Discovery of the Josephson Effect

B. D. Josephson
University of Cambridge, Cavendish Laboratory, Cambridge CB3 0HE, UK
bdj10@cam.ac.uk

The present chapter is closely based on an earlier article, "The History of the Discovery of Weakly Coupled Superconductors,"[a] in turn based on my response to an earlier request of the History of Physics Group of the Institute of Physics (Great Britain) on the historical background of the work I did in 1962 on weakly coupled superconductors [1, 2], and resulting in my talk on 12 November 1986 to the IOP History of Physics Group at University College, Cardiff, Wales.

1.1 The Background

I very much regret to say that, despite my personal involvement, the amount of historical accuracy I can give is not as high as I would wish, as although in the course of my searches I have found a certain

[a]Chapter 20 in *Physicists Look Back: Studies in the History of Physics*, ed. John Roche (Adam Hilger, Bristol, 1990), pp. 366–377. Used with permission.

Josephson Junctions: History, Devices, and Applications
Edited by Edward Wolf, Gerald Arnold, Michael Gurvitch, and John Zasadzinski

ISBN 978-981-4745-47-5 (Hardcover), 978-1-315-36452-0 (eBook)
www.panstanford.com

amount of material relating to that period of time, a lot of the ideas were worked out in my head, and seem not to have been written down on paper at all.

I should perhaps, as some readers may not be familiar with this work, say a little bit about what it is. At very low temperatures some materials become perfect conductors of electricity—this is called superconductivity—and we now understand that the superconducting state is a state of order which is very similar to that found in a laser, there being coherent electron waves where in a laser you have coherent electromagnetic waves, and the phenomenon involves something which is closely connected with the waves.

The basic equations which apply to superconducting barriers are shown below. A barrier between superconductors can be regarded as a two-dimensional system, whose behaviour (if dissipative processes are neglected) is governed by the following set of equations:

$$\frac{\partial H_y}{\partial x} - \frac{\partial H_x}{\partial y} = \frac{4\pi}{c} j_z + \frac{4\pi C}{c} \frac{\partial V}{\partial t} \tag{1.1}$$

$$\frac{\partial \varphi}{\partial x} = \frac{2ed}{\hbar c} H_y \tag{1.2a}$$

$$\frac{\partial \varphi}{\partial y} = \frac{-2ed}{\hbar c} H_x \tag{1.2b}$$

$$\frac{\partial \varphi}{\partial t} = \frac{2e}{\hbar} V \tag{1.3}$$

$$j_z = j_1 \sin \varphi \tag{1.4}$$

These equations apply for a barrier which occupies the *xy* plane. φ is the difference between the values of the phase of the superconducting order parameter on the two sides of the barrier, j_1 the critical current density of the barrier, C its capacitance per unit area, d its thickness plus the sum of the penetration depths in the superconductors on the two sides, and V the potential difference across the barrier. Gaussian units have been used. Equation 1.1 (essentially one of Maxwell's equations) describes the effect of the supercurrent through the barrier on the electromagnetic field, 1.2

and 1.3 describe the effects of the electromagnetic field on the superconducting order parameter, and finally 1.4 relates the barrier current to the order parameter [3].

To cut a long story short, Eq. 1.4 is the main result I discovered. Appearing on the right is a phase difference. The system which my work was involved with was where you have two superconductors that can exchange electrons, and the calculation I did showed that under those conditions you get a current given by that equation (1.4), which depends on the sine of the phase difference. So that was the effect, and now I will describe the background to it.

Surprisingly, perhaps, my official Ph.D. research project was not a theoretical one but an experimental one. My view at that time in my life was that I wouldn't find sitting at a desk spending most of my time thinking too congenial, although it's what I do now more or less, and so I decided I would do an experiment instead. I ultimately got my experiment to work, though it wasn't until 1974, about 10 years afterwards, that I got round to writing it up [4]. The experiment dealt with a phenomenon that had been of particular interest to my supervisor Professor Pippard (the late Sir Brian Pippard). He believed that second sound, a phenomenon that was well known in liquid helium, might also be found in superconductors (however, during the course of my Ph.D. work I managed to convince myself that a published calculation claiming to demonstrate the existence of second sound in superconductors [5] must be wrong, and at the present time no one seems to believe that second sound can actually occur in superconductors). There were anomalous results in the non-linear electrodynamics of superconductors which Pippard thought might be indicative of a resonance associated with second sound, and the proposed experiment involved looking in a new frequency range to see what happened there. It was because I was doing this experiment that I started looking into the theories of superconductivity.

As a consequence of the contacts between David Shoenberg, the director of the Mond Laboratory, and Soviet scientists, my first acquaintance with the theory of superconductivity was with a formidable book, *A New Method in the Theory of Superconductivity*, by N. Bogoliubov, V. Tolmachev and D. Shirkov [6], which it was suggested I try to understand. I found this book almost totally incomprehensible, which was hardly surprising in view of the way it assumed knowledge of topics such as many-body theory, second quantisation and diagram expansions, which neither I nor anyone else at the Cavendish at that time knew anything at all about. In the course of time through my browsings in the literature I came across explanations of these matters, and the mysteries of superconductivity theory started to become a little clearer. I found that there were a number of approaches to the problem, all of which were in some sense equivalent since they all implicitly assumed that a superconducting state was a state of pairing of electrons of equal and opposite momentum and spin like that hypothesised by J. Bardeen, L. N. Cooper and J. R. Schrieffer [7], who worked more or less directly with the actual wavefunction. The Bogoliubov theory [6] used a transformation to convert a situation that couldn't be treated by perturbation theory into one that could, while the Gor'kov theory [8, 9] used the very powerful Green function technique. Finally, there was the approach of P. Anderson [10], who described the superconducting state in terms of interacting pseudospins.

1.2 The Phase of a Superconducting Current

Both the Gor'kov and the Anderson theories contain something corresponding to a phase (in the Anderson theory the pseudospins lie in a plane through the relevant axis of symmetry whose orientation is otherwise arbitrary, and which is related to the phase in the Gor'kov picture), and these theories started me thinking about phases in superconductors. I have some scribblings in a notebook which suggested that I was attempting at that time to understand the structure of the superconducting state and the nature of its long-range order in terms of the kind of correlation functions that appeared in the Gor'kov theory. And there was something in the Gor'kov theory—the so-called anomalous Green function—which did have a phase attached to it. This phase allows one to understand certain properties of superconductors and in particular implied the phenomenon of flux quantisation, a phenomenon with a somewhat curious history. It is manifested first as a suggestion in the famous book of F. London [11], and later as a prediction in a paper by A.

Abrikosov [12] based on the 1951 Ginzburg-Landau theory [13], which prediction Landau himself had forbidden to be published at the time on account of the result being 'clearly unphysical', although he later changed his mind about this. The Ginzburg-Landau theory itself, despite its ability to give a simple picture of many of the features of the superconducting state, was virtually unknown in the West at that period because of the language problem and the poor communications between Eastern and Western scientists. Thus the experimental demonstration of the existence of flux quantisation (Deaver and Fairbank [14], Doll and Nabauer [15]) came as somewhat of a surprise.

In Gor'kov's theory the phase comes over as something more mathematical than physical: it is related to the phase of a matrix element between two states with different numbers of electron pairs and is essentially arbitrary because the two states could be chosen with arbitrary phases. Anderson made use of diagonal matrix elements instead, computed in states which did not have a definite number of electrons. In this approach the phase is a property of the specific state of the system and seems less of a mathematical thing, although it could well be argued that such states involving coherent combinations of states with different numbers of electrons are themselves unphysical. It would take some further analysis to sort this problem out, but in the meantime I found it very fruitful to think of the phase as something that was 'in principle real', real enough in fact to produce the phenomenon of flux quantisation.

If the phase was real, this implied that there might be a way of exploring its existence more directly. The conditions under which an effect could occur were regulated by the existence of symmetry principles. From these you could show that only phase differences could be observed, not the phases themselves. Furthermore, even phase differences could not be physically significant unless a process of electron exchange, which would have to be one that did not lose any phase information, occurred between the two regions concerned.

The ultimate answer to the conundrum about the reality of the phase (discussed in some detail in my Trinity College fellowship dissertation [2]) was that, whatever may be the case for the phases themselves, phase differences are perfectly respectable entities (i.e.

entities related to operators with non-trivial expectation values even for perfectly reasonable and physical kinds of quantum state).

It was not clear at that time that the possibility of setting up the kind of situation envisaged could be more than a dream, although in fact at around that time experimenters were actually busy creating such situations in experiments using thin films.

1.3 Boundaries and Junctions

A line of investigation that led me towards the actual discovery sprung from Pippard's interest in the question of what happened at the boundary between a normal metal and a superconductor. This was interesting because superconductors can be put into what is known as an 'intermediate state', where there are fairly macroscopic regions of normal metal and superconductor that alternate with each other. If you measure the electrical resistance and thermal resistance in such materials you find it is quite low, suggesting that there is not the amount of scattering at the boundaries that you would expect. That was a little bit mysterious. In fact it wasn't too clear even how the electric current could flow at all through such materials at low temperatures, because the current flow carried by the individual particles in a normal metal has somehow got to turn into current in a superconductor, and at very low temperatures, because of the energy gap, there are hardly any particles around to carry the current, unless there is a mechanism by which the ordinary electrons in a normal metal can be transformed into superconducting electrons.

Thinking about this and trying to apply some of the concepts that were around, I succeeded in producing a qualitative understanding of what went on at a normal-to-superconducting boundary. The detailed calculations would have been very complex and I did not pursue them, but subsequently Andreev [16] was to publish calculations based on exactly such concepts. He had enough energy to go through the calculations, which were somewhat messy, and I didn't. In fact, I don't think I've ever done any very messy calculations. I tend to go for problems that won't involve such calculations. In any event, I was as a result accustomed to using the self-consistent field method to think about this kind of problem. A notebook in which I wrote a bird's-eye view of my whole research project contained the following entry:

> Meanwhile we turned our minds to the question of transport properties of superconducting boundaries. A self-consistent field equation for quasi-particles, treating them as two-component electron-plus-hole wave functions, was developed and the consequences explored. However, curious difficulties arise when treating SIS junctions and SNS junctions. An approximate treatment was thought up and seems to imply oscillatory effects. It seems to be rather difficult to persuade experimentalists that these are more than a mathematical fiction, but we hope to publish a paper on the effects, and then someone may try the experiments . . . And so it came to pass that the cautiously entitled paper 'Possible new effects in superconductive tunnelling' was published in Physics Letters of July 1st 1962 [1].

Brian Pippard had a somewhat sceptical attitude to theoretical manipulations, and insisted on the incorporation of the word *possible* into the title of the paper.

1.4 Detailed Origin of the Theory of Weakly Coupled Superconductors

Let me now go a little into the details. The most important thing was that the various approximations that people introduced had the feature that not only could electrons scatter into electrons, but they could also scatter into holes (a hole is the absence of an electron). The apparent violation of charge conservation involved here is dealt with by treating the background of Cooper pairs as a reservoir which can supply or absorb pairs of electrons as needed without having to be put explicitly into the equations. You then find that to a first approximation all the electron–electron interactions that are important have become miraculously incorporated into the processes of the scattering of single particles already described. To calculate anything you then have to find out what the normal modes of this kind of system are, and these then correspond to mixtures of electrons and holes which you can then regard as being independent of each other.

There was, however, an awkward technical problem involved in a situation where you had two superconductors. The approximations that people introduced had the feature that not only could electrons be scattered from one state to another, but also electrons could be scattered into holes (a hole is the antiparticle to an electron in a metal) and vice versa, i.e. charge was apparently not conserved. The requirement that charge really should be conserved was taken into account by invoking the reservoir of superconducting electrons (a situation that was in due course clarified by Anderson [1958]). Now, in order to get this to work properly, you had to take the energy of the electrons in the reservoir, which is the Fermi energy, as your origin of energy (otherwise you would have problems with energy conservation, since the energy of the electrons in the reservoir would not have been taken into account). This creates no problem if you have only one region of superconductor, but what do you do if there are two regions and their Fermi levels are different? Fortunately for me, other people who were thinking about two-superconductor systems regarded this difficulty as something not really significant and failed to follow it up. In particular, at about this time Cohen, Falicov and Phillips [17] had applied perturbation theory to treat tunnelling involving superconductors by perturbation methods. They treated the case where only one side was superconducting correctly but omitted discussion of the two-superconductor case from their paper. I learnt afterwards that one of them had been asked to examine the two-superconductor case and had found terms equivalent to those I had found later, but couldn't understand their significance and had decided that they probably had no effect.

Because they had not dealt with the problem of the energy origin their formalism was inadequate to deal with the problem in any case. I managed to invent a rather esoteric formalism to handle the situation (which has rightly gone into oblivion now), involving putting in explicitly into the Bogoliubov operators the transfer of electron pairs from the background reservoir. Although that was a rather awkward method, I did it in preference to a Green function method because I wasn't familiar with Green functions and I thought

I'd have more chance of getting it right if I kept to the Bogoliubov operator method instead.

The interest in doing these calculations in the first place stemmed from the fact that Ivar Giaever, inventor of the tunnelling method for studying the properties of superconductors, had given a very simple-minded theory of the tunnel junctions that had the merit of agreeing with experiment (Giaever 1960 [18], Fisher and Giaever 1961 [19]). But it was not at all clear (as had been observed by Pippard) why Giaever's calculation should give the right answer, because it ignored the 'coherence factors' that normally have to be taken into account in a situation involving superconductors.

The Cohen et al. [17] calculation justified Giaever's formula for the single-superconductor case, which was, however, the case where the coherence factors turn out to cancel out anyway. What I expected would happen in the two-superconductor case was that there would be a non-zero coherence contribution which would have no DC component (so that Cohen et al. were in a sense correct to believe that the extra terms didn't contribute) because it would be multiplied by a factor depending on the phase difference which would average out to zero. The physical reality of this component was linked to the physical reality of the phase, and my physical intuition told me that both were real, although this was a controversial idea at the time.

My calculations confirmed my expectations, but they predicted an additional term that I had not expected. This term was the one that came from the delta-function contribution to the energy denominators, and had the unexpected feature of not vanishing when the applied voltage was set equal to zero. I checked my calculation several times trying to find a sign error whose existence would have prevented the terms from cancelling each other out, but had to conclude in the end that the contribution that did not vanish at zero voltage really did exist.

The presence of such a term at some order of approximation was not surprising, since it merely indicated the presence of a supercurrent. I had not expected it to occur in my calculation to second order because of the following argument due to Pippard [20]: a supercurrent will involve pairs of electrons tunnelling, so that the matrix element for the process will be proportional to the square of the matrix element for the tunnelling of a single electron. The rate at which the pairs will tunnel will be proportional to the square of this again, and so to the fourth power of the single-electron tunnelling matrix element. In due course I realised where this argument was in error: the tunnelling supercurrents are a coherent flow process, and so the usual rule that its rate is proportional to the square of the matrix element does not apply. I was then satisfied that my calculation was correct and went ahead and submitted the paper for publication. (Incidentally, my original calculation was not quite correct. In it I had managed to introduce two extra factors of two, through taking two integrals over the whole real line instead of just for positive values of the energy as should have been done. Thus the result given in my first paper was too large by a factor of four. I discovered the error when writing up the calculation for my Trinity College fellowship dissertation [2] a couple of months later, while a correct value was obtained also by Ambegaokar and Baratoff [21] who repeated the calculation using Green's functions).

If you look at my paper [1] you will see that there was considerable concern on my part with making a convincing case that the concepts I was using were valid ones. This raises an interesting issue. I think this is something that's liable to happen in the history of science when a different way of looking at things comes about. Now everyone accepts that you can just write down the phase of the wave in a superconductor because this is what all the experiments confirm, but then the experimenters were inclined just to think of it as just a bit of mathematics, and were very doubtful about it. Fortunately Phil Anderson was spending a year's sabbatical in Cambridge and he said, 'Yes, this is certainly right and an important and interesting result.' So I had some back-up.

1.5 Testing for Predictions

Although I found the theoretical arguments for the effects convincing, it was rather worrying that the theory made definite predictions of how much supercurrent there should be, and these predictions didn't seem to fit the experiments. Theory predicted that for an average sort of junction you should get something like some hundreds of microamperes of supercurrent; indeed the supercurrent should be of the same order of magnitude as typical normal currents. I noted with interest a published paper by Nicol et al. [22] that showed a clearly visible zero-voltage blip, but this was not typical and such events when they did occur were normally ascribed to superconducting shorts across the insulating barrier. I tried to think up coherence-destroying mechanisms for reduction of the supercurrent, such as the effects of magnetic impurities, so as to bring theory and experiment into line. If such effects were significant, it might mean that the effects would be unobservable.

One possible source of non-observation of the effects that could be investigated, which had been proposed by Anderson, was that of magnetic field, since the theory indicated that the junctions should be highly sensitive to such fields. An amount of flux in the junction corresponding to one flux quantum would have a drastic effect on the magnitude of the critical supercurrent. Anderson's suggestion was that only the flux in the insulating region would be effective, but using the Ginzburg–Landau theory to calculate the effects of the field I convinced myself that the flux penetrating into the superconducting regions also contributed, and this implied that the earth's field could be important. So I set up compensating coils to reduce the local fields to a few milligauss, and looked for zero-voltage currents using an electronic ammeter sensitive to a nanoampere, smaller by a factor of around 10^5 than the supercurrents predicted by the theory. Disappointingly, no supercurrents, even at this level, showed up, even though I was able to obtain junctions with good normal characteristics.

The predictions in my paper were quite controversial, and John Bardeen in particular asserted that my predictions were in error [23] because, according to him 'the pairing does not extend into the barrier'. I did not believe in this criticism, since it seemed to involve the assumption that the situation could be treated on the basis of local equations, which did not seem to me to be correct, but the fact that the supercurrents had not been observed argued in favour of Bardeen's theories. He and I debated the issue at a special session (not published in the proceedings) during the International Low Temperature Conference held at Queen Mary College later that year. My view was supported by a theoretician there, de Gennes I believe.

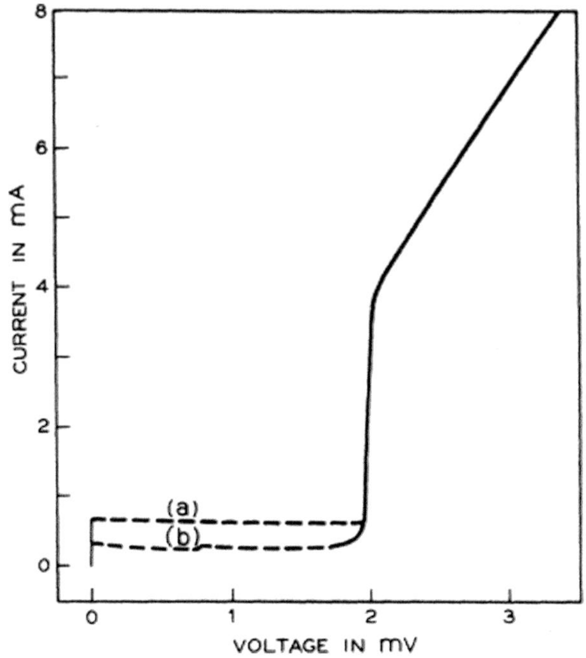

Figure 1.1 Experimental verification of the Josephson effect. Current–voltage characteristics for a tin–tin oxide–lead tunnel structure at ~1.5 K (a) for a field of 6×10^{-3} gauss and (b) for a field of 0.4 gauss. Reproduced with permission from Anderson, and Rowell [24].

It took nine months from the publication of my theoretical paper before convincing evidence for the reality of the phenomenon was obtained by Phil Anderson and John Rowell at Bell Labs [24] (Fig. 1.1). They used lower-resistance junctions made from lead instead of the aluminum that I had used, and observed zero-voltage currents whose maximum value before a voltage appeared was highly sensitive to magnetic fields. Such low-resistance junctions were liable to have metallic shorts through the barrier, which was why I had used aluminum instead, but the way the currents depended on magnetic fields provided very good evidence for the existence of a genuine effect.

Why had the Anderson–Rowell junctions showed an effect whereas mine had not? The explanation involved something that

I had clearly seen, but there was an unexpected twist as well. The coupling energy across the barrier is given by $-(\hbar/2e)\ I_c\cos(\Delta\varphi)$, and I had worked out that this was large compared with $k_B T$ if T is taken to be the sample temperature and concluded that thermal fluctuations would not be large enough to destroy the phase coupling process. But what I had failed to realise is that room-temperature electrical noise which is two orders of magnitude greater than the noise at helium temperature, comes down the leads to the specimen, and that for my samples with their low I_c this was enough to destroy the phase coupling and with it the zero-voltage supercurrents.

The Anderson–Rowell paper meant that the effect was real and its prediction could really be celebrated at last. It was only a matter of time before the other effects predicted in my original paper were confirmed experimentally, and practical applications of the effects were developed over the years as well.

For a comprehensive review of subsequent developments, see A. Barone and G. Paterno (1982), *Physics and Applications of the Josephson Effect* (New York: Wiley). Later chapters in the present book describe more recent developments that are primarily based on the refractory Josephson junctions typically made of Niobium.

References

1. Josephson, B. D. (1962). Possible new effects in superconductive tunneling. *Phys. Lett.* **1**, 251–253.
2. Josephson, B. D. (1962). The relativistic shift in the Mössbauer effect and coupled superconductors, *Dissertation for the Annual Election of Fellows*, Trinity College, Cambridge.
3. Josephson, B. D. (1966). Superconducting barriers: the plasma resonance and related properties, *Quantum Fluids. Proceedings of the Sussex University Symposium 16–20 August 1965* ed. D. F. Brewer (North-Holland, Amsterdam) pp. 174–175.
4. Josephson, B. D. (1974). Magnetic field dependence of the surface reactance of superconducting tin at 174 MHz. *J. Phys. F: Met. Phys.* **4**, 751–756.

5. Thouless, D. J., and Tilley, D. R. (1961). Collective modes in the theory of superconductivity. *Proc. Phys. Soc.* **77**, 1175–1181.
6. Bogoliubov, N. N., Tolmachev, V. V. A., and Shirkov, D. V. (1959). *A New Method in the Theory of Superconductivity* (Consultants Bureau, New York).
7. Bardeen, J., Cooper, L. N., and Schrieffer, J. R. (1957). Theory of superconductivity. *Phys. Rev.* **108**, 1175–1204.
8. Gor'kov, L. P. (1958). On the energy spectrum of superconductors. *Sov. Phys.-JETP* **34**, 505–508.
9. Gor'kov, L. P. (1959). Microscopic derivation of the Ginzburg-Landau equations in the theory of superconductivity. *Sov. Phys.-JETP* **36**, 364–367.
10. Anderson, P. W. (1958). Random-phase approximation in the theory of superconductivity. *Phys. Rev.* **112**, 1900–1916.
11. London, F. (1950). *Superfluids* (Wiley, New York).
12. Abrikosov, A. (1957). The magnetic properties of superconductors of the second group. *Sov. Phys.-JETP* **5**, 1174–1182.
13. Ginzburg, V. L., and Landau, L. D. (1965). *On the Theory of Superconductivity. Collected Papers of L D Landau* ed. D. Ter Haar (Gordon and Breach, New York) pp. 546–568.
14. Deaver, B. S., Jr., and Fairbank, W. M. (1961). Experimental evidence for quantized flux in superconducting cylinders. *Phys. Rev. Lett.* **7**, 43–46.
15. Doll, R., and Nabauer, M. (1961). Experimental proof of magnetic flux quantization in a superconducting ring. *Phys. Rev. Lett.* **7**, 51–52.
16. Andreev, A. F. (1964). The thermal conductivity of the intermediate state in superconductors. *Sov. Phys.-JETP* **19**, 1228–1229.
17. Cohen, M. H., Falicov, L. M., and Phillips, J. C. (1962). Superconductive tunneling. *Phys. Rev. Lett.* **8**, 316–318.
18. Giaever, I. (1960). Energy gap in superconductors measured by electron tunneling. *Phys. Rev. Lett.* **5**, 147–148.
19. Fisher, C., and Giaever, I. (1961). Tunneling through thin insulating layers. *J. Appl. Phys.* **32**, 172–177.
20. Pippard, A. B. (1961). Recent experiments on the electrical properties of superconductors, *Proc. Seventh Int. Conf. Low Temp. Phys.* ed. G. M. Graham and A. C. Hollis Hallett (North-Holland, Amsterdam) pp. 320–327.

21. Ambegaokar, V., and Baratoff, A. (1963). Tunneling between superconductors. *Phys. Rev. Lett.* **10**, 486–489 (and errata *Phys. Rev. Lett.*, **11**, 104).
22. Nicol, J., Shapiro, S., and Smith, P. H. (1960). Direct measurement of the superconducting energy gap. *Phys. Rev. Lett.* **5**, 461–464.
23. Bardeen, J. (1962). Tunneling into superconductors. *Phys. Rev. Lett.* **9**, 147–149.
24. Anderson, P. W., and Rowell, J. M. (1963). Probable observation of the Josephson superconducting tunnel effect. *Phys. Rev. Lett.* **10**, 230–232.

Chapter 2

Introduction to Refractory Josephson Junctions

E. L. Wolf
Department of Applied Physics, New York University Tandon School of Engineering, 6 MetroTech Center, Brooklyn, New York 11201, USA
ed.wolf@nyu.edu

As explained in the previous chapter by Dr. Josephson, the Josephson coupled-superconductor effect is inherent in any superconductor–insulator–superconductor tunnel junction if the insulating tunnel barrier is sufficiently thin to allow the coupling energy between superconducting pairs on the two sides to exceed thermal fluctuations.

2.1 Review of Physical Aspects

The Josephson effect entails a tunneling supercurrent density

$$\mathbf{J} = \mathbf{J}_0 \sin \varphi \tag{2.1}$$

at zero voltage; proportional to $\sin\varphi$, where

$$\varphi = \theta_1 - \theta_2 \tag{2.2}$$

is the difference between the phases of the pair states on each side. In this book we focus on refractory Josephson junctions, dating

Josephson Junctions: History, Devices, and Applications
Edited by Edward Wolf, Gerald Arnold, Michael Gurvitch, and John Zasadzinski

ISBN 978-981-4745-47-5 (Hardcover), 978-1-315-36452-0 (eBook)
www.panstanford.com

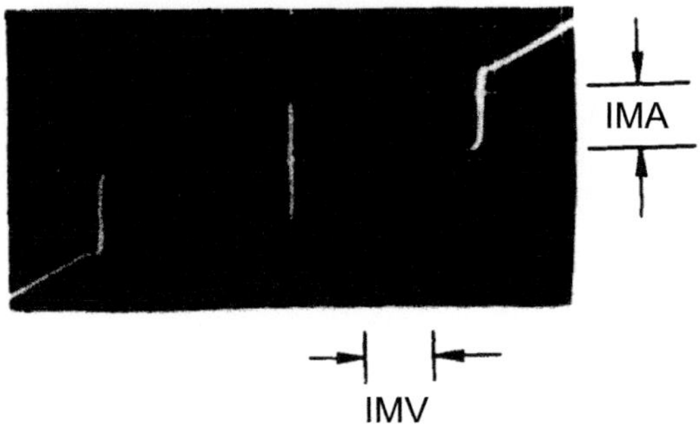

Figure 2.1 Current–voltage characteristic at 4.2 K of refractory Josephson junction of the form Nb/Al–Al-oxide–Nb shows supercurrent and quasi-particle branch at onset 2.95 mV. This junction of dimensions 8×8 μm^2 had nominal Al metal thickness on Nb/Al electrode of 50 A and gives a supercurrent density of 1300 A/cm^2. Reproduced with permission from Gurvitch et al. [1].

from around 1983, that now completely dominate applications. The dominance of refractory Josephson junctions comes from their robust nature, allowing frequent temperature cycling, and from the generally higher critical temperatures, exemplified by Nb (~9.2 K) and its compound NbN (16.6 K) compared to Pb (7.2 K). The physics is the same for both types of structure, and in the following we occasionally take examples from soft superconductor devices to illustrate relevant points of the Josephson junction physics. The materials-related technical difficulties in changing tunnel junction and Josephson junction fabrication from soft superconductors to the now widely applied refractory superconductors are a central theme in this book. The overall I–V characteristic of a Josephson junction is shown in Fig. 2.1 on a Nb/Al–oxide–Nb junction in the work of M. A. Gurvitch et al. [1] of 1983.

This curve approaches an ideal characteristic also from the point of view of the expected strength of the Josephson critical current. That value was given by Ambegaokar and Baratoff [2], who immediately confirmed the original Josephson result using a different theoretical method. Their expression for the product of the critical current times the junction resistance, for a symmetric junction, is

$$I_c R_n = \frac{\pi \Delta}{2}, \tag{2.3}$$

where I_c is the critical current, R_n the resistance in the normal state, and Δ is the gap parameter, half the energy gap of the superconductor. The measured value for the junction of Fig. 2.1 is $I_c R_n = 1.71$ mV [1], while the theory value, taking Δ as half the sum-gap voltage 2.95 mV, is nominally 2.31 mV, and in the range 1.95–2.1 mV in a more accurate estimate made by Gurvitch et al. based on [2].

From an experimental view, the Josephson effect is best distinguished from artifacts in the nature of short circuit paths by the dependence of the superconducting phase difference φ on the local magnetic vector potential **A**, such that $\mathbf{B} = \Delta \times \mathbf{A}$. This magnetic field sensitivity, of great importance in applications, arises from the generalized electron-pair momentum $\mathbf{p} + 2e\mathbf{A}$ in a pair wavefunction Ψ_P,

$$\Psi_P \approx \exp i(\mathbf{p} + 2e\mathbf{A}) \cdot \frac{\mathbf{r}}{\hbar}, \tag{2.4}$$

where, for a superconducting pair of charge $2e$, $\mathbf{p} = 0$. Requiring that such a wavefunction be single-valued in traversing a closed path in the superconductor gives a derivation of the flux quantum,

$$\begin{aligned} \Phi_0 &= \frac{h}{2e} = 2.068 \times 10^{-15}\ \text{Wb} \\ &= 2.068 \times 10^{-7}\ \text{gauss cm}^2 \end{aligned} \tag{2.5}$$

This is well verified experimentally, even though the value is exceedingly small, confirming the notion of phase coherence in the macroscopic paired state, as implied in Eq. 2.4. John Bardeen was quoted as commenting that (the phase) is coherent "over miles of dirty lead wire," suggesting what can be meant by a "macroscopic quantum state." As shown by Josephson [3] the variation of phase difference φ across the junction with magnetic field **B** in the barrier of effective thickness $d + 2\lambda$, with λ the superconducting penetration depth, satisfies

$$\mathbf{grad}\varphi = \left(\frac{2e}{\hbar}\right)(d + 2\lambda)(\mathbf{B} \times \mathbf{n}), \tag{2.6}$$

where **n** is a unit vector normal to the barrier, $\hbar = h/2\pi$, and d is the barrier thickness. This leads to a characteristic Fraunhofer-like magnetic field dependence of the maximum supercurrent, I_C, of the Josephson tunnel junction:

$$I_C = I_C(0)\left|\left[\sin\left(\frac{\pi\Phi}{\Phi_0}\right)\right] \Big/ \left(\frac{\pi\Phi}{\Phi_0}\right)\right|. \tag{2.7}$$

Here $I_C(0)$ is maximum current in absence of magnetic flux, and Φ and Φ_0, respectively, are the magnetic flux through the effective junction barrier thickness, and the flux quantum, $h/2e$. If the in-plane magnetic field $\mathbf{B}$ is parallel to dimension b of a rectangular junction of area ab, with insulating barrier thickness d, then $\Phi = \mathbf{B}b(2\lambda + d)$, where λ is the magnetic field penetration depth into the superconductor.

Beautiful data illustrating this characteristic dependence are shown in Fig. 2.2, due to R. C. Dynes and T. A. Fulton [4]. Note in the figure that the first data points from $\mathbf{B} = 0$ are plotted at 1/5

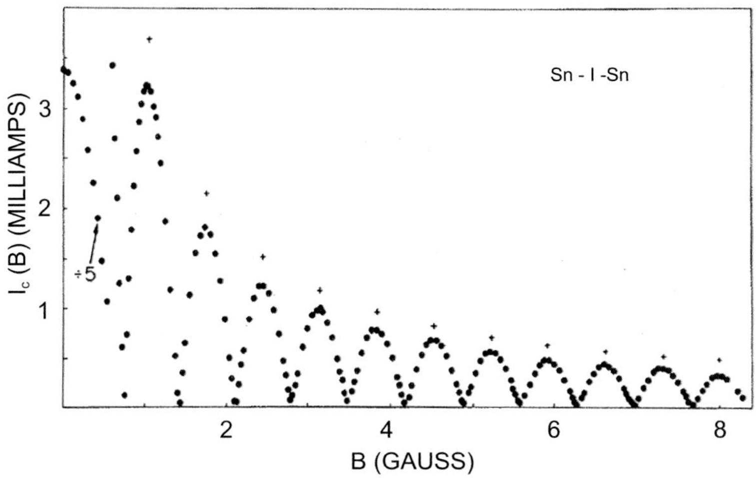

Figure 2.2 Typical Fraunhofer-like dependence of Josephson critical current on magnetic field, that underlies many applications. Experimental plot of critical supercurrent I_C vs. magnetic field $\mathbf{B}$ parallel to Sn–I–Sn Josephson junction at 1.43 K. The average spacing between nodes is 0.695 gauss, corresponding to 1140 A for $2\lambda + d$, where λ is the penetration depth of the magnetic field into the Sn and d is the barrier thickness. Between nodes one additional flux quantum flows through area set by junction length and the width of the tunnel barrier augmented by two penetration lengths, $(2\lambda + d)$. Reproduced with permission from Dynes, and Fulton [4].

height relative to later data points, and that the + symbols show the peak heights and positions predicted by the Fraunhofer function (2.7). For the junction shown, having dimensions $a = 0.26$ mm by $b = 0.41$ mm, the measured average spacing of minima, 0.695 gauss, implies $(2\lambda + d) = 1140$ A, a reasonable value for Sn.

As will be discussed later in this book, in a parallel connection of two Josephson junctions, a device known as a SQUID, or superconducting quantum interference device, the effective area for intercepting flux, described for the single junction as $b(2\lambda + d)$, with b the dimension of the rectangular junction parallel to the field, becomes the macroscopic area enclosed by the circuit loop containing the junctions. The corresponding critical current becomes, in the limit of small SQUID inductance, L,

$$I_C = 2I_C(0)\left|\cos\left(\frac{\pi\Phi}{\Phi_0}\right)\right|, \tag{2.8}$$

where $\Phi \approx \mathbf{B}A$ is the flux intercepted by the macroscopic circuit loop of area A containing the two junctions. This much larger intercepting area A makes the SQUID device, commonly now constructed with refractory Josephson junctions, a most sensitive detector of magnetic field, and by simple extensions, of electric current and voltage. The SQUID is the basis of several of the most important applications of refractory Josephson junctions, and this topic is reviewed below in Chapter 8 of this book by D. Drung and J. Beyer. (Equation 2.8 is formally valid in the limit $b_L \to 0$, where b_L is the inductance parameter according to Eq. 8.3 in the SQUID chapter).

The beautiful data of Dynes and Fulton in Fig. 2.2 was analyzed to provide, by an inversion method, an experimental mapping of the thickness d of the basic tunnel barrier. The tunnel barrier in soft superconductors such as Al, Sn, Mg, and Pb was easily provided by exposure of the soft superconductor metal to air or oxygen, perhaps in a plasma, via self-limiting growth of the native oxide. As Dynes and Fulton showed, the resulting tunnel barrier can be more than satisfactory.

A primary difficulty in extending fabrication of high-quality tunnel junctions to refractory superconductors, notably Nb, was in the failure of their grown oxides to provide similarly satisfactory tunnel barriers. Approaches to this technical difficulty, that finally led to the first high-quality refractory Josephson junction as shown in Fig. 2.1, from the work of Gurvitch et al., are a central theme of the early chapters in this book.

Returning to the nature of the full Josephson junction I–V characteristic in Fig. 2.1, one sees at once the basic bistability that offers a basis for application of the Josephson junction as a binary computer element. The supercurrent state gives way abruptly to the voltage state, when the supercurrent is destroyed by an overcurrent or a magnetic perturbation. The origin of the current–voltage characteristic in the voltage state, the state of quasiparticle tunneling, is suggested in Fig. 2.3 [5], where, for simplicity, the supercurrent has been omitted.

The upper portion of Fig. 2.3 schematically shows opened energy gaps, $2\Delta_1$ and $2\Delta_2$, and density-of-states peaks characteristic of most superconductors, as confirmed by methods of electron tunneling spectroscopy [5]. The lower portion of the figure shows the basic quasiparticle current onset at the "sum-gap" voltage $(\Delta_1 + \Delta_2)/e$. The difference gap feature is seldom of importance and will not appear in a symmetric junction as shown in Fig. 2.1. Comparing the current onset here to that in Fig. 2.1, there is a small "overshoot" feature at the sum gap in the experimental data that is associated with the actual Nb/Al bilayer structure. The origin of this small feature, not detrimental to applications, will be discussed in the following two chapters.

2.2 History of the Josephson Program

Before going into the issues in the technical development of refractory tunnel junctions, we mention that a large effort was made at the IBM Corporation to develop a computer technology using Josephson junctions based on Pb alloys, including Bi and Au. The supercurrent and voltage states of such junctions, similar to those shown in Fig. 2.1, were the binary zero and one of this technology. Review of this major effort [6] in 1980 was followed by announcement of its termination in September 1983 [7]. It was stated [7] that the IBM effort had been at the level of $20 million per year and employed 115 workers.

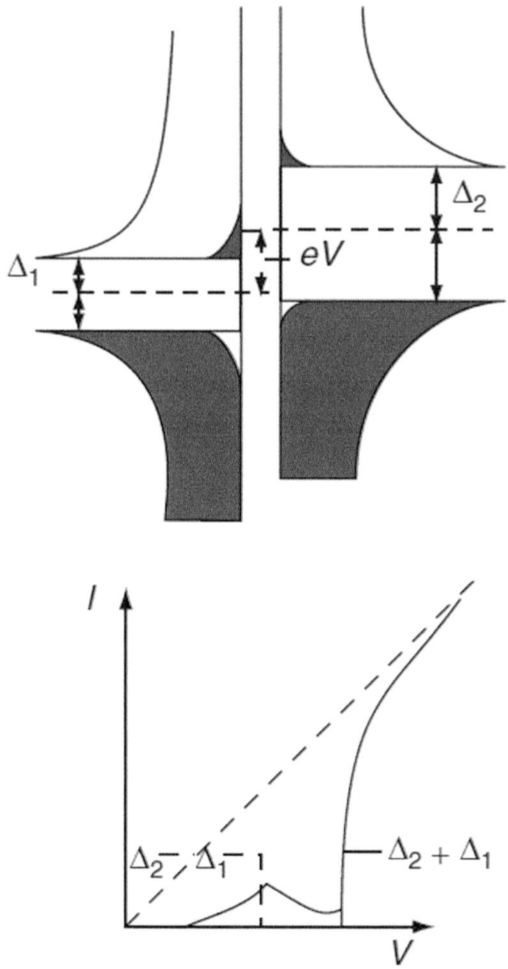

Figure 2.3 (Upper) Sketch of quasiparticle states and their occupation at finite temperature in junction at bias voltage V between superconductors of energy gap parameter values Δ_1 and Δ_2. (Lower) Current vs. voltage plot shows basic quasiparticle tunneling onset at sum gap voltage $(\Delta_1 + \Delta_2)/e$. Supercurrent at $V = 0$ is not shown. Reproduced with permission from Wolf [5].

By 1986, as reviewed by Hayakawa [8], important work in Japan on refractory Josephson junctions, based in part on work by Kroger et al. [9] in 1981 using amorphous Si barriers, and by Gurvitch et al. [1] in 1983 (the Nb/Al trilayer technology; see Chapter 5), showed improvements over the earlier soft superconductor technology, which was by then largely abandoned. Hayakawa's paper showed a good Josephson I–V characteristic from a refractory NbN/MgO/NbN junction, where the MgO is regarded as an artificial barrier, in extension of work by Kroger et al. [9], where amorphous Si (am-Si) was deposited on sputtered Nb as an artificial tunnel barrier. The fully refractory NbN/MgO/NbN Josephson junctions first reported in [8] have been extended to epitaxial form by Kawakami et al. [10]. Extensions of the work on NbN junctions are treated later in Chapter 6 below by J.-C. Villegier.

After the demise of the soft superconductor IBM Josephson program, Likharev and Semenov [11] suggested that a reason for the failure, beyond the fragility of the junctions, was that the $I(V)$ of the unshunted Josephson junction, as shown in Fig. 2.1, is inevitably of the *latching type*, meaning that the junction will remain in the switched state near $V = 2\Delta/e$ (that is determined by the *load line*, i.e., the outside resistor in series with the junction) unless reset by a pulse from the outside. The resetting is done by the clock in the digital Josephson technology. As was emphasized by Likharev and Semenov [11] the maximum clock rate in such a system is limited to the order of 1 GHz, because at higher rates the probability of spontaneous change was increased, rendering the logic unreliable. While 1 GHz was initially an acceptably high frequency, with later improvements in the semiconductor logic, the JJ technology limit on clock rate became a disadvantage. This speed difficulty was avoided in a shunted junction technology, known as RSFQ or SFQ [9], for rapid single flux quantum, that has survived, and now has almost completely replaced the initial latching logic technology. The original introduction of the RSFQ technology was an important advance, increasing the available frequency by a factor of 100. It is important to note that the RSFQ technology is now typically implemented in the Nb/Al trilayer technology. The early history of these developments in switching circuitry was summarized by Hayakawa [6] as well as by Likharev and Semenov [11].

2.3 Development of Tunnel and Josephson Junctions on Niobium

We first focus on the tunnel junction technology needed for the refractory metals, of which the most important is Nb, having the highest transition temperature, 9.26 K, among the elements. (This value is observed in clean bulk crystals and the best not-very-thin films, but in 100 nm films the observed value is often slightly lower, 9.1 K or 9.0 K.) The difficulty in making good oxidized tunnel barriers on Nb was suggested in 1975 in the work of Okaz and Keesom [12], who showed that NbO is not an insulator, but a superconducting metal of T_c 1.61 K. In exposure of Nb to air or oxygen one may find some NbO as well as the insulating oxide Nb_2O_5, as was summarized later by Halbritter [13]. Indeed, anomalous results from tunneling spectroscopy of oxidized Nb tunnel junctions were reported by Bostock and MacVicar [14] in 1976. They reported obtaining anomalous negative values of the parameter μ^* and in their paper questioned whether Nb could be treated by the conventional electron–phonon strong coupling theory of superconductivity. Their paper raised the question that Nb might represent an entirely different form of superconductivity. This idea was very controversial, and the research community finally agreed that the anomaly arose from the complicated thermal oxide grown on Nb rather than from an unusual form of superconducting pairing in that metal. In fact a resolution of this controversy, by preparing improved Nb tunnel junctions not containing the problematic niobium oxide, was one of the aims of the "Ames tunneling group," whose work is now described.

An extended effort to make excellent tunnel junctions on Nb in the "tunneling group" of the Ames Lab at Iowa State University from 1976 to 1985, starting in 1977 [15, 16] and summarized at length in [17, 18] 1980, focused on the use of Nb/Al bilayers, the Al subsequently being oxidized to provide a reliable insulating tunnel oxide. The use of Al as an overlayer, subsequently oxidized, was not new at that time; see the work of Hauser et al. [19] in 1966 and references therein, and p. 507 in Ref. [5]. The focus of the "tunneling group" was to make excellent tunnel junctions on

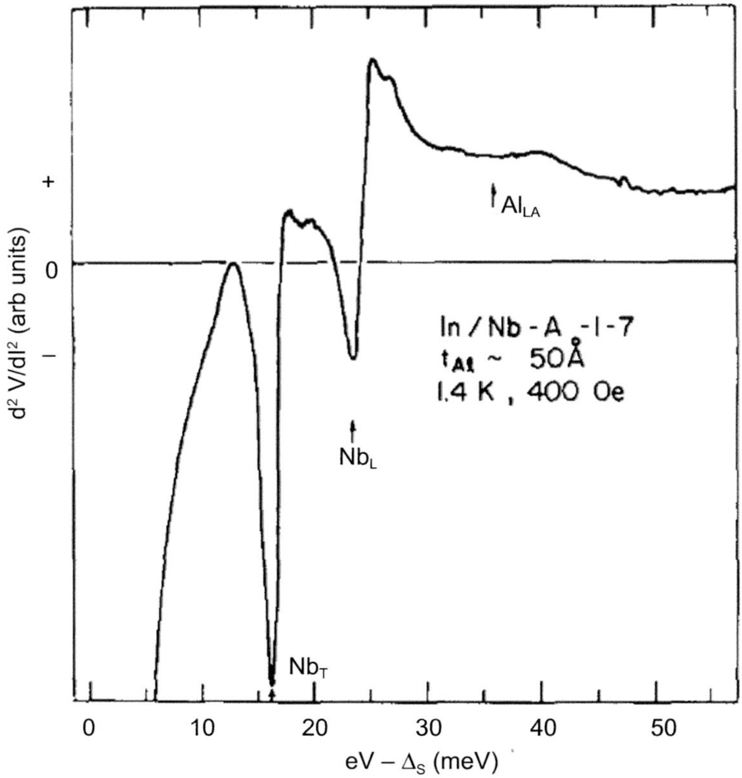

Figure 2.4 Tracing of d^2V/dI^2 spectrum for In/Al_2O_3/AlNb junction at 1.4 K in 400 Gauss parallel field, applied to simplify the spectrum by suppressing the superconductivity of In. Arrows mark positions of expected phonon peaks corresponding to transverse and longitudinal Nb modes, and to longitudinal Al mode. Reproduced with permission from Wolf, and Zasadzinski [15].

Nb (see Fig. 2.4) and use them to infer the detailed mechanism of superconductivity in that metal. The junctions were termed *proximity junctions* because the superconducting pairing in Nb induced pairing in the Al, as paired electrons flow from Nb to Al. The proximity junctions are described in detail in Ref. [18]. The Eliashberg theory of superconductivity, an extension of the successful BCS theory, shows in detail how metal vibration modes or phonons can couple electrons into pairs and leave a detailed picture of this interaction in the measured quantity dI/dV. This spectrum can be inverted, using the McMillan–Rowell computer program (as was done in Ref. [14]), to extract the effective phonon density of states. These technical aspects are described in Ref. [5] and the extension of the method to bilayers such as Nb/Al is the main topic of Ref. [18].

The Josephson effect is inherent in any tunnel junction between superconductors (although, as explained by Dr. Josephson in the previous chapter, it can be absent if there is external electrical noise), but the Josephson effect in itself was not the focus of the Ames Lab tunnel group. Single crystals of Nb were produced by recrystallizing Nb foils from 1800°C to near the Nb melting point (~2460°C) in ultra-high vacuum (UHV) 2.6×10^{-12} Bar or less. (The cryogenic UHV vacuum system with extremely high pumping speed was designed by L. Y. L. Shen at Bell Labs [20] and later donated to the Ames tunnel group. L.Y. L. Shen in his paper [20] thanks J. M. Rowell "for many helpful discussions" in the work). In these atomically clean conditions, after the foil had melted across, and abruptly cooled to room temperature, thin layers of Al were thermally evaporated, in variable thicknesses, across sections of the centimeter-wide foils. It was later found [21] that the recrystallized Nb foil typically had a (110) surface, with at least millimeter-size grains, and that the Al layer was (111), epitaxially related to the Nb. Under such atomically clean conditions the Al would be expected to smoothly spread out across the Nb to maximize its contact area, since the cohesive energy of Nb greatly exceeds that of Al.

The "wetting" of the Nb with the Al, avoiding any kind of "pinholes," was confirmed from the tunneling results, first shown in [15] (Fig. 2.4).

The tunneling results that clearly show the wetting of the Al on the Nb are the feature near 37 mV bias in Fig. 2.4 and Figs. 7, 8, 9, and 10 of [17], on junctions for 27 A and 50 A deposited Al thickness. The feature near 37 mV comes from the longitudinal acoustic LA phonon of metallic Al. The wetting of the Nb was also indicated by surface analysis. Page 36 of [17], published in 1980, states that "above 30 A deposited thickness of Al the film is continuous. This conclusion is supported by Auger analysis of freshly coated Nb/Al sandwiches which for $t_{Al} > 30$ A showed Al Auger peaks but no lower energy Nb peaks." Wolf and Zasadzinski [15] (see also p. 48 of [17]) state, "We also thank ... L. Y. L. Shen and J. M. Rowell and Bell Laboratories for the donation of a vacuum system." There were in fact two cryogenic UHV systems, designed by and built for L. Y. L. Shen at Bell Laboratories and shipped to Ames, IA, the first in 1975, with the approval of John Rowell, after Lawrence Shen was unable to use them at Bell.

In fact, John Rowell visited the Ames Lab tunneling group in the fall of 1979, and on that day, before giving a seminar talk, intently inspected direct $\mathrm{d}V/\mathrm{d}I{-}V$ chart recorder curves for the indication of the Al phonon at 37 mV, which all knew indicated the wetting of the Nb by Al under atomically clean conditions, down to the thinnest layers. Also discussed on that day was the ongoing work on tantalum foils that was published in 1981 [22] as IV in the series in the *Journal of Low Temperature Physics* on proximity electron tunneling spectroscopy. An earlier connection of the Ames Lab tunnel program with John Rowell, beyond donation of the vacuum systems, was the visit in 1976 to Ames of Bennett Robinson, whose Ph.D. thesis in physics at Stanford University, dated 1976, under the supervision of T. H. Geballe, was titled "Tunneling into Quench Condensed Transition Metals" and was in fact partially supervised by John Rowell, who was on sabbatical at Stanford in 1976. Bennett Robinson and Ames tunnel group members compared his data on Nb to the Ames results, and the consensus was that the Ames data (see Fig. 2.4) were definitely superior. Bennett Robinson agreed that the proximity tunnel method, based on single-crystal bilayers, and using the first vacuum system donated by Bell Labs, slightly rebuilt and already extensively in use, was superior to quench condensing, to learn about the superconductivity in Nb.

As noted by Dr. Josephson in Fig. 1.1, J. M. Rowell, with P. W. Anderson, first experimentally confirmed the Josephson effect, and J. M. Rowell obtained a U.S. patent [23] for such devices. Shortly after fall in 1979, when J. M. Rowell visited the Ames Lab tunneling group, three experimental papers, [24] in 1981, [25] in 1982 and [1] in 1983, were published from Bell Laboratories, in a new Josephson junction effort. (The new Bell effort, while encouraged by J. M. Rowell, was partially in a different laboratory from his, as is explained by M. A. Gurvitch in Chapter 5. These works, culminating in 1983 with the Josephson characteristic shown in Fig. 2.1, utilized DC sputtering systems with Nb and Al targets, so that deposition of Nb could be quickly followed by deposition of Al. (The first encouraging all-refractory $I(V)$ appeared in 1981, as explained in Chapter 5 by M. A. Gurvitch.) The paper by Rowell, Gurvitch, and Geerk [24] confirmed that the same wetting of clean Nb by Al that was shown in the published Ames Lab works [15–17], starting in 1977, could be obtained using sequentially sputtered thin films. This was good news from the point of view of making Josephson junctions, but the observed "wetting" could hardly have been "surprising" (see p. 2278 of [24]), at least to the lead author. This paper failed to reference any of the published Ames Lab works on the directly relevant Nb/Al, including the personal communication that had occurred. In the second paper [25] devoted to oxidation of Al on Nb/Al bilayers, again based on sputtered bilayers using the same parameters as employed in [24], the authors Kwo et al. thanked J. M. Rowell for suggesting their research project.

Kwo et al. [25] found that with the sputtered Nb films under clean conditions the Al layer could flow around the grains, leaving a diminished amount on the front surface, and that by heating the substrate to enlarge to size of the Nb grains, this effect could be diminished. But enough Al was typically left wetted to the Nb to allow good oxidation for the tunnel barriers. This paper referenced [17] of 1980 as indicating that Al would spread out over the surface of clean Nb, as evidenced by strong proximity effect tunneling, as shown in Figs. 7–10 of [17]. The 1983 paper [1] of Gurvitch et al. shows excellent Josephson junction tunnel characteristics (Fig. 2.1) for devices made in a similar but separate DC sputtering system, now with three sputtering targets. This paper refers only to Ref. [24] as a reason to expect wetting of the Nb by Al, with no reference to the prior Ames work.

The structures in this paper [1] are of the form Nb/Al-AlOx-Nb, made by sequential sputtering and thermal oxidation, and the process is now referred to as the Nb/Al trilayer Josephson junction process, and is widely used to the present. See, for example, Berggren et al. [26] describing arrays of junctions of critical current density 1.7 kA/cm^2. The term SNEP (selective niobium etching process) following the terminology of Kroger et al.

[9] was used by Gurvitch et al. These authors referenced earlier work by Laibowitz and Mayadas, using 800 A Al layers on Nb to make "trilayer" Josephson junctions [27] with critical current density about 12 A/cm^2. This early Nb/Al "trilayer" process, using sequentially electron-beam-evaporated films of Nb and Al, clearly did not allow full coverage of the Nb below 800 A thickness of Al, and one speculates that the vacuum conditions, even though conventionally competent, did not leave a sufficiently clean Nb surface for the Al to flow around the grains as was observed in the Ames work starting in 1977 and in the later, 1981 work of Rowell, Gurvitch, and Geerk [24] and Kwo et al. [25]. Gurvitch et al. [1] acknowledged advice from, and a reading of their manuscript by, J. M. Rowell. The developments at Bell Labs, starting shortly after 1980, leading to the important 1983 paper [1] are described in detail by M. A. Gurvitch in Chapter 5. Improvements in the performance of the sputtered Nb/Al trilayer junctions, including lower subgap current and critical current densities up to 4600 A/cm^2, were reported by Morohashi et al. [28] in 1985. The several applications of the Nb/Al trilayer junctions, and of other refractory junctions such as NbN/MgO/NbN [8] and $Nb_{0.62}Ti_{0.38}N/Ta_xN/Nb_{0.62}Ti_{0.38}N$ devices [29] are summarized in later chapters of this book.

2.4 Development of Tunnel and Josephson Junctions on Niobium Nitride

Following the work of Gurvitch et al. on Nb junctions, the next important advance in refractory junctions may well have been the work of Shoji et al. [30], which was the key reference in [8]. Junctions of the form NbN/a-MgO/NbN were made by sequential RF sputtering from Nb and MgO targets.

The NbN films were RF sputtered from a Nb target in a 6% N_2–Ar gas mixture at 1.1 Pa, while the MgO layer was RF sputtered from an MgO target in 1.3 Pa of argon gas. The MgO barrier film was determined to be amorphous by appearance of halo rings in reflection high-energy electron diffraction RHEED measurement and an absence of any peak in x-ray diffraction. The sum gap voltage of these devices was typically 5.1 mV, suggesting a gap parameter value 2.55 meV. Critical current densities in the range 600–1400 A/cm^2 were observed, with $I_c R_n$ product values of 3.15–3.25 mV, compared to an ideal value of 4 mV based on the Ambegaokar–Baratoff relation, using $\Delta = 2.55$ meV [2]. The subgap leakage parameter V_m is in the range 30–50 mV. The specific capacitance was inferred to be in the range 7–8 μF/cm^2. The advantage of using NbN is the higher transition temperature. On the basis of the strong coupling relation

$$2\Delta = 4.0 k_B T_C \text{ (Thouless)} \tag{2.9}$$

rather than the BCS relation

$$2\Delta = 3.52 k_B T_C \tag{2.10}$$

(see Thouless [31], also p. 100 in [5]), one predicts that $T_C = 14.8$ K for the NbN fabricated by Shoji et al., very close to the observation. The actual temperature dependence of the critical current for the junction shown in Fig. 2.5 is shown in Fig. 2.6. This dependence is very close to that predicted by Ambegaokar and Baratoff, ([2] in the Erratum) namely,

$$I_C(T) = \pi/2 \; R_N^{-1} \Delta(T) \tanh[\Delta(T)/2k_B T]. \tag{2.11}$$

This expression leads [2], near T_C, to

$$I_C/I_C(0) = 2.67(T_c - T)/T_c, \; T \approx T_C \tag{2.12}$$

assuming the BCS dependence

$$\Delta(T) = 1.8\Delta(0)\left(1 - \frac{T}{T_C}\right)^{1/2}, \text{ near } T_C \tag{2.13}$$

and noting that $\tanh(x) \approx x$ for small x. The predicted linear slope with prefactor $2.66 \approx 2.67$ was observed in weak-coupling Sn/I/Sn Josephson junctions by Fiske [32].

The linear dependence near T_C is also nicely shown in Fig. 2.6, but these data for the strong coupling superconductor NbN near T_C fit more closely to

$$I_C/I_C(0) = 2.19(T_C - T)/T_C, \; T \approx T_C. \text{(observed, see Fig. 2.6)} \tag{2.14}$$

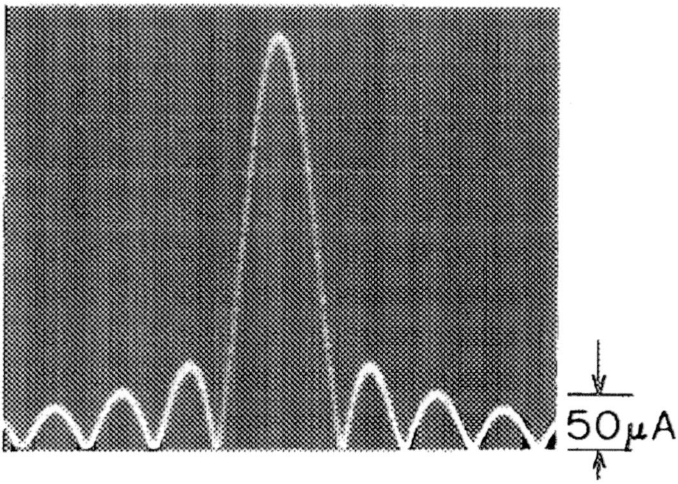

Figure 2.5 Dependence of Josephson critical current at 4.2 K on parallel magnetic field for the first NbN Josephson tunnel junctions, of the form NbN/MgO/NbN from the work of Shoji et al. in 1985. These nearly ideal Josephson junctions were fabricated by RF sputtering of Nb and MgO targets (see text), and the Josephson current persisted up to 14.5 K. Reproduced with permission from Shoji et al. [30].

This observed slope value (2.18) is in acceptable agreement with the slope, 2.26, obtained from the Thouless strong coupling theory [31] that gives

$$\Delta(T) = 1.73\Delta(0)\left(1 - \frac{T}{T_C}\right)^{1/2} \quad \text{near } T_C, \tag{2.15}$$

and $2\Delta(0) = 4\ k_B T_C$. So these refractory NbN tunneling Josephson junctions [30] behave consistently as expected for a strong coupling superconductor.

The higher critical temperature of the NbN devices is important to applications, in that it makes possible cooling by refrigerators rather than by liquid helium, but their fabrication may be more difficult than the trilayer Nb/Al junctions described earlier.

The route to these superior NbN/MgO/NbN junctions, with directly deposited barriers, perhaps started in the 1979 work of Kroger et al. [33] (see also [9]) on oxidized amorphous Si, followed by work of Rudman and Beasley [34] on the same subject. Rudman and Beasley conclude that the oxidized amorphous Si barriers

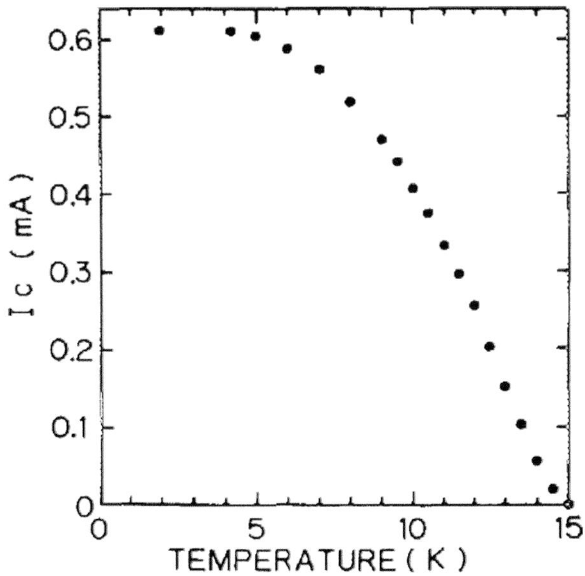

Figure 2.6 Temperature dependence of the maximum critical current for a NbN/a-MgO/NbN junction similar to that shown in Fig. 2.5. The critical temperature for the device is near 14.8 K. Reproduced with permission from Shoji et al. [30].

work well provided the "native oxide of the base electrode has a sufficiently low conductance to block pinholes." So the am-Si process lacked any "wetting" character to fully cover the base electrode, such as the Nb/Al process that was being worked on at the same time [15–18]. "Pinhole" defects could lead to excessive subgap leakage in such devices. The am-Si process, nonetheless, led in 1980 to workable early NbN/am-Si/NbN junctions [35]. These devices were surpassed in 1985 in the same group by the excellent NbN devices described in [30].

The next advance in this area appears to be the discovery that entire NbN/MgO/NbN epitaxially crystalline structures could be sequentially grown using sputtering methods on single-crystal MgO substrates, culminating around 2001 in excellent junctions as described in [10]. The epitaxial NbN Josephson junction development may have started with the work [36] of Kerber et al. in 1989. These researchers reported that if NbN were sputtered onto a thin layer of MgO, deposited on an oxidized Si substrate, rather than directly onto the oxidized Si, the NbN T_C rose from about 14.8 K to 15.7 K. The NbN layer in this case was found to be oriented. This work was extended as described in 1991 by Shoji [37] who reported, using single-crystal NbN electrodes with MgO barriers, sum gap voltages 5.6–5.8 mV, and by Wang et al. [38] in 1997, who reported Josephson current densities up to 54 kA/cm^2 using AlN barriers.

Finally, Kawakami in 2001 [10] described fully epitaxial NbN/MgO/NbN structures grown on single-crystal MgO substrates with critical current densities in the range 0.2–70 kA/cm^2. The NbN electrode T_C was reported as 15.7 K and gap voltages in the range 5.6–5.9 mV were reported for those junctions with critical current densities up to 15 kA/cm^2.

The same authors explained the motivation of their work in application of these excellent, fully epitaxial, devices as Josephson oscillators and SIS mixers for use above 700 GHz in [38]. They performed experiments with these devices to indicate low losses up to 1.2 THz. Pioneering work on NbN films and devices grown on sapphire Al_2O_3 is described by Villegier et al. [39]. A review of work on SIS mixers and oscillators based on refractory Josephson junctions is given by Dmitriev, Filippenko, and Koshelets in Chapter 7.

2.5 Non-hysteretic Josephson Junctions and Generalized Josephson Devices

Fabrication of non-hysteretic shunted Josephson junctions is of interest to exploit the RSFQ computing circuitry mentioned above [11] as a successor to the latching circuitry investigated earlier at IBM. The bulk of RSFQ circuits remove the bi-stable nature of the $I(V)$ of Nb/Al trilayer junctions, by applying an external shunt resistor [11]. To make a smaller overall device, efforts have been made to incorporate the shunt into the Josephson junction itself.

The details of the $I(V)$ of Josephson junctions with shunt resistor R were investigated by Stewart [40] and by McCumber [41]. These descriptions are found to apply to a wider class of weakly coupled superconductor junctions, where the original tunnel barrier coupling in an SIS structure may be replaced with a resistive layer, for example, to make an SNS junction [42], or even SINIS structures [43, 44], found to exhibit supercurrents and other Josephson phenomena. These generalized structures are now fabricated in refractory forms, for use in contemporary Josephson junction applications including Josephson voltage standards. The degree of shunting needed to give a single-valued $I(V)$ for a Josephson junction is established by the Stewart–McCumber parameter,

$$\beta_c = 4\pi e I_c R^2 C/h, \tag{2.16}$$

where R is the shunt resistance, C the capacitance, and h is Planck's constant. The overdamped non-hysteretic regime corresponds to $\beta_c \leq 1.0$.

For example, the paper of Senepati and Barber [45] (see also references therein) describes internally shunted NbN/MgO/NbN devices. A typical set of $I(V)$ curves is shown in Fig. 2.7, illustrating the behavior predicted by Stewart and McCumber. The maximum supercurrent I_C for the device is 0.37 mA at 4.2 K and the product.

$I_C R_N \approx 0.15$ mV at 4.2 K. The shunt resistance itself is estimated as 0.45 Ω, while the R_N value measured on similar, but unshunted, junctions is ~3.5 Ω. The estimated capacitance C is 1.3 pF and $\beta_c = 0.23$ at 4.2 K.

The $I_C R_N \approx 0.15$ mV is much reduced from what is available in unshunted junctions, e.g., values $I_C R_N = 3.15$–3.25 mV for the NbN/MgO/NbN junctions of Kawakami et al. [30]. The authors [42] point out that in the RSFQ logic the inherent speed is proportional to $I_C R_N$, based on [11]. Nonetheless, the authors suggest that their shunted devices remain workable for RSFQ logic. Since their device has a parallel current path, the shunt resistor, to the supercurrent path, the authors have been at pains to establish that the total current is dominated by the Josephson supercurrent. This is established as shown in the inset to Fig. 2.7, where the total measured supercurrent is plotted vs. temperature, and shown, by comparison to the theory curve based on the Ambegaokar–Baratoff relation (2.11), to be in excellent agreement. In their application of the Ambegaokar–Baratoff formula (2.11) the authors [45] have used an interpolation formula for temperature dependence of the BCS gap

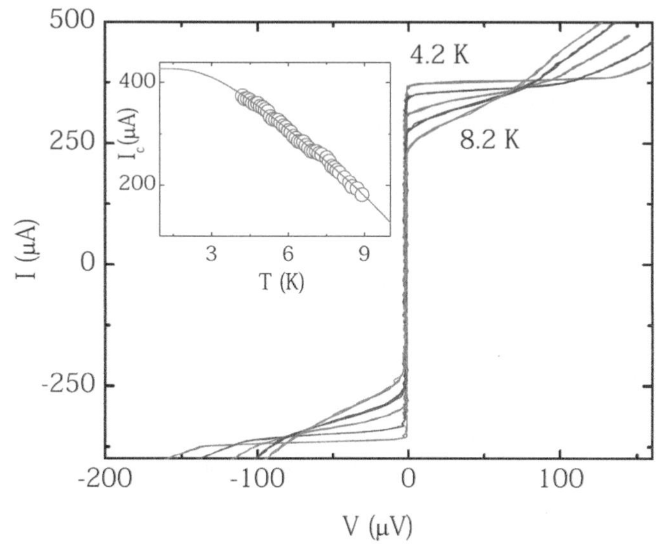

Figure 2.7 $I(V)$ measured at temperatures 4.2, 5.2, 6.2, 7.2, and 8.2 K for sidewall-shunted NbN/MgO/NbN overdamped Josephson junction of dimensions 30 μm^2 and estimated Stewart–McCumber parameter β_c = 0.23 [40]. $I(V)$ is single-valued as consistent with reduced value of β_c. Inset shows temperature dependence of the critical current as compared to theory. Reproduced with permission from Senapati, and Barber [45].

$\Delta(T)$ that is not limited to the region near T_C: this formula [46] is

$$\Delta(T) = \Delta(0)\tanh[4.8(T_C - T)/T]^{1/2}. \tag{2.17}$$

Using this formula excellent agreement to the data is found, note the offset zero on the current scale in the inset to Fig. 2.7.

2.5.1 Shapiro Steps, Highly Hysteretic Junctions, and Josephson Voltage Standards

As described by Dr. Josephson in Chapter 1, and in his first published paper, when the junction is biased at voltage V the usual DC current appears, but there is also an AC supercurrent of frequency

$$f = 2eV/h(483{,}597.9 \text{ GHz/V}), \tag{2.18}$$

where h is Planck's constant. If a junction is exposed to microwave rf radiation it is predicted by Josephson and observed [47] that the $I(V)$ will be interrupted by constant voltage (Shapiro) steps occurring at quantized voltages

$$V_{\mathrm{n}} = n\left(\frac{h}{2e}\right) f, \text{ where } n = 1, 2, 3\ldots \quad (2.19)$$

In the current range of a given step, the phase of the Josephson oscillating current is locked to the phase of the applied rf field. There has been a longstanding and well-published international effort to use this relation to establish voltage standards, making use of agreed high precision frequency standards. Chapter 10 of this book, by Johannes Kohlmann, is devoted to this effort, which has reached an extremely high level of sophistication. In the present chapter we are primarily concerned with the underlying refractory Josephson junctions that have dominated the voltage standard field since perhaps 1990 [48]. An advance in the voltage standard field, which involves series arrays of Josephson junctions, was the prediction, based on accurate modeling at $\beta_{\mathrm{c}} = 500$, of Levinson et al. [49], that high β_{C} junctions provide "zero crossing steps," where each Shapiro step extends from negative to positive current. Such steps are more amenable to forming a voltage standard. This is illustrated in Fig. 2.8, taken from the review in 2000 of Clark A. Hamilton [50].

Zero-crossing Shapiro steps of the type illustrated in Fig. 2.8b were clearly shown in the 1990 work of Popel et al. [48], demonstrating 1 V and 10 V standards, based, respectively, on series arrays of 2000 and 20,160 Nb/Al trilayer Josephson junctions, to produce stable quantized-voltage steps up to 1.5 and 14.5 V. In describing their method, Popel et al. point out that with their input frequency f of 70 GHz, the first step will appear at about 0.145 mV. A single junction will provide about seven steps, up to about 1 mV, in their mode of operation. To achieve 1 V, the entire series array must be biased on the 7000th step, and to achieve 10 V the entire series array is biased onto the 70,000th step. The method of controlling the bias voltage and the bias impedance to select a particular quantized step was earlier described by Hamilton et al. in 1987 [51] in connection with a soft-superconductor array. The capability of reaching any one of the 70,000 steps in the range 14.5 V, using refractory Josephson junctions at 70 GHz, was demonstrated in the Popel paper [48] of 1990.

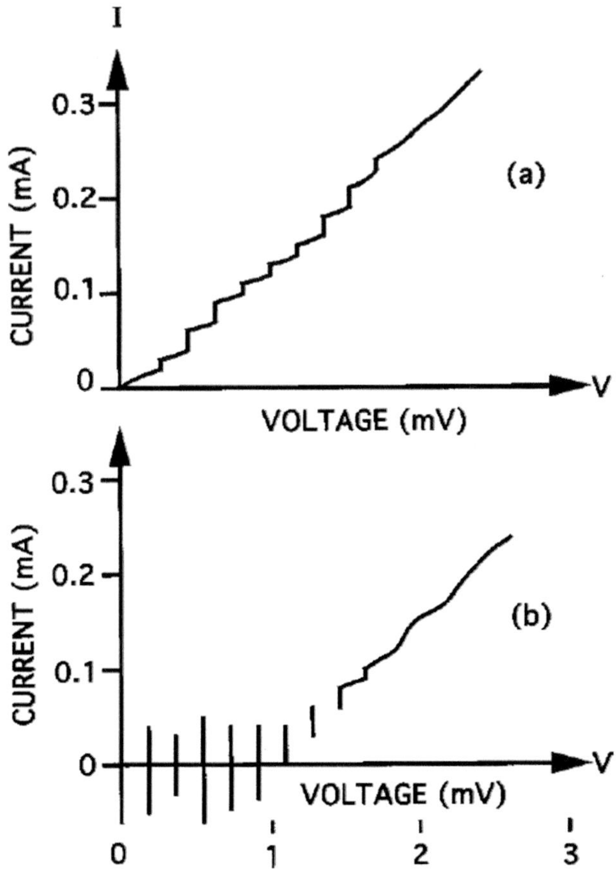

Figure 2.8 (a) Shapiro steps for a non-hysteretic, low β_c, junction. (b) Zero-crossing Shapiro steps, desired for voltage standard arrays, appear in high β_c high-capacitance junctions, as predicted by Levinson et al. [49]. After Hamilton [50].

2.5.2 Josephson Junction Arrays

A general feature of the Josephson arrays is design such that the linear array of junctions also serves as a stripline for the microwave propagation to the junctions. This is suggested in Fig. 2.9. In such striplines, according to Hamilton [50], the junction's capacitive impedance (about 1 mΩ) is so small compared to the stripline impedance, about 3 Ω, that each junction has only a minor effect

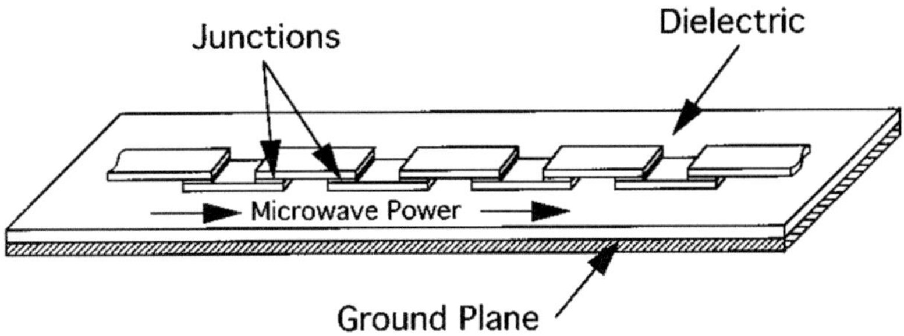

Figure 2.9 Schematic layout of series array of Josephson junctions on insulator above a ground plane, thus serving as a low-loss stripline propagating microwave radiation. After Hamilton [50].

on the propagation of the microwave power. Typically each junction absorbs 0.02–0.04% of the power propagating through it. Thus it is possible to have an array of several thousand junctions with a power uniformity of ± 1.5 dB. Strings of up to 4800 junctions have been used in this way [50].

According to the review of Hamilton in 2000, voltage standard systems of the general type demonstrated by Popel et al. were installed by year 2000 in about 50 standards laboratories around the world, and such DC systems were commercially available.

An alternative advanced design of the voltage standard arrays was described by Hamilton et al. in 1995 [52], with the important advantage that access times to individual voltage steps could be reduced to microseconds. Detailed specifications on the required junctions for the rapid-access standards were given, now allowing shunted junctions, rather than highly hysteretic junctions as were earlier advocated by Levinson et al., shown in Fig. 2.8b and incorporated in the working 1 and 10 V refractory standards described by Popel et al. (1990). The more rapid access time is clearly desirable for many applications.

Following the suggestion of [52] for rapid access arrays, SNS structures were demonstrated for use in the rapid access Josephson voltage standards, for example, Nb/PdAu/Nb junctions as reported by Benz [53]. These junctions, referred to as "trilayer SNS" devices, were deployed in arrays up to 8192 junctions, and were judged suitable for voltage standards using the alternative advanced design described by Hamilton et al. [52] in 1995 (and also described in the review by Hamilton [50]).

Returning to novel types of refractory Josephson-like junctions, suitable for the rapid access voltage standards of the alternative advanced design, stacked SNS structures using $MoSi_2$ as normal metal N, with inherently shunted $I(V)$ curves, have been studied by Chong et al. [54]. Five-layer structures of the type $Nb/MoSi_2/Nb/MoSi_2/Nb$, where the middle Nb layer varied in thickness from 5 nm to 80 nm, were fabricated, each unit considered as two stacked Josephson junctions. Doubly stacked junctions mean that the stripline need only be half as long if two junctions can be incorporated at each location marked "Junctions" in Fig. 2.9. In the voltage standard array format, series arrays of the doubly stacked devices were fabricated by Chong et al., containing 4100 stacks (8200 junctions in series).

Figure 2.10 shows the $I(V)$ of the series array of 4200 stacks (8400 junctions) in the absence of irradiation and with microwave

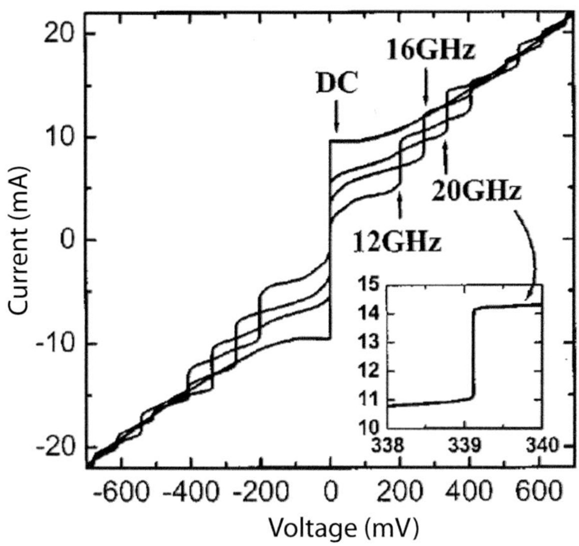

Figure 2.10 Characteristic $I(V)$ of series array of 4100 stacks (8200 Josephson junctions, with middle Nb electrode thickness 20 nm) under variable irradiation and simple current bias. Inset shows detail of $n = 1$ step at 339.2 mV with 20 GHz microwaves, stable in current range 11–14 mA; see text. After Chong et al. [54].

radiation at 12, 16, and 20 GHz and under simple current bias. The inset to the figure shows, at 339 mV across the whole array, the first step obtained over a current range of 3 mA with 20 GHz irradiation. The voltage across the whole array, for current in the range 11–14 mA, can be predicted as 8200 $\times$ 20 GHz/483.598 GHz/mV $=$ 339.12 mV, in agreement with the data in the inset of the figure. More elaborate bias conditions, the alternative advanced rapid access Josephson array design, as explained in [52], would be needed to quickly stabilize and display individual examples of the full complement of voltage steps inherent in such an array.

The present state of the sophisticated voltage standard field is described by Johannes Kohlmann in Chapter 10.

2.5.3 Josephson Devices for RSFQ Computing

The default devices for RSFQ computing have been the Nb/Al trilayer tunnel junctions with external shunt resistors, with some attention to devices based on NbN. Research has been reported aimed at internally shunted devices. One suggestion for a replacement for an externally shunted Josephson junction is shown in Fig. 2.11 [55]. This proposed structure is in the category of an SNS junction based on NbN, where the electrodes are $Nb_{1-x}Ti_xN$, an alloy with a slightly larger energy gap and transition temperature than NbN. The $Ta_{1-x}N_x$ middle layer is resistive depending on the composition

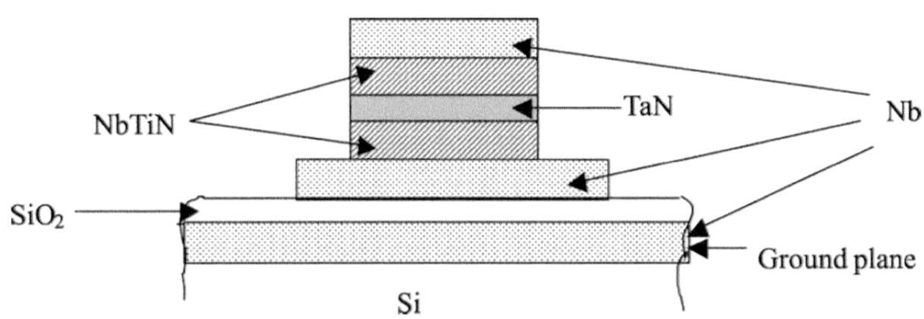

Figure 2.11 Pentalayer NbN-based SNS junction design for "drop-in replacement" of shunted Nb/Al trilayer tunnel junctions in Nb integrated circuits. Reproduced with permission from Van Duzer et al. [55]. This type of structure can be conveniently deposited on a conventional Si wafer using sequential sputtering methods.

chosen. This layer is made by sputtering from a Ta target in an atmosphere of argon with a small percentage of nitrogen. The authors display the 4.2K I(V) of a $2 \times 2\ \mu m^2$ junction of this type with $I_C = 2$ mA, $I_C R_N = 2.3$ mV, $J_C = 50$ kA/cm^2 and a slightly hysteretic characteristic. The design is intended for "drop-in replacement" of shunted Nb/Al trilayer junctions in Nb integrated circuits.

Internally shunted Josephson junctions with barriers tuned near the metal–insulator transition for RSFQ logic applications have been reported [29] in an extension of the above-mentioned work [55]. A full summary of work based on NbN is contained in Chapter 6 by J.-C. Villegier.

References

1. Gurvitch, M., Washington, M. A., and Huggins, H. A. (1983). High quality refractory Josephson tunnel junctions utilizing thin aluminum layers. *Appl. Phys. Lett.* **42**, 472–474.
2. Ambegaokar, V., and Baratoff, A. (1963). Tunneling between superconductors. *Phys. Rev. Lett.* **10**, 486–489 (and erratum *Phys. Rev. Lett.* **11**, 104).
3. Josephson, B. D. (1964). Coupled superconductors. *Rev. Mod. Phys.* **36**, 216–220.
4. Dynes, R. C., and Fulton, T. A. (1971). Supercurrent density distribution in Josephson junctions. *Phys. Rev. B* **3**, 3015–3023.
5. Wolf, E. L. (2012). *Electron Tunneling Spectroscopy*, 2nd Ed. (Oxford University Press, Oxford).
6. Anacker, W. (1980). Josephson computer technology: an IBM research project. *IBM J. Res. Dev.* **24**, 107–111.
7. Robinson, A. L. (1983). IBM drops superconducting computer project. *Science* **222**, 492–494.
8. Hayakawa, H. (1986). Josephson computer technology: recent advances on the road to superconducting computers include novel operating designs and memory circuits as well as stable and reliable devices made entirely from refractory materials. *Phys. Today* **39**, 46– 57.
9. Kroger, H., Smith, L. N., and Jillie, D.W. (1981). Selective niobium anodization for fabricating Josephson tunnel junctions. *Appl. Phys. Lett.* **39**, 280–282.

10. Kawakami, A., Wang, Z., and Miki, S. (2001). Fabrication and characterization of epitaxial NbN/MgO/NbN Josephson tunnel junctions. *J. Appl. Phys.* **90**, 4796.
11. Likharev, K. K., and Semenov, V. K. (1991). RSFQ logic/memory family: a new Josephson-junction technology for sub-terahertz-clock-frequency digital systems. *IEEE Trans. Appl. Supercond.* **1**, 3–28.
12. Okaz Ali, M., and Keesom, P. H. (1975). Specific heat and magnetization of the superconducting monoxides: NbO and TiO. *Phys. Rev. B* **12**, 4917–4928.
13. Halbritter, J. (1987). On the oxidation and on the superconductivity of niobium. *Appl. Phys. A* **43**, 1–28.
14. Bostock, J., Diadiuk, V., Cheung, W. N., Lo, K. H., Rose, R. M., and Mac Vicar, M. L. A. (1976). Does strong-coupling theory describe superconducting Nb?. *Phys. Rev. Lett.* **36**, 603–606.
15. Wolf, E. L., and Zasadzinski, J. (1977). On the observation of superconducting phonon structure in proximity tunneling. *Phys. Lett. A* **62**, 165–167.
16. Arnold, G. B., Zasadzinski, J., and Wolf, E. L. (1978). A resolution of the controversy on tunneling in Nb. *Phys. Lett. A* **69**, 136–138.
17. Wolf, E. L., Zasadzinski, J., Osmun, J. W., and Arnold, G. B. (1980). Proximity electron tunneling spectroscopy. I. Experiments on Nb. *J. Low Temp. Phys.* **40**, 19–49.
18. Arnold, G. B., Zasadzinski, J., Osmun, J. W., and Wolf, E. L. (1980). Proximity electron tunneling spectroscopy. II. Effects of the induced N-metal pair potential on calculated S-metal properties. *J. Low Temp. Phys.* **40**, 225–246.
19. Hauser, J. J., Bacon, D. D., and Haemmerle, W. H. (1966). Energy gap of V_3Si. *Phys. Rev.* **151**, 296–298.
20. Shen, L. Y. L. (1972). Design and performance of a high-speed economical ultrahigh vacuum chamber utilizing titanium sublimation. *Rev. Sci. Instrum.* **43**, 1301–1306.
21. Tao, H. J., Gibson, E. D., Verhoeven, J. D., and Wolf, E. L. (1984). Combined tunnelling and SEM channelling analysis of metallic bilayers Al(111) epitaxed on Nb(110). *Philos. Mag. B* **50**, L55–L61.
22. Wolf, E. L., Burnell, D. M., Khim, Z. G., and Noer, R. J. (1981). Proximity electron tunneling spectroscopy. IV. Electron-phonon coupling and superconductivity of tantalum. *J. Low Temp. Phys.* **44**, 89–118.
23. Rowell, J. M. (1966). Cryogenic supercurrent tunneling devices. *U. S. Patent* 3,281,609.

24. Rowell, J. M., Gurvitch, M., and Geerk, J. (1981). Modification of tunneling barriers on Nb by a few monolayers of Al. *Phys. Rev. B* **24**, 2278–2281.
25. Kwo, J., Wertheim, G. K., Gurvitch, M., and Buchanan, D. N. E. (1982). X-ray photoemission spectroscopy study of surface oxidation of Nb/Al overlayer structures. *Appl. Phys. Lett.* **40**, 675–677.
26. Berggren, K. K., Macedo, E. M., Feld, D. A., and Sage, J. P. (1999). Low T_c superconductive circuits fabricated on 150-mm-diameter wafers using a doubly planarized Nb/AlOx/Nb process. *IEEE Trans. Appl. Supercond.* **9**, 3271–3274.
27. Laibowitz, R. B., and Mayadas, A. F. (1972). Josephson junctions with Nb/Al composite electrodes. *Appl. Phys. Lett.* **20**, 254–256.
28. Morohashi, S., Shinoki, F., Shoji, A., Aoyagi, M., and Hayakawa, H. (1985). High quality Nb/Al-AlOx/Nb Josephson junction. *Appl. Phys. Lett.* **46**, 1179–1181.
29. Yu, L., Gandikota, R., Singh, R., Gu, L., Smith, D. J., Meng, I., Zeng, X., Van Duzer, T., Rowell, J. M., and Newman, N. (2006). Internally shunted Josephson junctions with barriers tuned near the metal-insulator transition for RSFQ logic applications. *Supercond. Sci. Technol.* **19**, 719–731.
30. Shoji, A., Aoyagi, M., Kosaka, S., Shinoki, F., and Hayakawa, H. (1985). Niobium nitride Josephson tunnel junctions with magnesium oxide barriers. *Appl. Phys. Lett.* **46**, 1098–1100.
31. Thouless, D. J. (1960). Strong-coupling limit of the theory of superconductivity. *Phys. Rev.* **117**, 1256–1260.
32. Fiske, M. D. (1964). Temperature and magnetic field dependences of the Josephson tunneling current. *Rev. Mod. Phys.* **36**, 221–222.
33. Kroger, H, Potter, C. M., and Jillie, D. W. (1978). Niobium Josephson junctions with doped amorphous silicon barriers. *IEEE Trans. Magn.* **MAG-15**, 488–489.
34. Rudman, D. A., and Beasley, M. R. (1980). Oxidized amorphous–silicon superconducting tunnel junction barriers. *Appl. Phys. Lett.* **36**, 1010–1013.
35. Shinoki, F., Shoji, A., Kosaki, S., Takada, S., and Hayakawa, H. (1980). Niobium nitride Josephson tunnel junctions with oxidized amorphous silicon barriers. *Appl. Phys. Lett.* **38**, 285–288.
36. Kerber, G. L., Cooper, J. E., Morris, R. S., Spargo, J. W., and Toth, A. G. (1989). NbN/MgO/NbN Josephson tunnel junctions fabricated on thin underlayers of MgO. *IEEE Trans. Magn.* **25**, 1294–1297.

37. Shoji, A. (1991). Fabrication of all-NbN Josephson tunnel junctions using single crystal NbN films for the base electrodes. *IEEE Trans. Magn.* **27**, 3184–3187.
38. Kawakami, A., Wang, Z., and Miki, S. (2001). Low-loss epitaxial NbN/MgO/NbN trilayers for THz applications. *IEEE Trans. Appl. Supercond.* **11**, 80–83.
39. Villegier, J. C., Bouat, S., Cavalier, P., Setzu, R., Espiau de Lamaestre, R., Jorel, C., Odier, P., Guillet, B., Mechin, L., Chauvat, J. P., and Ruterana, P. (2009). Epitaxial growth of sputtered ultra thin NbN layers and junctions on sapphire. *IEEE Trans. Appl. Supercond.* **19**, 3375–3378.
40. Stewart, W. C. (1968). Current-voltage characteristics of Josephson junctions. *Appl. Phys. Lett.* **12**, 277–280.
41. McCumber, D. E. (1968). Effect of ac impedance on dc voltage-current characteristics of superconductor weak-link junctions. *J. Appl. Phys.* **39**, 3113–3118.
42. Barnes, S. E. (1980). Validity of the resistively shunted Josephson junction model for small area superconductor-normal-superconductor junctions in a magnetic field. *J. Appl. Phys.* **51**, 6438–6440.
43. Kuprianov, M. Yu., and Lukichev, V. F. (1997). Influence of boundary transparency on critical current in "dirty" SS'S structures. *Sov. Phys.-JETP* **70**, 3603–3605.
44. Maezawa, M., and Shoji, A. (1997). Overdamped Josephson junctions with Nb/AlO_x/Al/AlO_x/Nb structures for integrated circuit applications. *Appl. Phys. Lett.* **70**, 3603–3605.
45. Senapati, K., and Barber, Z. H. (2009). Sidewall shunted overdamped NbN-MgO-NbN Josephson junctions. *Appl. Phys. Lett.* **94**, 173511-173511-3.
46. Sergeenkov, S. A. (2002). Electric field dependence of the thermal conductivity of a granular superconductor: giant field-induced effect predicted. *JETP Lett.* **76**, 170–174.
47. Shapiro, S. (1963). Josephson currents in superconducting tunneling: the effect of microwaves and other observations. *Phys. Rev. Lett.* **11**, 80–82.
48. Popel, R., Niemeyer, J., Fromknecht, R., Meier, W., and Grimm, L. (1990). 1- and 10- V series array Josephson voltage standards in Nb/Al_2O_3/Nb technology. *J. Appl. Phys.* **68**, 4294–4303.
49. Levinson, M. T., Chiao, R Y., Feldman, M. T., and Tucker, B. A. (1977). An inverse ac Josephson effect voltage standard. *Appl. Phys. Lett.* **31**, 776–778.

50. Hamilton, C. A. (2000). Josephson voltage standards. *Rev. Sci. Instrum.* **71**, 3611–3623.
51. Hamilton, C. A., Kautz, R. L., Lloyd, F. L., Steiner, R. L., and Field, B. F. (1987). The NBS Josephson array voltage standard. *IEEE Trans. Instrum. Meas.* **36**, 258–261.
52. Hamilton, C. A., Burroughs, C. J., and Kautz, R. L. (1995). Josephson D/A converter with fundamental accuracy. *IEEE Trans. Instrum. Meas.* **44**, 223–225.
53. Benz, S. P. (1995). Superconductor-normal-superconductor junctions for programmable voltage standards. *Appl. Phys. Lett.* **67**, 2714–2716.
54. Chong, Y., Dresselhaus, P. D., Benz, S. P., and Bonevich, J. E. (2003). Effect of interlayer thickness in $Nb/(MoSi_2/Nb)_N$ stacked Josephson junctions. *Appl. Phys. Lett.* **82**, 2467–2469.
55. Van Duzer, T., Zheng, L., Meng, X., Loyo, C., Whiteley, S. R., Yu, L., Newman N., Rowell, J. M., and Yoshikawa, N. (2002). Engineering issues in high-frequency RSFQ circuits. *Physica C* **372–376**, 1–6.

Chapter 3

Tunnel Junctions on Niobium Using Aluminum: Experiment

J. F. Zasadzinski
Department of Physics, 3101 S. Dearborn Street, Illinois Institute of Technology, Chicago, Illinois 60616, USA
zasadzinski@iit.edu

Modern Josephson electronics [1] (e.g., high-speed digital logic circuits, qubits, SQUIDS, photon detectors) utilize multilayer superconductor–insulator–superconductor (SIS) thin-film, tunnel junctions of the form Nb/Al-Al_2O_3/Nb or Nb/Al-Al_2O_3/Al/Nb where the self-limiting Al oxide serves as a high-quality, reproducible, insulating tunnel barrier. The Al overlayer approach has been the key advance to Josephson technology, allowing thermally and mechanically stable tunnel junctions with the refractory metal Nb, while at the same time mitigating the various problems associated with the native Nb oxides [2]. These problems, discussed in more detail later, include defects, conducting sub-oxides, and magnetism. The residual, non-oxidized Al, which wets the Nb, serves as a capping layer, protecting the Nb base layer surface from the deleterious effects of air exposure. Quantum tunneling involves electron states within a Fermi wavelength of the barrier (approximately a few angstroms) and thus it is the local density of electron states (DOS) at

Josephson Junctions: History, Devices, and Applications
Edited by Edward Wolf, Gerald Arnold, Michael Gurvitch, and John Zasadzinski

ISBN 978-981-4745-47-5 (Hardcover), 978-1-315-36452-0 (eBook)
www.panstanford.com

the surface of the Al film adjacent to the oxide layer that determines the properties of the Josephson junction. At the typical temperatures of operation the Al is in the normal state and the Nb/Al bilayer is of the superconductor/normal metal (S/N) type and therefore the junctions should more accurately be described as SNIS or SNINS. To support a Josephson current the Al must be induced into the superconducting state, allowing Cooper pair tunneling, and therefore the junction properties relevant for device performance depend critically on the detailed nature of the proximity effect between Nb and Al.

In this chapter we focus on the experimental quasiparticle (or single electron) characteristics from tunneling into the N side of such S/N bilayers which have elucidated the proximity effect. Specifically, junctions of the type Nb/Al-Al_2O_3/C are reviewed, where C is a convenient counterelectrode such as In, which can be switched easily between superconducting and normal states via a small magnetic field with minimal effects on the Nb superconductivity. These junctions, developed at Ames Laboratory (at Iowa State University) between 1976–1979 [3–6], were the first to reveal what might be called a technologically ideal proximity effect, i.e., the normal Al is induced into a superconducting state with an *effective* gap for quasiparticle excitations, $\Delta(E)$, and a Cooper pair density, n_S, that closely mimic the underlying Nb. The following chapter by G. B. Arnold focuses on the theoretical proximity effect model used in the analysis of these tunneling conductance data. The combined experimental and theoretical approach opened a field known as proximity electron tunneling spectroscopy (PETS). The PETS technique (in some cases just the theoretical analysis) was used to unveil quantitative, spectral details about the electron–phonon interaction in a variety of superconductors, e.g., Nb, V, Ta, NbZr, Nb_3Sn, V_3Ga, and normal metals including Al and Mg [7].

It might seem surprising that this monograph on Josephson junctions (devices that explicitly involve dynamics of Cooper pairs) would have two chapters dedicated primarily to the quasiparticle tunneling characteristics, including of non-Josephson SIN junctions. However, such junctions demonstrated the feasibility and utility of the Al overlayer method and are thus of historical significance. But more importantly, the quasiparticle tunneling conductance, dI/dV

vs. V, in SIN junctions, appropriately normalized, is a measure of the local, tunneling DOS, $N_T(E)$, and such spectral information, combined with the appropriate theoretical analysis, provides a detailed physical description not only of the proximity effect between Nb and the Al overlayer but of the tunnel junction itself. The relevant parameters, such as Al thickness and uniformity, electron mean-free-path, tunneling momentum selectivity, possible spin-flip channels, as well as N/S interface transmission and scattering, play a critical role in the performance and reproducibility of the Josephson junction. These physical quantities can be deduced from the quasiparticle dynamical conductance.

It should be noted that during the initial, exploratory period of proximity effect tunneling in the 1970s there was no way to know, a priori, whether this approach would yield satisfactory results, either for determination of the electron–phonon interaction in the superconductor or for Josephson junction devices. Early attempts at IBM in 1971 to fabricate Nb/Al Josephson junctions [8] did not seem promising. The Al thickness, d_N, required for "pinhole-free" overlayers was larger than expected; i.e., reported values were ~800 Å, the energy gap was much smaller than the bulk value of Nb, and subgap quasiparticle currents were relatively large (~25% of the normal state at $V = 2.0$ mV) and of unknown origin. The prevailing model of the proximity effect at that time was due to McMillan (1968), and it assumed, for practical reasons, that the interface between S and N was inherently a disordered, nonconducting layer [9]. For example, it might be imagined that even a brief exposure of a fresh Nb film to ambient pressure prior to Al deposition, under typical thin-film vacuum conditions, might lead to patches of poorly conducting sub-oxide on the surface. As a rule of thumb, at a pressure of 10^{-6} torr, a monolayer of ambient air strikes a surface each second. The McMillan model thus solved the proximity effect problem within perturbation theory, assuming the N/S interface allowed only a diffuse, weak, tunneling-type coupling between electron states in the S and N layers. As will be shown, had the Nb/Al fabrication process been constrained to be in the McMillan limit, it is safe to say that the present Josephson junction industry likely would not exist.

Here we review selected, representative spectra of the Ames Lab junctions [3–5] which showed, convincingly, that the Nb/Al bilayer could be fabricated in the "Arnold limit" where the N/S interface is clean and the coupling between N and S is strong. A principal assumption of the Arnold model (highlighted in Chapter 4), which allows an exact, one-dimensional solution, is that the N/S interface is specularly transmitting and reflecting. This leads to an important effect that is quite relevant for Josephson junction circuits. Quasiparticle excitations above the induced gap in the N layer, $\Delta_N \ll \Delta_S$, are prohibited by coherent Andreev reflections up to a bound state energy, E_0, which is very close to Δ_S. This provides low subgap quasiparticle currents, which, for Josephson junctions, are a critical performance criterion. The early Ames Lab junctions provided the necessary "proof-of-principle" and laid the groundwork for the successful extension of the Al overlayer method to all-Nb, SNIS, Josephson junctions developed by Gurvitch and Rowell (see Chapter 5). A focus here is on experimental results that indicate the presence of this Andreev bound state. Another result of the Arnold model is that in the limit where d_N is relatively small, i.e., less than characteristic coherence lengths (ξ_S, ξ_N). All terms containing $\Delta_N(E)$ either have a very small prefactor or cancel out exactly for $E \gg \Delta_S(E)$. Thus in the region where one expects to observe phonon structure ($E = 10$–40 meV) due to strong coupling effects, as described by Eliashberg theory and first observed in Pb [10], it will be shown that the influence of Al phonons can be made negligible by decreasing d_N, again a confirmation of the Arnold model. This was another discovery of the Ames Lab proximity effect studies, that Al overlayers as thin as $\sim$30 Å could be made continuous and pinhole free. The ability of Al to wet the Nb surface has been verified by high-resolution TEM studies [11].

In the final section of this chapter we return to the issue of the Nb oxides and their deleterious effects on the quasiparticle tunneling conductance, showing some recent results. While the Al overlayer mitigates such problems in the Josephson junction, there are still free Nb surfaces in Josephson devices, e.g., SQUID loops, which may be affected by the Nb oxides. Also, for SNIS trilayer Josephson junctions, the top Nb film is in contact with the Al oxide and any diffusion of oxygen into the Nb may lead to sub-oxide formation with potential consequences. There is also evidence that Nb oxides may be affecting large-scale superconducting devices such as superconducting radiofrequency (SRF) cavities. Thus, there has been a resurgence of interest in the properties of the Nb oxides. A tunnel junction is a natural way to probe this layer. We report key results of extensive point contact tunneling (PCT) studies of SIN-type junctions on Nb using the native oxide as the barrier and a gold tip [12, 13]. These junctions reveal that Nb oxides are sometimes paramagnetic, leading to zero bias conductance peaks in SIN tunnel junctions due to spin-flip, Kondo tunneling. Such magnetic moments may be the source of the two-level systems that produce $1/f$ noise in SQUIDS as well as enhanced decoherence in superconducting qubits.

3.1 Fabrication Methods

The approach at Ames Lab was to attempt to make Nb/Al bilayers with a near-perfect interface, i.e., atomically flat, abrupt, and clean, without any oxidation of the Nb. Commercial foils of the highest-purity grade of Nb were resistively heated to the melting point in a specially built, ultrahigh vacuum chamber that maintained a pressure of low 10^{-9} torr during annealing [5]. A large fraction of the stainless steel chamber was immersed in liquid nitrogen, preventing outgassing during the foil annealing, and vacuum was maintained by a titanium sublimation pump coating the large surface area of the chamber. After annealing, the Nb foils exhibited an RRR of $\sim$385. Significant recrystallization of the Nb foil occurred, with grains up to a few millimeters in diameter, a more recent example of which is shown in the inset of Fig. 3.2b. As revealed by $\Theta-2\Theta$ x-ray diffraction, high-symmetry orientations (100), (110), and (111) were found but also other orientations, e.g., (210), suggesting no clear evidence of preferential orientation. Given the large grains, tunnel junctions were often confined to a single-crystal region, allowing an investigation of gap anisotropy in Nb. No clear evidence of gap anisotropy was observed.

After the foil broke, interrupting the current, the external Cu leads were quickly cooled by contact to attached Cu braid in a liquid nitrogen bath. Base pressure quickly returned to low 10^{-9} torr or mid 10^{-10} torr. Vapor deposition of high-purity Al was achieved with a nearby Ti wire basket boat. After a reasonable deposition rate (of approximately a few Å/s) was established, as monitored by a calibrated quartz crystal microbalance, a shutter, connected to a linear motion feedthrough, was opened in sequential steps. In a typical run, various Al thicknesses from 20 to 250 Å were obtained over discrete, micrometer-size sections of the Nb foil. The chamber was first vented with dry, high-purity N_2 gas, and then the foils were removed and exposed to laboratory air for several hours, oxidizing the Al surface. The discrete regions of the Nb foil coated with Al layers of varying thickness were visible by optical inspection. The foil, after masking with collodion, could be oriented onto a metal mask, and coated with an In cross-strip counterelectrode at a base pressure of 10^{-7} torr, in a way that at least two junctions could be made on each Al thickness. Up to 10 junctions were obtained on each of the separate sections of the broken Nb foil. In this way, Nb/Al junctions, prepared in identical fashion, could be studied with a single adjusted variable, the Al thickness. Current–voltage (I–V) measurements were obtained by d.c. methods, while dynamic resistance, $\mathrm{d}V/\mathrm{d}I$ vs. V, was obtained by standard a.c. harmonic detection methods, utilizing a Kelvin bridge and lock-in amplifier. Typical junction area was $\sim$0.5 mm^2 with d.c. resistance $\sim$1–100 Ω.

3.2 Gap Region Spectra of the Ames Lab Nb/Al Tunnel Junctions

Before reviewing the S/N/I/C data of the Ames Lab junctions, it is worth discussing the predicted tunneling DOS, $N_T(E)$, from the McMillan and Arnold models, which span the extremes of weak and strong N/S coupling, respectively. The basic geometry of the N/S bilayer is shown in the inset to Fig. 3.1a, where an assumed abrupt change in the pairing gap parameter is indicated. Leakage of Cooper pairs from S leads to superconducting off-diagonal order and resulting pair density function, $F(x)$, in the N layer and an induced gap parameter $\Delta_N(x) = \lambda_N F(x)$, where λ_N is the electron–phonon coupling strength in the N layer. For $d_N \ll (\xi_S, \xi_N)$ it can be assumed that Δ_N is constant, and enhanced from its bulk value; however, both gap parameters, Δ_S and Δ_N, must be determined self-consistently. For the McMillan model, the presumed delta function tunnel barrier at the N/S interface leads to a discontinuity in $F(x)$ and thus the induced gap $\Delta_N(x)$ will in general be smaller than obtained in the Arnold model. Quasiparticle excitations above Δ_N are inside a potential well of height $\Delta_S - \Delta_N$ on one side and a normal barrier given either by the work function of the N layer for a vacuum interface or the tunnel barrier in a SNIC junction on the other side. At the vacuum interface, in addition to the reflection off the real potential, there can be an Andreev reflection off the superconducting step potential, Δ_N. This leads to two independent types of oscillatory terms mixing the quasiparticle wave functions [5].

In the Arnold model, only those excitations are allowed which satisfy the boundary conditions imposed by the phase-preserving reflections off the two potential steps. This leads to coherent bound states, similar to the standard quantum well problem; however, the reflections off the superconducting potential step involve Andreev processes whereby an electron-like quasiparticle gets reflected back as hole-like excitation. For thin N layers there is only one Andreev bound state at energy $E_0 \cong \Delta_S$. In the McMillan model the N/S interface barrier provides diffuse, incoherent scattering. Thus there are no coherent bound states and all quasiparticle excitations above $\Delta_N(E)$ are allowed. The McMillan DOS exhibits a pronounced BCS singularity at $\Delta_N \ll \Delta_S$ and this type of Nb/Al proximity junction would be unsuitable for Josephson devices. The differences in $N(E)$ between the two models are considerable, as shown in Fig. 3.1.

It turned out that the Ames Lab junctions were completely described by the Arnold theory, not only in the gap region but also in the higher-bias phonon region. An example is shown in Fig. 3.2, where one of the junctions with $d_N = 27$ Å is presented from the very first processed Nb foil. The I–V characteristic reveals a single gap feature near 2.0 mV when the In counterelectrode is superconducting ($\Delta_{In} = 0.53$ meV) consistent with a gap parameter close to that of bulk Nb. Subgap currents with In normal are small, leading to a d.c. conductance I/V at zero bias $< 0.2\%$ of the normal state value observed for $V > 2$ mV. The self-consistent solution [6] leads to $\Delta_N = 0.78$ meV, but there is no direct observation

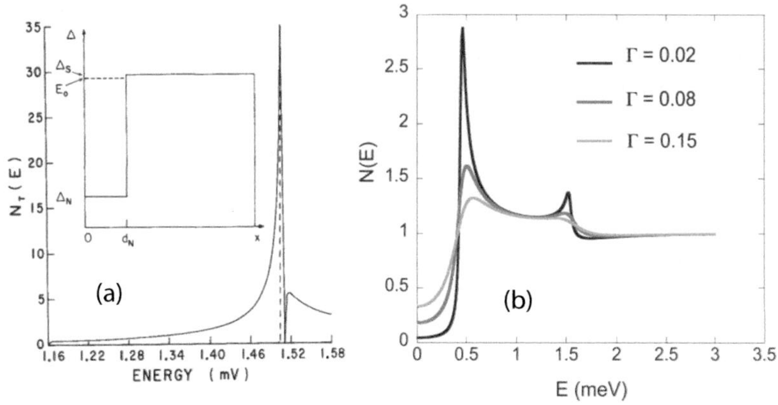

Figure 3.1 (a) Arnold DOS (see Chapter 4, Fig. 4.3) and (b) McMillan DOS for tunneling into the N side of an N/S proximity bilayer with phenomenological quasiparticle scattering rate term, Γ. In both cases the S layer gap parameter chosen is close to that of Nb ($\Delta_S = 1.51$ meV). For the Arnold DOS the induced gap parameter, $\Delta_N = 0.78$ meV (determined self-consistently), is not observed. Only the Andreev bound state E_0, indicated as dashed line in (a) satisfies the boundary conditions inside the potential well of the N layer. For the McMillan DOS in (b) the induced gap Δ_N is chosen to be 0.46 meV, where a pronounced BCS singularity is observed. In (a) the Arnold model displays the bound state as a vertical dashed line, essentially a delta function peak in DOS at E_0. The is what would be found in the tunneling DOS, $N_T(E)$, assuming 1-d tunneling perpendicular to the barrier. With thermal smearing the bound state, being so close to Δ_S, appears similar to a BCS singularity. If the tunnel junction supports random tunneling then a continuum of bound states is observed below E_0 (solid line).

of this Al gap, a consequence of the coherent Andreev reflections, as explained above. Figure 3.2b shows the normalized dynamical conductance dI/dV, compared to a BCS model and the Arnold model. The reasonable agreement with a BCS model with $\Delta_N = 1.49$ meV shows that with small d_N and thermal smearing, there is little difference between the Arnold theory and BCS.

More striking evidence of the Andreev bound state is found for thicker N layers where E_0 is farther below Δ_S. The resulting gap in $N_T(E)$ between E_0 and Δ_S leads to a dip feature in the tunneling conductance, dropping below the normal state conductance, a feature not found in the McMillan DOS. To minimize the effects

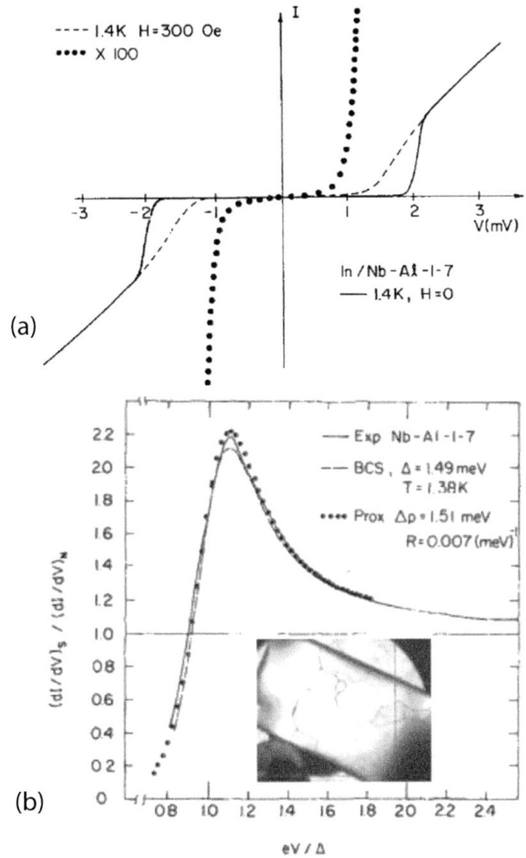

Figure 3.2 (a) I–V characteristics of Nb/Al-Al_2O_3/In junction with the In in the superconducting ($H = 0$) (solid line) or normal state ($H = 300$ Oe) (dashed line). (b) Normalized dynamic conductance (solid line) compared to BCS and Arnold fit. Inset of (b) shows typical recrystallization pattern of UHV annealed foils with grains up to a few mm diameter. Reproduced with permission from Cao [16].

of thermal smearing and observe the dip feature, it is required that the counterelectrode be superconducting. In this case the BCS singularity of the counterelectrode sharpens the thermal convolution function arising from Fermi factors [7]. In Fig. 3.3 are shown normalized dynamic resistance data for a V/Al-Al_2O_3/In junction, where the Al thickness is 230 Å [14]. A pronounced dip

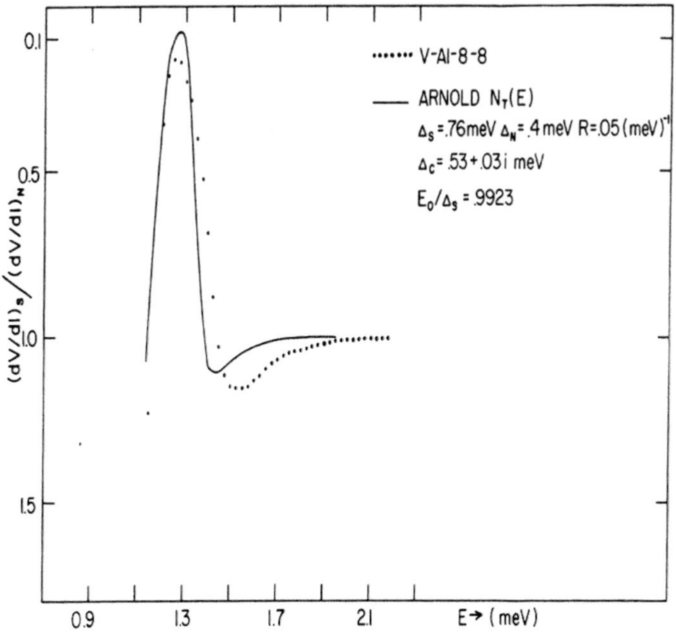

Figure 3.3 Normalized dynamic resistance dV/dI of V/Al-Al_2O_3/In junction (dots) and Arnold fit (solid line). The spectral dip drops below unity conductance, indicating a gap in the DOS, $N(E)$, between E_0 and Δ_S.

feature (reduced conductance) is observed and a reasonable fit using the Arnold theory can be found with the parameters indicated in the figure legend. Here $R = \frac{2d_N}{\hbar v^*_{FN}}$ where v^*_{FN} is the renormalized Fermi velocity in the N layer. While a full examination of the parameter space in the Arnold theory was not undertaken [14], it is nevertheless clear that the model successfully accounts for a dip feature of characteristic size and shape that drops below unity in the normalized conductance. Here a small imaginary term to the gap parameter in the superconducting In counterelectrode has been used to account for any gap inhomogeneity.

The presence of a gap in the Arnold DOS between E_0 and Δ_S has an effect on the I–V quasiparticle characteristics of Nb trilayer Josephson junctions, giving rise to the well-known "knee" feature, most easily observed at very low temperatures [15], as displayed in Fig. 3.4. This can be explained as follows. The jump in current close

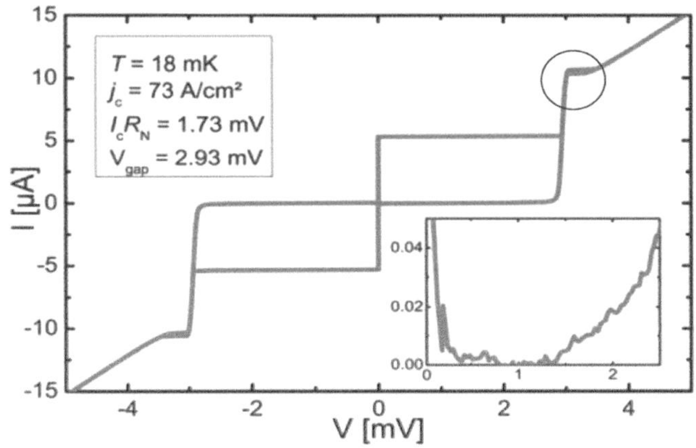

Figure 3.4 Current–voltage (I–V) characteristic of $Nb/Al/Al_2O_3/Nb$ Josephson junction [15] displaying the characteristic "knee" feature indicated by the circled region. Inset shows subgap currents of $\sim$0.02 μA at $V = 2.0$ mV, significantly below the 10 μA current near the gap voltage of $\sim$3 mV.

to $2\Delta_{Nb}$ (often referred to as the Nb gap) corresponds to a voltage where the two $N(E)$ singularities of each Nb electrode, one for filled quasiparticle states and the other for empty states, are matched in energy. An increase in voltage will correspond to a relative shifting of the $N(E)$ peaks, such that a peak of one electrode is matched to the gap in the $N(E)$ between E_0 and Δ_S in the other electrode. This gives rise to a fixed, or even decreasing current (negative dI/dV), with increasing voltage, with an associated hysteresis. The ability of the Arnold model to explain this commonly observed phenomenon in Nb trilayer Josephson junctions demonstrates that it is the appropriate model for junction analysis.

At this point a reasonable question to ask is, What is the quasiparticle gap, generally referred to as 2Δ, in SIS proximity Josephson junctions? Figure 3.4 indicates that $V_{gap} = 2.93$ mV, determined from the sharp upturn in current, and the relevant value for this particular Josephson device. But this is below the expected value based on the bulk gap of Nb, which should be closer to 3.1 mV. In the Arnold model, there are three parameters relevant to this issue: the bound state, E_0, the superconducting gap parameter at the N/S interface, Δ_S, and the bulk energy gap parameter of

Nb, Δ_S^{bulk}. The bound state, $E_0 < \Delta_S$, Δ_S^{bulk} is the minimum energy for quasiparticle excitations and that corresponds to the $V_{\text{gap}}/2$ of Fig. 3.4, and explains why it is reduced from the expected gap. The value of Δ_S can be obtained directly from fitting of the dynamic conductance using the Arnold theory and is in the range where the "knee" is found, above 3 mV in Fig. 3.4. However, Δ_S is reduced from the bulk value due to the presence of the N layer.

Using a straightforward correction, the theory can predict the expected bulk value of the gap of Nb. This correction has the form $\Delta_S^{\text{bulk}} \cong \Delta_S(1 + \pi R \Delta_S)$ For the Ames Lab Nb/Al junctions [5] this correction leads to a bulk gap parameter of Nb, $\Delta^{\text{bulk}} = 1.57$–1.61 meV.

3.3 High-Bias Spectra of the Ames Lab Nb/Al Tunnel Junctions

Another success of the Arnold proximity effect model is its ability to quantitatively fit the measured high-bias tunneling conductance where strong-coupling effects (phonon structures) from both gap functions, $\Delta_S(E)$ and $\Delta_N(E)$, enter the DOS. Details are presented elsewhere [5, 6]. Here we are interested in a particular limit, relatively small d_N, where an important result is obtained. For typical phonon energies (10–40 meV), one is in the limit where $E \gg \Delta_S(\text{E})$ and the Arnold $N_T(\text{E})$ becomes

$$N(E) = 1 + \frac{1}{2}\text{Re}\frac{\Delta_N^2(E)}{E^2} + \frac{1}{2}\text{Re}\frac{[\Delta_S(E) - \Delta_N(E)]^2}{E^2}\exp(2i\Delta K d_N)$$
$$+\text{Re}\Delta_N(E)\frac{[\Delta_S(E) - \Delta_N(E)]}{E^2}\exp(i\Delta K d_N) \qquad (3.1)$$

where $\Delta K d_N = \frac{2d_N E}{\hbar v_{\text{FN}}^*} + i d_N/l$.

This form highlights the two oscillatory terms arising from Andreev reflections off the potential step, $\Delta_S(E)$–$\Delta_N(E)$, described earlier. The first term of $\Delta K d_N$ can be written as *RE*, where, as shown in Fig. 3.2b, *R* is relatively small for $d_N = 27$ Å, such that even for quasiparticle energies $E = 50$ meV, the term RE $\approx 0.35 \ll \pi$. Thus the Re part of $\Delta K d_N$ leads to $\exp(i\Delta K d_N) \cong 1$. The Im part is d_N/l where l is the quasiparticle mean free path in N. For clean N layers, d_N/l is also $\ll 1$. Taking the leading terms of the expansion, one finds that terms containing Δ_N either cancel exactly or have small prefactors of the form [1–exp(–d/l)]. Thus in the small d_N limit, Eq. 3.1 becomes

$$N(E) = 1 + \frac{1}{2}\mathrm{Re}\frac{\Delta_S^2(E)}{E^2}e^{-d/l} \tag{3.2}$$

Equation 3.2 implies that for thin N layers, the tunneling DOS in the N layer at characteristic phonon energies is that of the underlying superconductor, scaled by $e^{-d/l}$, a term close to unity. Equation 3.2 is the expansion of the strong coupling version of the BCS DOS [10] in the limit $d/l = 0$. Evidence for this effect has been shown in Fig. 2.4 for the second derivative spectrum d^2V/dI^2 showing minimal contribution from Al phonons. In Fig. 3.5a we show the high-bias reduced normalized conductance for the same junction shown in Fig. 3.1.

The fitting of the data using Eq. 3.2 and the McMillan inversion procedure leads to the dashed line in Fig. 3.5a and the resulting $\alpha^2F(\omega)$ for Nb is shown in Fig. 3.5b. This indicates clearly that the principle structures observed are due to the transverse and longitudinal phonon modes of Nb. Corrections to the fit can be obtained to refine this analysis using the calculated $\Delta_N(E)$ [6]. This confirms the suitability of the Arnold theory to explain the entire tunneling conductance spectrum.

3.4 Point Contact Tunneling Studies of Nb

While the Al overlayer method has played a pivotal role in the development of Josephson junction technology, there is still a need to understand in more detail the problems of the native Nb oxide layer, as it enters the free Nb surfaces in such devices, including SQUID loops. There is also evidence that the Nb oxides are playing a role in the RF surface impedance of superconducting radiofrequency (SRF) cavities [12, 13] and thus may be related to decoherence in superconducting qubits.

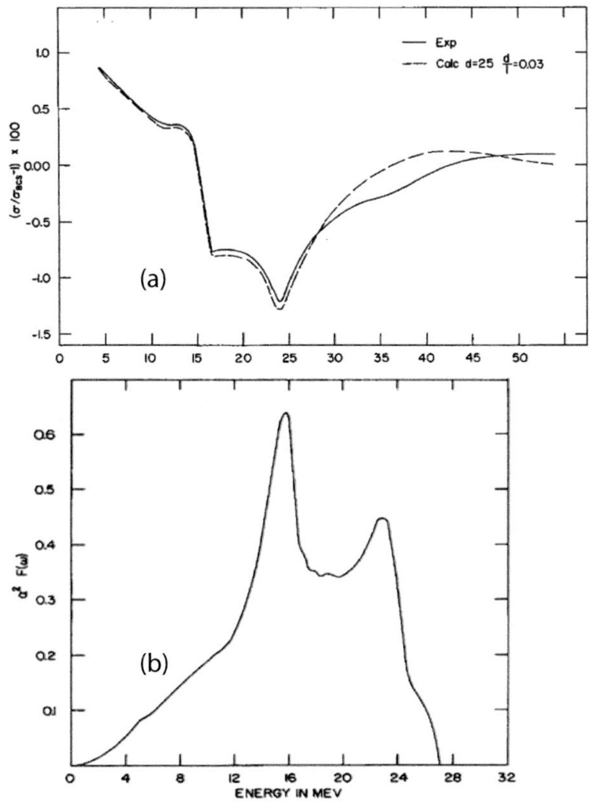

Figure 3.5 (a) Reduced normalized conductance at high bias (solid line) along with Arnold model fit (dashed line) using $d_N = 25$ Å and $d/l = 0.03$. Weak phonon feature between 35 and 40 meV arising from the Al acoustic phonon mode peak as a strong coupling effect in $\Delta_N(E)$ has been ignored in the initial fitting. (b) Resulting $\alpha^2F(\omega)$ for Nb from the fitting in (a).

A pristine Nb surface exposed to air develops a complex set of oxides including NbO, NbO_2, and Nb_2O_5. Each of these oxides is thermodynamically stable with substantial off-stoichiometry. The topmost Nb_2O_5 layer is an ordinary band insulator with an energy gap > 4 eV in bulk. However, tunneling studies utilizing the native Nb oxide typically display barrier heights of 300–400 mV and such a low value may originate in the amorphous, defective nature of this

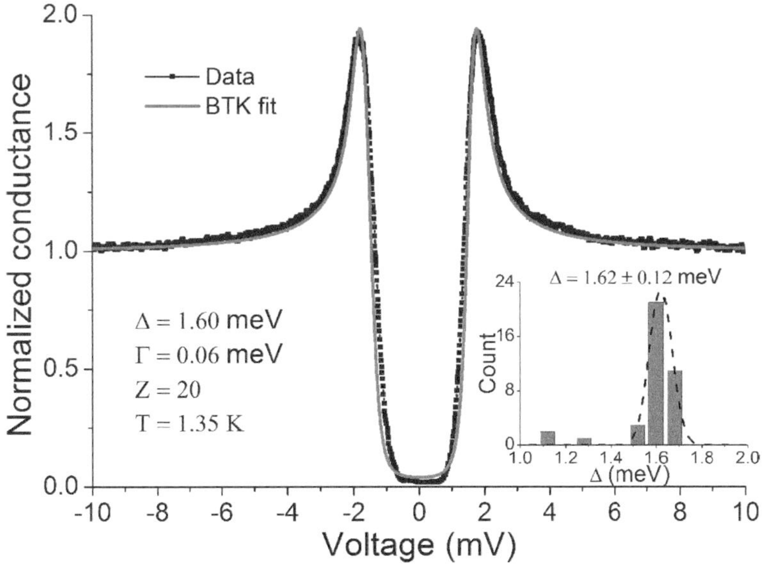

Figure 3.6 PCT normalized conductance for a particular region of a Nb SRF cavity cold-spot. The fit indicates a near-ideal BCS DOS. The fit uses the BTK model, and the high value of the barrier strength $Z = 20$ indicates the Nb oxide is a good insulator. Inset shows a statistical distribution of gap values peaked near 1.6 meV. Reproduced with permission from Cao [16].

oxide [2]. Point contact tunneling (PCT) utilizing a Au tip offers a chance to study junctions of area $<1\ \mu m^2$ over a wide area of a Nb surface utilizing the native oxide. Such surface mapping has been particularly useful for the study of processed Nb samples and SRF cavity cutouts [14]. By varying the tip pressure one can change from tunneling to ohmic regimes and thus the analysis utilizes the theory of Blonder–Tinkham–Klapwijk (BTK) (Sections 3.8 and 5.2 of [7]). An example of tunnel junction from a cold spot of an SRF cavity, i.e., a region with no excess RF losses, is shown in Fig. 3.6.

The normalized dynamic conductance data display a near-ideal fit to a BCS DOS at 1.35 K with a $\Delta_S = 1.60$ meV consistent with a low RF dissipation region. This gap value is consistent with Δ_S^{bulk} obtained from the Nb/Al proximity junctions suggesting a minimal native proximity layer. BTK analysis leads to $Z = 20$, a high barrier strength. Such behavior indicates that locally the Nb oxide may be sufficiently insulating and thick to provide a suitable tunnel barrier. The processing of Nb SRF cavities involves deep drawing, chemical etching, and annealing. The final step of the standard processing is a mild bake at 120 C for 24–48 hours. This final step is similar to some of the early tunneling studies of single-crystal Nb by Bostock and MacVicar (see [4] and references therein). However, while gap region spectra may not display any obvious evidence of a proximity layer, Eq. 3.2 shows that the phonon structure can be reduced even with a small value of d/l and this feature of the Arnold theory helped resolve a controversy about the electron–phonon interaction in Nb [4].

The spectra on SRF cavity cold spot regions, e.g., Fig. 3.6, are not typical of processed Nb. Spectra from hot spot regions of an SRF cavity generally displayed smaller gaps and broader gap features [12, 14]. Cavity grade Nb crystals subjected only to the electropolishing and mild baking processing steps revealed considerably broader spectra, and the gap region analysis indicated that magnetic pair breaking, including Shiba states, might be responsible [13]. More striking evidence of magnetism on the surface of hot spot regions is shown in Fig. 3.7. Numerous spectra showed a zero bias conductance peak (Fig. 3.7a) that displayed a logarithmic T-dependcnce (Fig. 3.7b) and Zeeman splitting in a magnetic field (Fig. 3.7c) that identified this feature with a Kondo tunneling channel [16]. This clearly indicates the presence of localized magnetic moments in the Nb oxide layer.

The origin of magnetic moments Nb oxide is likely due to oxygen vacancies. In the work of Cava et al. [17], it was shown that the introduction of 3% oxygen vacancies into Nb_2O_5 changes this material from an insulator into a conductor with Curie–Weiss ($1/T$) magnetic susceptibility. This is not surprising. The pentavalent Nb in Nb_2O_5 has filled core levels and an empty d-band and is nonmagnetic. An oxygen vacancy introduces two electrons into the formula unit allowing a d-orbital and spin moment to the Nb. The presence of such moments in an oxide region with higher conductivity means that Cooper pairs can communicate with such moments and this leads to pairbreaking and quasiparticle states near the Fermi level even at $T = 0$ and intrinsic RF losses [12].

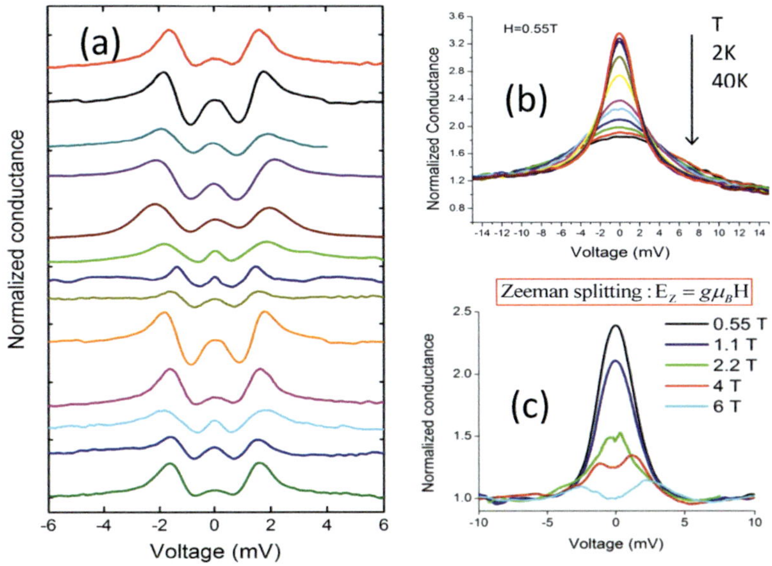

Figure 3.7 (a) Representative PCT spectra from SRF cavity hot spot region revealing a zero bias conductance peak (ZBCP). (b) T dependence of ZBCP, which is logarithmic in $1/T$. Reproduced with permission from Proslier et al. [12]. (c) Zeeman splitting of ZBCP indicative of a Kondo tunneling channel. Reproduced with permission from Cao [16].

More recent studies of Nb tunnel junction SQUIDS by McDermott et al. revealed the presence of magnetic moments [18]. They characterized the temperature dependence of the flux threading dc SQUIDs cooled to millikelvin temperatures. The flux increases as $1/T$ as temperature is lowered; moreover, the flux change is proportional to the density of trapped vortices. The data are compatible with the thermal polarization of surface spins in the trapped fields of the vortices. In the absence of trapped flux, they observe evidence of spin-glass freezing at low temperature (Fig. 3.8). These results suggest an explanation for the longstanding problem of universal $1/f$ flux noise in SQUIDs and superconducting qubits as due to a two-level system originating from the localized moment. The PCT measurements strongly suggest such moments originate in the defective Nb oxide.

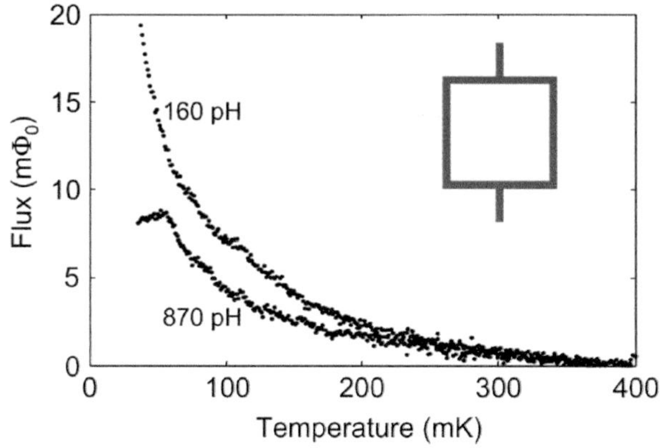

Figure 3.8 Temperature dependence of the flux threading narrow (2 μm) linewidth Nb/Al oxide/Nb SQUIDs with inductances 160 and 870 pH. SQUID loop geometry is shown in the inset. Curie–Weiss behavior indicates surface spins [18].

References

1. Fourie, C. (2014). Superconducting electronics. *Wiley Encyclopedia of Electrical and Electronics Engineering*. pp. 1–18.
2. Halbritter, J. (1987). On the oxidation and on the superconductivity of niobium. *Appl. Phys. A* **43**, 1–28.
3. Wolf, E. L., and Zasadzinski, J. (1977). On the observation of superconducting phonon structure in proximity tunneling. *Phys. Lett. A* **62**, 165–167.
4. Arnold, G. B., Zasadzinski, J., and Wolf, E. L. (1978). A resolution of the controversy on tunneling in Nb. *Phys. Lett. A* **69**, 136–138.
5. Wolf, E. L., Zasadzinski, J., Osmun, J. W., and Arnold, G. B. (1980). Proximity electron tunneling spectroscopy. I. Experiments on Nb. *J. Low Temp. Phys.* **40**, 19–49.
6. Arnold, G. B., Zasadzinski, J., Osmun, J. W., and Wolf, E. L. (1980). Proximity electron tunneling spectroscopy. II. Effects of the induced N-metal pair potential on calculated S-metal properties. *J. Low Temp. Phys.* **40**, 225–246.
7. Wolf, E. L. (2012). *Electron Tunneling Spectroscopy,* 2nd Ed. (Oxford University Press, Oxford).

8. Laibowitz, R. B., and Mayadas, A. F. (1972). Josephson junctions with Nb/Al composite electrodes. *Appl. Phys. Lett.* **20**, 254–256.
9. McMillan, W. L. (1968). Tunneling model of the superconducting proximity effect. *Phys. Rev.* **175**, 537–542.
10. McMillan, W. L., and Rowell, J. M. (1969). Tunneling and strong-coupling superconductivity, in *Superconductivity*, ed. Parks, R. D. (Marcel Dekker, New York), Vol. I, pp. 561–614.
11. Imamura, T., and Hasuo, S. (1991). Cross sectional transmission electron microscopy observation of Nb/AlO$_x$-Al/Nb Josephson junctions. *Appl. Phys. Lett.* **58**, 645–647.
12. Proslier, T., Kharitonov, M., Pelin, M., Zasadzinski, J., and Ciovati, G. (2011). Evidence of surface paramagnetism in niobium and consequences for the superconducting cavity surface impedance. *IEEE Trans. Appl. Supercond.* **21**, 2619–2622.
13. Proslier, Th., Zasadzinski, J. F., Cooley, L., Antoine, C., Moore, J., Norem, J., Pellin, M., and Gray, K. E. (2008). Tunneling study of cavity grade Nb: possible magnetic scattering at the surface. *Appl. Phys. Lett.* **92**, 212505–212508.
14. Zasadzinski, J. F. (1980). Superconducting tunneling study of V and V_3Ga, Iowa State University Retrospective Theses and Dissertations. Paper 7361.
15. Kaiser, Ch., Bauch, T., Lombardi, F., and Siegel, M. (2011). Quantum phase dynamics in an LC shunted Josephson junction. *J. Appl. Phys.* **109**, 093915-1–093915-6.
16. Cao, C. (2014). Surface characterization of niobium for superconducting RF cavities, Chicago, IL: Illinois Institute of Technology, PhD thesis, THESIS.ST3h792eb.
17. Cava, R. J., Batlogg, B., Krajewski, J. J., Poulsen, H. F., Gammel, P., Peck, W. F., and Rupp, L. W. (1991). Electrical and magnetic properties of Nb_2O_5 crystallographic shear structures. *Phys. Rev. B* **44**, 6973–6981.
18. Sendelbach, S., Hover, D., Kittel, A., Muck, M., Martinis, J. M., and McDermott, R. (2008). Magnetism in SQUIDS at Millikelvin temperatures. *Phys. Rev. Lett.* **100**, 227006–227009.

Chapter 4

Tunnel Junctions on Niobium Using Aluminum: Theory

Gerald B. Arnold
Department of Physics, University of Colorado, Boulder, Colorado, 80309, USA
gbarnold46@gmail.com

As explained in the previous chapter, superconducting tunnel junctions which utilized the oxidation of niobium to form a tunnel barrier presented a variety of problems. When an aluminum overlayer was evaporated onto the niobium and that layer was oxidized before evaporating the counter-electrode, however, the tunneling characteristics were markedly improved. Ideally, one would like to oxidize all of the aluminum, thereby attaining a good S'IS tunnel junction. In practice, it was found that some aluminum was not oxidized, leading to an S'INS junction, where N signifies the thin layer of aluminum. Due to the superconducting proximity effect, superconductivity is induced in the aluminum by the S layer of niobium. This leads to the theoretical problem of describing the tunneling characteristics of S'INS junctions wherein a thin layer of N metal is induced to be superconducting by a thick layer of superconducting S metal. In this chapter, I will summarize the theoretical facts concerning these tunnel junctions.

Josephson Junctions: History, Devices, and Applications
Edited by Edward Wolf, Gerald Arnold, Michael Gurvitch, and John Zasadzinski

ISBN 978-981-4745-47-5 (Hardcover), 978-1-315-36452-0 (eBook)
www.panstanford.com

Much of the research which I report on in this chapter was done in connection with the development of proximity electron spectroscopy [1], which is concerned with extracting detailed information on the electron–phonon interaction in the N and S layers of an N'INS junction. Of primary interest in that research was the energy dependence of the gap functions arising from this interaction, which produces the so-called phonon structure in the tunneling density of states. In this work, I focus instead on those aspects of NS bilayers which affect the gap region in the tunneling density of states, the region most important for Josephson tunneling.

4.1 Proxity Effect in Thin N Layers on Thick S Layers

Consider a thin (in a sense which will be defined) N metal layer in ideal (epitaxial) contact with a semi-infinite S metal layer. In this section, I will also assume that there is negligible elastic scattering in the N metal. Later on, I will discuss the effects of elastic scattering in the N metal. The N metal will be presumed to be thin if its thickness is very much less than the characteristic length scale for spatial variations in the induced N metal pair density. If this were a bulk N metal at zero temperature, then this length scale would be $\xi_N = \hbar v_F/\pi \Delta_N^{(bulk)}$, the BCS coherence length. What is the appropriate length scale for the NS proximity effect sandwich?

First, note that the self-energy for the NS sandwich is spatially dependent. There are two contributions to this self-energy: a pair function $\phi(x)$ and a renormalization function $Z(x)$. The equations for these are given in Ref. 1 as

$$\phi(x) = \int_0^\infty dE' \mathrm{Re}[f(E', x)] K_+(E, E' \cdot x) \tag{4.1}$$

$$Z(x) = \int_0^\infty dE' \mathrm{Re}[N(E', x)] K_-(E, E' \cdot x) \tag{4.2}$$

The "pair density function," $\mathrm{Re}[f(E', x)]$, and the "normalized density of states", $\mathrm{Re}[N(E', x)]$, are given by

$$f(E,x) = \int_0^1 d\cos\theta \left(-\frac{1}{2}\hbar v_F \cos\theta\right) \mathrm{Im}[G(x,x,E)_{12}] \quad (4.3)$$

$$N(E,x) = \int_0^1 d\cos\theta \left(-\frac{1}{2}\hbar v_F \cos\theta\right) \mathrm{Im}[G(x,x,E)_{11}] \quad (4.4)$$

where $G(x, x, E)$ is the Nambu matrix Green function for the NS bilayer, as obtained in Ref. 1, and v_F is the bare Fermi velocity (assumed, for simplicity, to be the same in N and S). The electron–phonon interaction kernels, $K_+(E, E', x)$ and $K_-(E, E', x)$ are given by

$$K_+(E,E',x) = \int_0^\infty d\nu\, \alpha^2(\nu,x)F(\nu,x)[(E+E'-\nu)^{-1} + (E-E'-\nu)^{-1}] - \mu^*(x)\Theta(E_c(x)-E') \quad (4.5)$$

$$K_-(E,E',x) = \int_0^\infty d\nu\, \alpha^2(\nu,x)F(\nu,x)[(E+E'-\nu)^{-1} - (E-E'-\nu)^{-1}] \quad (4.6)$$

The interaction kernels, $K_+(E, E', x)$ and $K_-(E, E', x)$, change in a step-like way at the NS interface, since the interactions are spatially local. However, the pair density function and the normalized density of states vary smoothly across the NS interface, because the Green's function varies smoothly across the interface. In the S metal, the Green's function (and therefore the pair density function) varies significantly only over the BCS coherence length in the S metal, $\xi_S = \hbar v_F/\pi\Delta_S$. This variation carries over into the N metal, by continuity, implying that the N metal gap function also varies over the BCS coherence length of the S metal. Hence, the thin N metal limit is that in which the N metal thickness, d, is much less than the BCS coherence length in the S metal. The relevant length scale for the NS system in the thin N metal limit is thus ξ_S.

The spatial dependence of the pair potential, $\Delta(x) = \phi(x)/Z(x)$, is sketched in Fig. 4.1. In Ref. 2 it is shown that the depression of the pair potential is approximately equal to the ratio of the N metal thickness to the BCS coherence length in the S metal times the bulk value of the pair potential in the S metal, $(d/\xi_S)\Delta_S$. When the ratio is negligibly small, as in the thin N metal limit, the pair potential at the NS interface is equal to the bulk S pair potential. In addition, the pair density function is equal to the bulk pair density

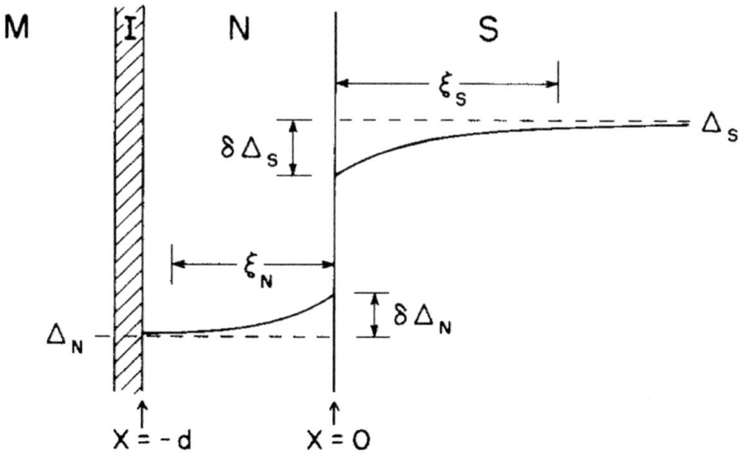

Figure 4.1 Sketch of spatial dependence of $\Delta(x)$.

function in the S metal in this limit. Therefore, from the theory of bulk superconductivity, in the thin N metal limit we find ($\Omega_S = (E^2 - \Delta_S^2)^{1/2}$ and $\Omega_S' = ((E')^2 - \Delta_S^2)^{1/2}$)

$$\lim_{d\to 0} \mathrm{Re}[f(E, x)] = \mathrm{Re}[\Delta_S/\Omega_S] \tag{4.7}$$

$$\lim_{d\to 0} \mathrm{Re}[N(E, x)] = \mathrm{Re}[E/\Omega_S] \tag{4.8}$$

$$\begin{aligned}\lim_{d\to 0} \Delta_N(E) &= \lim_{d\to 0} \phi_N(E)/Z_N(E)\\ &= (1/Z_N(E)) \int_0^\infty dE' \, \mathrm{Re}[\Delta_S/\Omega_S'] K_+(E, E')_N \end{aligned}\tag{4.9}$$

$$\lim_{d\to 0} Z_N(E) = E - \int_0^\infty dE' \, \mathrm{Re}[E'/\Omega_S'] K_-(E, E')_N \tag{4.10}$$

$$\begin{aligned}\lim_{d\to 0} \Delta_S(E) &= \lim_{d\to 0} \phi_S(E)/Z_S(E)\\ &= (1/Z_S(E)) \int_0^\infty dE' \mathrm{Re}[\Delta_S/\Omega_S'] K_+(E, E')_S \end{aligned}\tag{4.11}$$

$$\lim_{d\to 0} Z_S(E) = E - \int_0^\infty dE' \, \mathrm{Re}[E'/\Omega_S'] K_-(E, E')_S \tag{4.12}$$

These equations describe the proximity effect in thin N metal NS bilayers. They are strictly correct in the limit as the N metal thickness approaches zero. In this limit, the induced pair potential in the N metal, Δ_N, is spatially constant, as is the pair potential in S.

Corrections to these pair potentials are smaller by a factor equal to the ratio of the N metal thickness to the S metal BCS coherence length, (d/ξ_S).

4.2 Tunneling Density of States in NS Bilayers

In this section, I will continue to assume that the N metal is clean. Consistent with the thin N metal limit, I will also assume that the pair potentials in N and S are constant. The tunneling density of states will simply be the normalized density of states presented above. In the thin N metal limit, this is simply the normalized density of states in the bulk S metal. In this section, I will discuss this function in the regime wherein the N metal thickness is not necessarily negligible relative to the BCS coherence length in S, so that I can consider the nature of corrections to the thin N metal limit. I will continue to assume spatially constant pair potentials—the "step model," shown in Fig. 4.2.

At the NS interface, the pair potential has a step of height $\Delta_S - \Delta_N$. Quasiparticles in N having energies greater than Δ_N but less than Δ_S are confined in an unusual bound state by means of Andreev scattering from the pair potential step. Because of the induced gap in N, there are no quasiparticle states with energies below Δ_N. Between Δ_N and Δ_S, however, one can have one or more "Andreev bound states." Imagine a quasiparticle propagating perpendicular to the NS interface in N at an energy E (>0) between Δ_N and Δ_S. As it reaches the NS interface, it forms a Cooper pair propagating into the S metal by pairing with a quasiparticle state of energy $-E$ from the N metal, leaving behind a hole which propagates back to the N metal surface, reflects from that, thereafter reaching the NS interface taking half of a Cooper pair from the S metal, leaving a quasiparticle at E propagating back to the N metal surface where it interferes constructively provided the energy has the appropriate bound state energy value, E_0. In Fig. 4.2 I show the result for an N metal with thickness much less than the BCS coherence length in S. For quasiparticles not propagating perpendicular to the NS interface, the distance traveled in these processes before constructive interference occurs is greater, hence the energy of the bound state is *lower.*

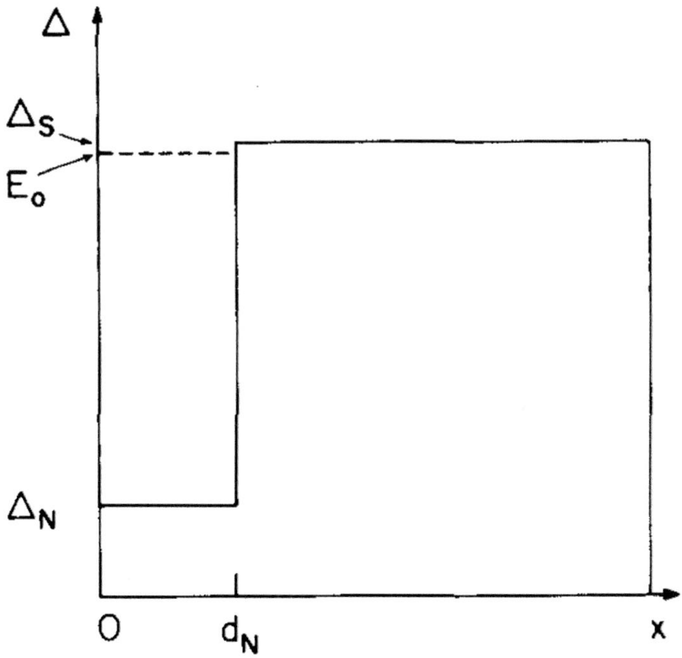

Figure 4.2 The step model for the thin N metal limit, indicating the single quasiparticle bound state at E_0.

In Ref. 2 I calculated the exact Nambu matrix Green function for the step model illustrated in Fig. 4.2. I found that the quasiparticle density of states is ($\varepsilon_N(E) = E/(E^2 - \Delta_N^2)^{1/2}$, $\delta_N(E) = \Delta_N/(E^2 - \Delta_N^2)^{1/2}$, with corresponding definitions for $\varepsilon_S(E)$ and $\delta_S(E)$):

$$(-1/\pi)\,\mathrm{Im}\,G(0,0)_{11} = (-2/\pi\hbar v_{Fx})\,\mathrm{Im}(\eta(E)/D(E)) \quad (4.13)$$

where

$$\begin{aligned}\eta(E) &= \varepsilon_N[i(\varepsilon_N\varepsilon_S - \delta_N\delta_S)\cos(\Delta K_N d) + \sin(\Delta K_N d)] \\ &\quad + i\delta_N(\varepsilon_N\delta_S - \delta_N\varepsilon_S) \\ D(E) &= i(\varepsilon_N\varepsilon_S - \delta_N\delta_S)\sin(\Delta K_N d) - \cos(\Delta K_N d) \\ \Delta K_N d &= 2dZ_N(E^2 - \Delta_N^2)^{1/2}/\hbar v_{Fx}\end{aligned} \quad (4.14)$$

In Fig. 4.3, I show the resulting density of states in the N metal.

The density of states relevant for tunneling will be different from that shown in Fig. 4.3 because the insulating tunnel barrier

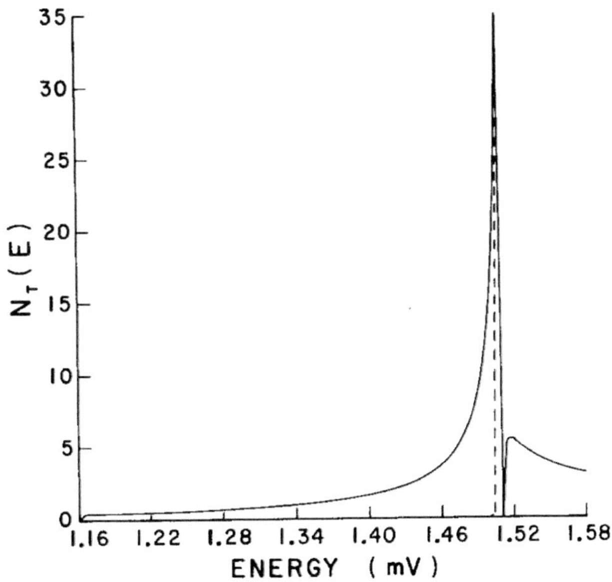

Figure 4.3 Density of states in N for a thin N metal in an NS bilayer for $\Delta_S = 1.51$ (Nb), and $d/\xi_S = 0.1$. The dashed curve is the *tunneling* density of states in N. Note that only the bound state region is affected by the selection of quasiparticles moving nearly perpendicular to the tunnel barrier.

is preferentially traversed by quasiparticles with wave vectors perpendicular to the barrier-N metal interface, exponentially decreasing the contribution of wave vectors not perpendicular. This has the effect of exponentially attenuating the bound state contributions from wave vectors not perpendicular to the barrier-N metal interface. The normalized (to the normal state) tunneling density of states is therefore given by

$$N_T(E) = \int_0^1 d(\cos\theta)(-1/\pi)\,\mathrm{Im}\,G(0,\,0)_{11}\,D(\cos\theta)(\pi\hbar v_{Fx}/2) \quad (4.15)$$

where $D(\cos\theta)$ is the "normalized probability distribution of tunneling electrons," which is very sharply peaked at $\cos\theta = 1$.

This probability distribution noticeably affects only the bound state region, resulting in the dashed curve in Fig. 4.3. In Fig. 4.4, I show the tunneling density of states on a much finer energy scale (with some smoothing applied to simulate the effect of temperature).

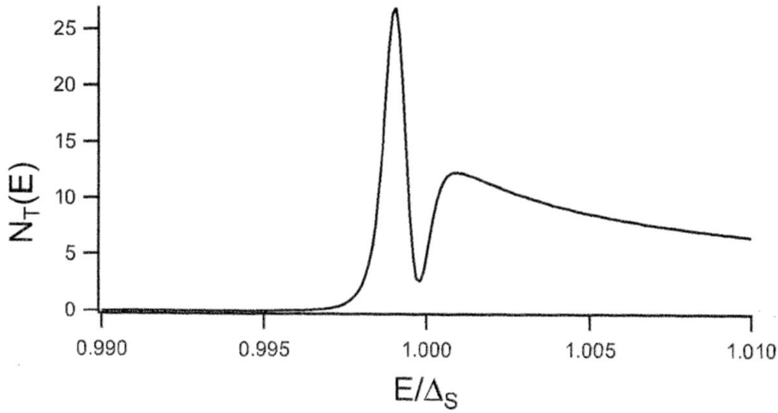

Figure 4.4 Tunneling density of states for an NS bilayer with $d/\xi_S = 0.05$ and $\Delta_N/\Delta_S = 0.2$, and normal counter-electrode. The results were smoothed to simulate the effect of finite temperature.

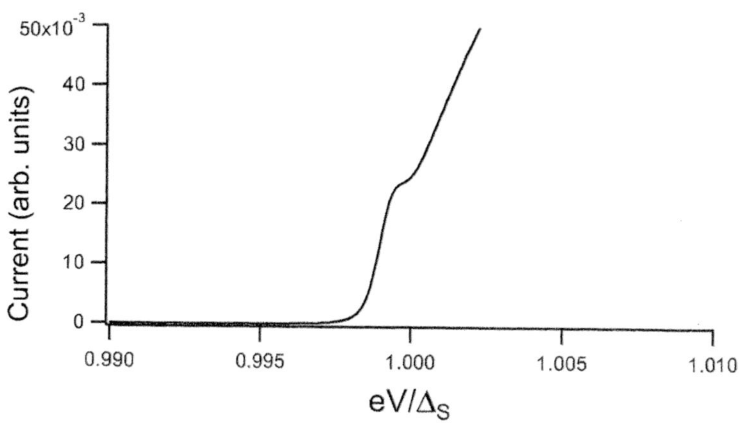

Figure 4.5 The current vs. voltage for tunneling into the NS bilayer of Fig. 4.4.

In Fig. 4.5, I show the current vs. voltage for tunneling into an NS bilayer for the NS bilayer of Fig. 4.4. This is just the integral of the tunneling density of states in Fig. 4.4. Note the "ski jump" for the finite N metal thickness case. The presence of this "ski jump" is indicative of a thin N metal layer and is a useful diagnostic, as such.

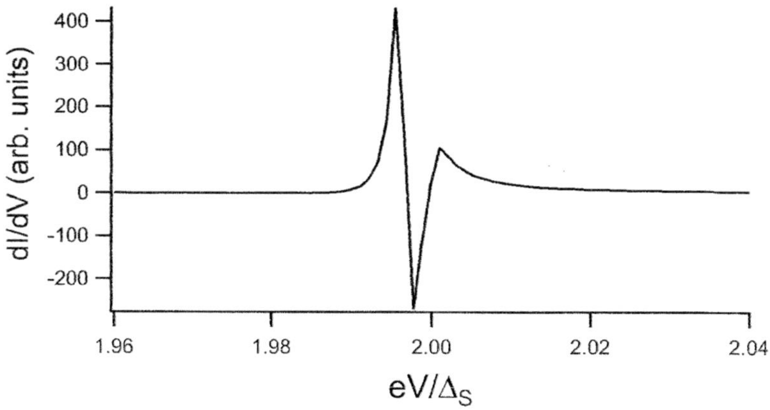

Figure 4.6 The tunneling density of states for a superconducting counter-electrode for an NS bilayer with $d/\xi_S = 0.1$ and $\Delta_N/\Delta_S = 0.2$, and $k_B T = 0.0001\Delta_S$.

For energies large compared to Δ_S, the tunneling density of states may be expanded to yield the expression (cf. Ref. 2)

$$N_T(E) \approx 1 + \Delta_N^2/2E^2 + (\Delta_S - \Delta_N)^2/2E^2 I((4d/L(E)) + \Delta_S(\Delta_S - \Delta_S)/E^2 I((2d/L(E)) \quad (4.16)$$

where

$$L(E) = \hbar v_F/Z_N(E)E \quad (4.17)$$

where $I(x)$ is a damped oscillatory function of x, with $I(0) = 1$. It is easy to see that for $d = 0$, this reduces to the density of states for tunneling into bulk S metal, as it must. The oscillatory functions give rise to the Tomasch–Rowell oscillations [2]. In the region of energies where phonon structure in the gap functions is present and $E \gg \Delta_S$, this shows that the phonon structure observed in tunneling will be dominated by that of the S metal provided that the thickness of the N metal is such that the integrals are near unity. This provides a foundation for proximity electron tunneling spectroscopy, as noted in the previous chapter.

In Figs. 4.6 and 4.7 I show the tunneling density of states for a superconducting counter-electrode, and the corresponding current vs. voltage curve, respectively, for an NS bilayer. With a superconducting counter-electrode, the current vs. voltage curve

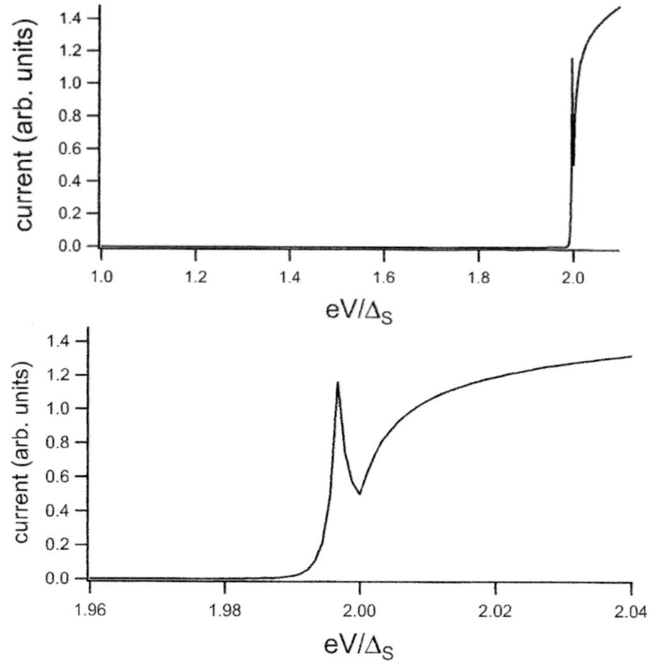

Figure 4.7 The current vs. voltage for tunneling into an NS bilayer for the NS bilayer of Fig. 4.6. The graph on the right gives an expanded view of the "gap region."

strongly resembles the tunneling density of states in the NS bilayer with a *normal* electrode, in the bound state region.

4.3 Effects of Elastic Scattering in the N Metal Layer

In a bulk superconductor, the Anderson theorem [3] states that elastic scattering has no effect on the gap function. In the N layer of an NS sandwich, however, I showed [4] that such scattering does affect the induced gap in the N metal. I will now briefly summarize the argument.

We will assume that the N metal region is smaller than the BCS coherence length in the S metal region, so that the pair function and the renormalization function can be replaced by their averages over the thickness of the N metal. This yields the equations for the phonon contributions to the pair density and renormalization function in N:

$$\begin{aligned} < \phi^{\text{ph}}(x) >_{\text{N}} &= \phi_{\text{N}}^{\text{ph}} \\ &= \int_0^\infty dE' < \text{Re}[f(E', x)] >_{\text{N}} K_+(E, E')_{\text{N}} \end{aligned} \quad (4.18)$$

$$\begin{aligned} < Z^{\text{ph}}(x) >_{\text{N}} E &= Z_{\text{N}}^{\text{ph}} E \\ &= E - \int_0^\infty dE' < \text{Re}[N(E', x)] >_{\text{N}} K_-(E, E')_{\text{N}} \end{aligned} \quad (4.19)$$

where $< >_{\text{N}}$ indicates the spatial average over the N metal layer. The self-energies $\phi_{\text{N}}^{\text{ph}}$ and Z_{N}^{ph} are the electron–phonon interaction self-energies, which are obtained from Eqs. 4.1–4.6 above. The elastic scattering contribution is obtained by replacing the kernels $K_{\pm}(E, E')_{\text{N}}$ in Eqs. 4.1 and 4.2 by

$$K_{\pm}(E, E')_{\text{N}} \rightarrow \pm(i\hbar/2\tau)\delta(E - E') \quad (4.20)$$

This yields the equations for the full pairing self-energy and renormalization functions:

$$\phi(E)_{\text{N}} = \phi_{\text{N}}^{\text{ph}}(E) + (i\hbar/2\tau) < f(E, x) >_{\text{N}} \quad (4.21)$$

$$Z(E)_{\text{N}} E = Z_{\text{N}}^{\text{ph}}(E) E + (i\hbar/2\tau) < N(E, x) >_{\text{N}} \quad (4.22)$$

The induced energy gap function in N is then given by

$$\begin{aligned} \Delta_{\text{N}} = [\phi_{\text{N}}^{\text{ph}} &+ (i\hbar/2\tau) < f(E, x) >_{\text{N}}]/[Z_{\text{N}}^{\text{ph}}(E) \\ &+ (i\hbar/2\tau) < N(E, x) >_{\text{N}} /E] \end{aligned} \quad (4.23)$$

The Anderson theorem states that in a bulk superconductor

$$\Delta_{\text{N}} = \phi_{\text{N}}^{\text{ph}} / Z_{\text{N}}^{\text{ph}} \quad (4.24)$$

By definition

$$\Delta_{\text{N}} = \phi_{\text{N}} / Z_{\text{N}} \quad (4.25)$$

so that, using Eqs. 4.4 and 4.5, the Anderson theorem would require that

$$\Delta_{\text{N}} = E < f(E, x) >_{\text{N}} / < N(E, x) >_{\text{N}} \quad (4.26)$$

This will hold, according to Eq. 4.6, only in the extreme limit wherein $\tau \to 0$, i.e., the limit of infinitely strong scattering.

In the limit as $d \to 0$, in Ref. 4, I found that the solution for the induced gap is

$$\lim_{d\to 0} \Delta_{\mathrm{N}}(E) = [\Delta_{\mathrm{N}}^{\mathrm{ph}} + i\Gamma_{\mathrm{N}}(E)\Delta_{\mathrm{S}}/\Omega_{\mathrm{S}}]/[1 + i\Gamma_{\mathrm{N}}(E)/\Omega_{\mathrm{S}}] \quad (4.27)$$

where

$$\Gamma_{\mathrm{N}}(E) = \hbar/2Z_{\mathrm{N}}^{\mathrm{ph}}(E)\tau \quad (4.28)$$

This result is the same as that found by McMillan [5] in his treatment of an NS bilayer *with a tunnel barrier separating the N and S layers.* In McMillan's theory, Γ_{N} is $\hbar/2Z_{\mathrm{N}}^{\mathrm{ph}}(E)\tau_{\mathrm{N}}$, where τ_{N} is the average time spent by a quasiparticle in the N layer. In our case, τ is the *elastic scattering lifetime* of a quasiparticle in the N layer. This similarity between the two different theories is due to the fact that both assume the relaxation of tunneling wave vector selection rules due to scattering in the N layer.

4.4 Tunneling Density of States in NS Bilayers with Elastic Scattering in the N Layer

In Ref. 4 I have calculated the tunneling density of states in NS bilayers in the limit (ℓ is the mean free path in N)

$$\hbar v_{\mathrm{F}}/(2Z_{\mathrm{N}}^{\mathrm{ph}}\Delta_{\mathrm{N}}) \gg \ell \text{ or } d/\ell \gg (2d/\hbar v_{\mathrm{F}})Z_{\mathrm{N}}^{\mathrm{ph}}\Delta_{N} \quad (4.29)$$

which is the "dirty limit" for the N layer.

In Fig. 4.8 I plot the real and imaginary parts of the solutions for Δ_{N} for four different values of $R = 2d/\hbar v_{\mathrm{F}}$ (in inverse milli-eV). Note that Δ_{N} is independent of mean free path in the dirty limit. In Fig. 4.9 I plot the tunneling density of states for these four cases, taking $d/\ell = 3$.

In Fig. 4.9, the solid curve which has an energy gap just below $E = 0.4$ is reminiscent of tunneling densities of states obtained in the McMillan tunneling model, while that corresponding to the thinnest N metal layer resembles those obtained in the previous section.

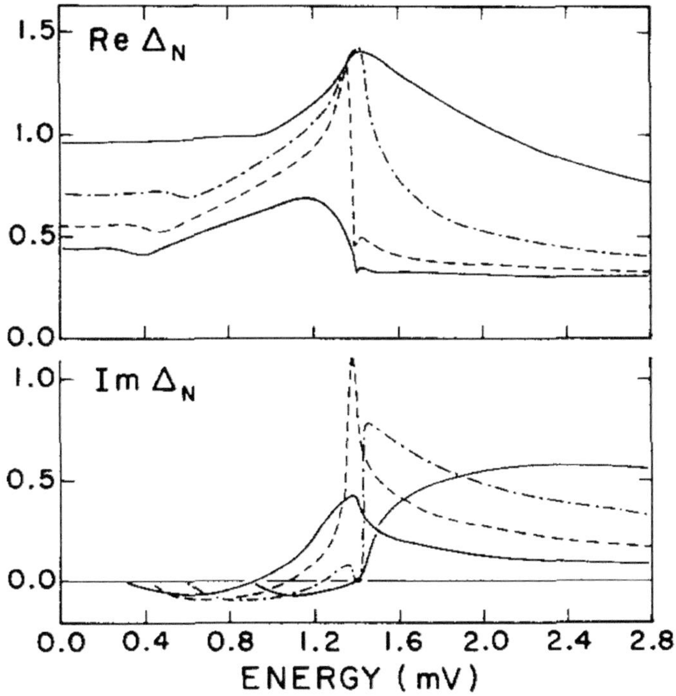

Figure 4.8 Real and imaginary parts of $\Delta_{\rm N}$, for $\Delta_{\rm N}^{\rm ph} = 0.3$, $Z_{\rm N}^{\rm ph} = 1.2$, $\Delta_{\rm S} = 1.4$, and $d/\ell = 3$. The larger of the two solid lines at $E = 1.4$ corresponds to $d/\xi_{\rm S} = 2.2$, the other solid line to $d/\xi_{\rm S} = 0.22$. The dashed lines are for $d/\xi_{\rm S} = 1.1$, the dash-dotted lines for $d/\xi_{\rm S} = 0.55$.

4.5 The Josephson Current in S'INS Tunnel Junctions with Thin N Metal Layers

In Ref. 6 I derived the following expression for the zero bias Josephson current (in the tunneling Hamiltonian limit):

$$IR = -2ek_{\rm B}T\,\Sigma_n B(i\omega_n)\delta_L(i\omega_n)\sin(\varphi) \qquad (4.30)$$

where R is the junction resistance, φ is the Josephson phase difference across the tunnel barrier, and $(\varepsilon_{\rm N}(i\omega_n) = \omega_n/(\omega_n^2+\Delta_{\rm N}^2)^{1/2}$, $\delta_{\rm N}(i\omega_n) = -i\Delta_{\rm N}/(\omega_n^2 + \Delta_{\rm N}^2)^{1/2}$, with corresponding definitions for $\varepsilon_{\rm S}(i\omega_n)$ and $\delta_{\rm S}(i\omega_n)$ and $\delta_L(i\omega_n)$):

$$B(i\omega_n) = Q(i\omega_n)/P(i\omega_n) \qquad (4.31)$$

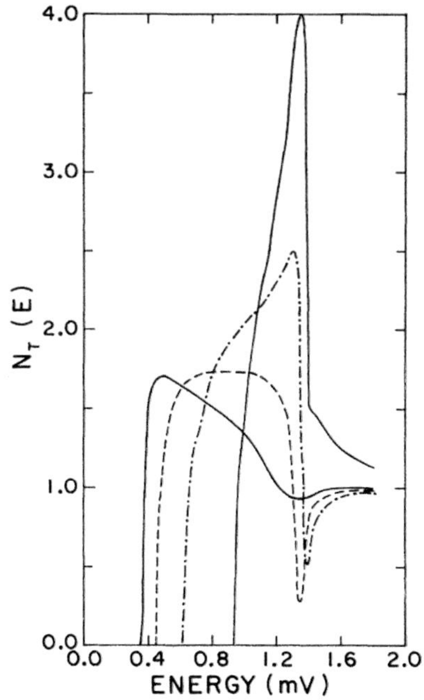

Figure 4.9 Tunneling density of states for the four cases in Fig. 4.8. The solid line which is largest near $E = 1.4$ corresponds to $d/\xi_S = 0.22$. All other lines are as given in the caption of Fig. 4.8.

where

$$\begin{aligned} Q(i\omega_n) &= \delta_N[i(\varepsilon_N\varepsilon_S - \delta_N\delta_S)\cos(\Delta K_N d) \\ &\quad + \sin(\Delta K_N d)] + i\varepsilon_N(\varepsilon_N\delta_S - \delta_N\varepsilon_S) \\ P(i\omega_n) &= i(\varepsilon_N\varepsilon_S - \delta_N\delta_S)\sin(\Delta K_N d) - \cos(\Delta K_N d) \\ \Delta K_N d &= i2d/(\pi\xi_S\Delta_S)(\omega_n^2 + \Delta_N^2)^{1/2} \end{aligned} \tag{4.32}$$

It is easy to see that in the limit as d approaches 0, $B(i\omega_n) = -i\Delta_S/(\omega_n^2 + \Delta_S^2)^{1/2}$ as expected. The new information which can be obtained from this result is the dependence of the Josephson critical current on the thickness of the N metal layer, d.

In the graph of Fig. 4.10, one can see that the critical current is relatively rapidly depressed with increasing N metal thickness. The asymptotic value of this ratio as $d \to \infty$ is just the value for the ratio of the critical current in an SIN junction to that for an SIS junction. The actual curves will decrease even faster because as the N metal

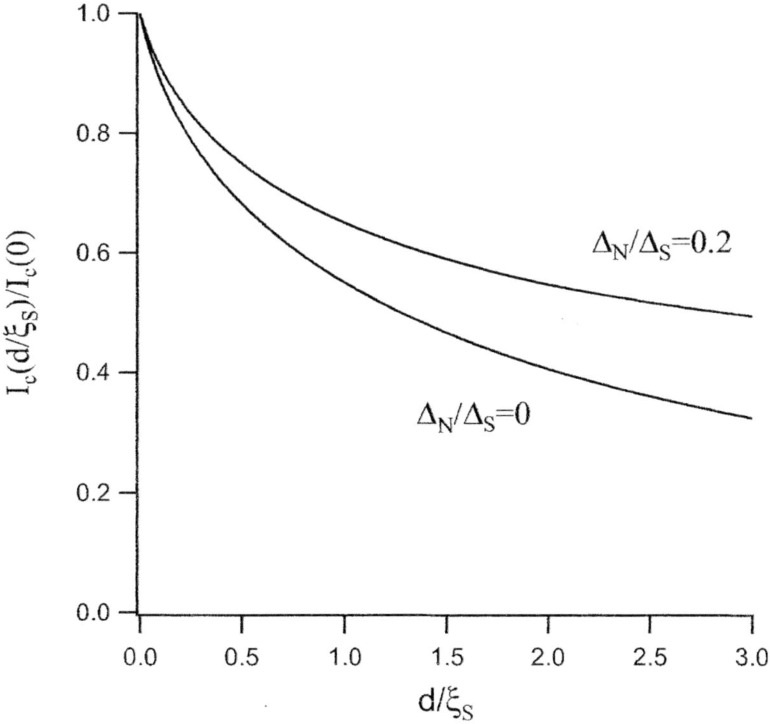

Figure 4.10 Ratio of Josephson critical current for a SINS tunnel junction with N thickness d to that for a SIS tunnel junction, as a function of d/ξ_S.

thickness increases the value of the S metal gap at the NS interface will also decrease. For $d/\xi_S \leq 0.1$, however, the junction is in the thin N metal limit and the calculation should be accurate.

References

1. Wolf, E. L., and Arnold, G. B. (1982). Proximity electron tunneling spectroscopy. *Phys. Rep.*, **91**, 31–102.
2. Arnold, G. B. (1978). Theory of thin proximity effect sandwiches. *Phys. Rev. B*, **18**, 1076–1100.
3. Anderson, P. W. (1959). Theory of dirty superconductors. *J. Phys. Chem. Solids*, **11**, 26–38.

4. Arnold, G. B. (1981). Theory of thin proximity effect sandwiches. II. Effects of s-wave elastic scattering. *Phys. Rev. B*, **23**, 1171–1178.
5. McMillan, W. L. (1968). Tunneling model of the superconducting proximity effect. *Phys. Rev.*, **175**, 537–542.
6. Arnold, G. B. (1985). Superconducting tunneling without the tunneling Hamiltonian. *J. Low Temp. Phys.*, **59**, 143–183.

Chapter 5

The Trace That Launched a Thousand Chips: Development of Nb/Al–Oxide–Nb Technology

Michael Gurvitch
Department of Physics and Astronomy, Stony Brook University, Stony Brook, NY 11794-3800, USA
michael.gurvitch@stonybrook.edu

Having been asked to write about the emergence of the Nb/Al–oxide–Nb technology, I have no choice but to put on a historian's hat. Becoming such an *accidental historian*, I will try to describe how refractory Josephson junction (JJ) technology was born, where it was born, of what parents, and why it was conceived in the first place, attempting to make a detailed "case study" of this particular innovation. As almost all new technologies in our time, it was a collective undertaking not limited to a single individual, to a single group, or even to a single institution. I will "name names" of those who were most involved in it at Bell Labs, and of those who most influenced our work from the outside of Bell, and will try to assign credit. In that, I may be subjective, although I will try to curb my subjectivity as best I can. Relevant technical detail without which the story would be less clear will be provided. To make it interesting on

Josephson Junctions: History, Devices, and Applications
Edited by Edward Wolf, Gerald Arnold, Michael Gurvitch, and John Zasadzinski

ISBN 978-981-4745-47-5 (Hardcover), 978-1-315-36452-0 (eBook)
www.panstanford.com

a human level, I will venture into a bit of reminiscence about the people I knew and the happy years at Bell Labs when this work was done. This inevitably requires extensive use of pronoun "I," which is normally avoided in scientific papers, and for which we (that is, I) apologize. The story would be incomplete without saying a few words about Bell Labs, where it took place, and that is where we will start.

5.1 Bell Labs Empire

Prior to Bell Labs, I was working on my thesis at Brookhaven National Lab. The Ph.D. program was at the Physics Department at Stony Brook, but the actual work, mainly on A-15 superconducting compounds, I did at Brookhaven, where I had a good fortune to learn from a scientist of the highest caliber and of the greatest devotion to physics that I have ever seen—my thesis advisor, the late Myron Strongin (1936–2013). I graduated in 1978 and started looking for a job. In the meantime, Stony Brook professor and well-known solid state theoretical physicist Phil Allen said something nice about me to his former coauthor, Bell Labs physicist and one of the stars of superconductivity world, Bob Dynes. As a result, in late summer or early fall of 1978, I was invited for an interview to Bell Labs in Murray Hill, New Jersey. It was the first time I saw that place, and it impressed me very much.

At that time "Ma Bell" was still at the height of her powers, this being six years before the fatal legislative breakup of 1984. Even though mortally wounded in 1984, Bell Labs kept its scientific traditions for a few more years. I was lucky to work there from 1979 through 1989, and not longer. That Orwellian year, 1984, marked *the beginning of the end*, the real end arriving a quarter century later, in 2008, when the owners of Bell Labs, Alcatel-Lucent, announced that they were pulling out of scientific research altogether. By then, research at Bell Labs had been reduced to a trickle anyway.

When I came for my interview, Bell Labs was undisputedly the best solid state physics lab in the world. It consumed only about 2% of the total American Telephone and Telegraph (AT&T) budget, but this 2% constituted 2 billion dollars a year, which in 1979 was still a lot of money. There were about 25,000 employees in all of the Labs spread around New Jersey and Pennsylvania, with about 3000 of them stationed at Murray Hill, the site of the leading lab and the seat of the top management. The total number of AT&T employees at that time was, I believe, close to 600,000, constituting the largest private enterprise in the world. The Bell empire had its own ocean fleet, its own communication satellites, its own manufacturing facility (Western Electric), and, of course, its world-famous scientific laboratories first established by the inventor of the telephone, Alexander Bell, in 1880. The labs were christened Bell Telephone Laboratories in 1925, when the business became part of AT&T. The word *telephone* was later dropped, the name becoming AT&T Bell Labs. Simultaneously with that (it may have happened in 1984), the company's logo was changed from the depiction of a bell (a nice play on the name of the founder) to a striped circular symbol that was supposed to look like a globe, but, according to Tony Fiory, my friend and collaborator at Bell, looked a lot like a "*finished* Death Star" of the then appearing *Star Wars* movies, in which, as the reader may recall, the Death Star was unfinished.

I was told that AT&T organization had 13 levels of management, and that salaries in the adjacent levels increased on the average by a factor of $\sqrt{2}$. At Bell Labs the lowest of those managerial levels consisted of Members of Technical Staff (MTS). MTS were scientists and engineers of all denominations, mostly with Ph.D. degrees: physicists, chemists, pure and applied mathematicians, materials scientists, computer scientists, electrical engineers, even some biologists and astrophysicists. In the basic research area of Bell Labs, Area 11, each MTS supervised a personal technician called Senior Technical Assistant, or STA, thus earning the title of a "manager," albeit at the lowest level (this concerned only experimentalists; theorists did not have technicians, and so were not managers). The second managerial level consisted of *supervisors* or *group leaders*, groups consisting of a few MTS (group leader level existed mostly in Area 52, the *development* area of Bell Labs). Then there were *department heads*, a department in Area 11 containing a dozen MTS and approximately the same number of STAs, then *lab directors*, labs consisting of five to seven departments, then *executive directors* in charge of subdivisions (Physics, Chemistry).

The executive director in charge of Area 111, Physics Division, was a solid state theorist and a whip-clever man, Bill Brinkman; above him was Arno Penzias, *vice president of research*. Penzias assumed that high position after getting in 1978 a Nobel Prize, which he shared with another Bell Labs man, Robert Wilson, the prize being awarded for their 1964 discovery of cosmic microwave background radiation from the Big Bang—in retrospect, one of the greatest scientific discoveries of the 20th century (and, therefore, of all times). Above Penzias was Bell Labs president, in my time, Ian Ross.

At Bell Labs, good, often truly distinguished scientists were promoted to the managerial positions. So far we counted seven levels, so, in Dante's classification, we are not out of Hell yet. The top six levels of AT&T management were invisible from Murray Hill; they were somewhere in Manhattan (I assume), and generally much closer to St. Peter than to us.

By 1979, the number of Nobel Prize winners who worked at various times at Bell Labs was eight, three of them still working there during my tenure (Anderson, Penzias, and Wilson). Before the end, this number increased by six, bringing the total number of Nobel laureates at Bell Labs to fourteen, those fourteen sharing in eight physics Nobel prizes awarded from 1937 through 2014.[a] Few countries can claim such statistics. A good chunk of the 20th century's high technology was invented at Bell Labs, and even more technologies were perfected there. To name a few highlights from different fields, without any claim on completeness:

- Practical vacuum tube amplifiers (improved De Forest's triodes) used as *repeaters* in long telephone lines (Harold Arnold, 1913–1915)
- Wave nature of matter (diffraction of electrons on crystals) (C. Davisson and L. Germer, 1927; Davisson received the Nobel Prize in physics in 1937)

[a]One of these Nobel prizes, for fractional quantum Hall effect, was awarded in 1998 to Dan Tsui, who at the time of his prize-winning research was MTS in our department 11151. He shared the prize with H. Stormer (department head of 11134) and R. Laughlin. Another Nobel Prize to someone we knew well and sometimes had lunch together was awarded to Steven Chu in 1997.

- Concept of negative feedback in electronics (Harold Black, 1927)
- Thermal noise in resistors (J. Johnson and H. Nyquist, 1928)
- Motion pictures with sound and high-fidelity (hi-fi) mono and stereo recordings (Arthur Keller, 1925–1932)
- Radio Astronomy (radio waves from the Milky Way; Karl Jansky, 1933)
- Identification of n and p conduction in semiconductors; a notion of positive *holes* (S. Scaff, H. Theurer, R. Ohl, 1939)
- First p-n junction in Silicon (R. Ohl, 1941)
- Patent on solar cell (R. Ohl, 1946)
- First mobile (cellular) wireless telephone system (from 1946)
- The first (point-contact) transistor, the crown jewel of all Bell Labs achievements (J. Bardeen and W. Brattain, 1947; Nobel Prize in 1956)
- Bipolar junction transistor, BJT (a much more practical transistor; W. Shockley, 1948; Nobel Prize in 1956, shared with Bardeen and Brattain)
- Information theory (Claude Shannon, 1948)
- Digital transmission of information (PCM, or pulsed code modulation) (B. Oliver, C. Shannon, J. Pierce, 1946–1948)
- Reliable solid-state diodes (late 1930s–1950)
- Efficient silicon solar battery (photovoltaic, or the solar cell) (C. Fuller, G. Pearson, D. Chapin, 1954)
- Zone-refining of semiconductor crystals (W. Pfann and H. Theurer, 1950–1955)
- High critical current superconducting wires (B. Matthias and T. Geballe, 1954)
- Replacing Ge with diffusion-doped Si in bipolar junction transistor (M. Tanenbaum, 1955)
- Principle of a laser (C. Townes, A. Schawlow, 1958; Townes shared a Nobel Prize with Basov and Prokhorov in 1964)
- Metal–oxide–semiconductor field-effect transistor, MOSFET (D. Kahng and M. Atalla, 1959)
- Experimental discovery of Josephson effect (P. W. Anderson and J. M. Rowell, 1963)
- CO_2 laser (K. Patel, 1964)

- Molecular beam epitaxy, MBE (R. Arthur and A. Cho, late 1960s – early 1970s)
- UNIX operating system and C programming language (D. Ritchie and K. Thompson, 1970–1972); C++ programming language (B. Stroustrup, 1986)
- Charge-coupled devices, CCDs (W. Boyle and G. Smith, 1969; Nobel Prize in 2009)
- Modulation doping (H. Stormer, 1980)
- Fractional quantum Hall effect (H. Stormer and D. Tsui, 1981–1983; the 1998 Nobel Prize was shared by Stormer and Tsui with R. Laughlin)
- Refractory Josephson junction Nb/Al–oxide–Nb technology (M. Gurvitch and J. M. Rowell, 1981–1983)
- Laser cooling of atoms (Steven Chu, 1985; 1997 Nobel Prize shared with W. D. Phillips and C. Cohen-Tannoudji)

I immodestly included in that list our refractory JJ technology, also known as the trilayer process, the workhorse of all He-cooled superconducting electronics to this day, and the subject of this book. Here I listed only two names of the principal inventors of that technology. Significant contributions of others, both inside and outside of Bell Labs, are detailed in subsequent sections. And, no—I do not think that Nb/Al technology can be compared in importance to the experimental discovery of waves of matter, or to the invention of a transistor. Still, it is a nice technology.

5.2 Starting at Bell Labs: Thin-Film Deposition System

As I said, in late summer or early fall of 1978, I came to Bell Labs for an interview, knowing little about its history. The Murray Hill building situated at the back of a front lawn was large, modern, and impressive. One entered through the high and spacious lobby housing the small but interesting Bell Labs museum, an exhibit of artifacts relating to some of the major Bell Labs achievements: a replica of the first transistor (the one we see reproduced in so many books); pages from Phil Anderson's notebook containing the description of one of his famous theories (it was either the theory of dirty superconductors or the localization theory); disks with first stereo recordings, and many other things. On the other side of the lobby, at the bottom of a wide staircase, there was a control post with guards checking Bell Labs passes which contained employees' names and photographs. Some years later, the future Nobel laureate Horst Stormer fixed a picture of a monkey over his own photograph, but was eventually caught by the guards and admonished for this little practical joke. The central building also contained two cafeteria, regular and special (the latter with waiter service, for top management and for treating lab's visitors), the offices of the highest company brass (Bell Labs president, vice president of research), and the great technical library rivaling best university libraries. On the sides of the central building were older long wings; as you faced the main entrance, Research Area 11 was located in the right wing, and Applied Area 52 in the left wing. I learned most of this local geography later. In the first day, I remember seeing the great staircases, mile-long corridors, and the special service cafeteria to which my hosts took me for lunch as a candidate. That fancy cafeteria, with tall slanted windows overlooking an inner garden, resembled, I thought, one of the Wonders of the World, the Hanging Gardens of Semiramis.

My two-day interview was hosted by Ted Fulton, who, as I found out, was heading a small group working on Josephson junction technology (the group was formed in Area 11, which, as I said, did not have the group leader position, so Ted, while heading the group, was a normal MTS). Ted explained to me that, should I be hired, I would have to devote at least some of my time to improving the existing Josephson junction (JJ) technology. The rest of the time, said Ted, I could use for anything I wanted to do. The exact allocation of time was not specified: Bell Labs were famous for allowing scientists considerable freedom in choosing the subjects of their research; at least it was so in Area 11, at the end of 1970s. I remember Ted explaining to me how a Josephson junction, when supercurrent exceeds the critical value, switches in about a picosecond from the supercurrent branch to the normal branch of its $I-V$ characteristic. He drew $I-V$ curves on a little blue notepad the stack of which, as I found later, was on every desk at Bell, used mostly for scrap paper. Ted Fulton, who had a great sense of humor, once said, "Boy, will the management regret the day they gave us these little blue pads." I knew something about Josephson effect, of course, but nothing about its applications. Nonetheless, I remember praising the elegance of this switching scheme (what can I say; I really wanted that job). I gave a talk based on our work at Brookhaven, and later I was passed around the building, on a typical exact-to-the-minute Bell Labs schedule, talking for exactly 30 minutes to each person, including Bob Dynes and John Rowell. John is the famous expert in tunneling and experimental discoverer, with P. W. Anderson, of the Josephson effect (see Dr. B. D. Josephson's chapter and Dr. E. L. Wolf's chapter). It all went remarkably well, and at the end of the second day I was offered a permanent MTS position in Fulton's unofficial group, which was part of the Electronics and Photonics Materials Research department #11151, headed at that time by Dick Slusher, a well-known physicist, expert in the so-called *squeezed* optical states. This department was part of Lab #1115, at that time headed by director Joe Giordmaine (an expert in nonlinear optics). I was also offered, as an alternative, a postdoctoral position with J. M. Rowell, who headed Solid State and Physics of Materials Research department 11112. I was happy about that, too, because at Bell, as I learned, the closer were all the digits in the departmental number to 1's, the more fundamental was the research. The most basic of them all, Theoretical Physics Research department, which had P. W. Anderson as a member, was assigned number 11111. Notwithstanding higher digits, I wisely chose a permanent position in 11151.

One of the people I talked to in that interview (or possibly later, when I was already hired) was Gerry Dolan, whom I met before at Stony Brook, where he was a postdoc with Professor Jim Lukens. Gerry was either in Bob Dynes' or John Rowell's department (I do not remember which); he said to me something like "Take seriously the request to improve Josephson technology, and if you will succeed, you will do well for yourself here." This was good advice—except that, as will be detailed below, we succeeded in improving JJ's just at the time when IBM abruptly stopped its supercomputer project in 1983, this decision taken, at least in part, in view of the very problems that we just resolved so successfully!

Bell Labs followed suit in winding down its Josephson effort as well. This was most unfortunate for me: I did not get a chance to rest on my laurels. In fact, my laurels were moving to Japan at that time, where large-scale projects using our JJ technology were just starting. Bell, like Moscow in the old Russian saying, *did not believe in tears*; at Bell past achievements did not mean much. So, after 1983, I was under the increasing pressure to leave superconductivity altogether, and to do something "useful," such as dreadful II–VI semiconductors, this pressure ending only at the end of 1986, when high-T_c superconductivity broke out. That started the second happy and exciting period at Bell, from the end of 1986 to 1989, of which I will not write here.

At the Research Area of Bell Labs, Area 11, every experimentalist MTS had a separate lab room, a personal assistant, and an unspecified budget to fill this room with whatever was needed and desired, from screwdrivers to MBE systems. You could ask for anything, in principle, if you had a good proposal for a new project, and it would be up to the management to say yes or no. However, when planning to ask for expensive items, one would do well by first finding out what is available in-house. More often than not, instead of asking for millions of dollars, you ended up asking for the room and telephone number of a colleague who had what you wanted. This promoted collaborations, which were the order of the day at Bell Labs. The ability to mingle with all kinds of experts and to find among them potential collaborators, usually the best in the world at what they did, was probably the strongest point of being at Bell Labs, famously envisioned and implemented by pre-WWII research director and post-WWII president of Bell Labs, the legendary Mervin Kelly. It is said that he insisted on designing Murray Hill buildings, those older wings with their incredibly long corridors, individual scientist's labs positioned along them, to promote chance interactions between people. It is true that you could not walk that corridor on your way to the library or to the cafeteria without meeting somebody well worth conversing with. In the cafeteria, people joined established or semi-established lunch groups, each large round table fitting up to perhaps six or eight people. So, at lunch, you would be sometimes sitting with your friends and colleagues, and sometimes with people you did not know before. If you could interest a potential collaborator with your project, you got instant access to the unique expertise, equipment, and materials. Sometimes it took only an hour or two from the moment you asked, "Who do I talk to about . . ." to the moment when some valuable advice was received or a new collaboration was established, and something be actually tried in the lab. The same was true if you had a theoretical question. The number of very clever people surrounding you at Bell was just incredible. When all else failed, people would say, "Go talk to P. W.," meaning Nobel laureate P. W. Anderson, the sage of solid state physics.

I had never been able to ask for a lot of money for a research project, in part because a large commitment ties you down (if you own an MBE system which costs a million dollars, you are married to it, and if it can make II–VI semiconductors, then you are in real trouble). At Bell, my excuses for not asking for a big new investment were (a) readily available good collaborations and (b) the existence of a large basement at Murray Hill, which was full of unwanted equipment—something a university experimentalist would sell his or her soul to have access to. I got there a few things for my lab, including a giant electromagnet, so heavy that the floor had to be reinforced with a steel plate to place it in my room. To my shame, I never used it at all, and at some point, it was removed from my lab by a team of workers with dollies, while I was trying to look the other way and not to think about it.

The room I was assigned was 1D-321, which meant building 1 (Basic Research Area 11), section D, third floor. The location was great, some of the best in Murray Hill. My lab happened to be one floor down and perhaps some 50 meters to the side from 1E-455, the old Walter Brattain's lab, where the first transistor had been famously demonstrated in December 1947. History apart, it was close to the labs of the people I collaborated with, as well as to the offices of the theorists from 11111. In that room I found an old vacuum system for making thin films and without a delay started retrofitting it with three DC sputter guns made by the company Basic Sputtering Inc. ("US guns," taking 2 inch diameter flat targets), and the rotating table that would pass a substrate under these guns. I decided that the table had to be water-cooled to prevent deposited films from overheating and thus from unwanted interdiffusion of metallic layers. This indeed proved to be useful, but to make such a rotating and simultaneously water-cooled table inside a high-vacuum chamber, its movements precision-controlled by a stepping motor, was not a trivial engineering problem.

Soon after I started, my department head, Dick Slusher, told me that I could hire an assistant. I hired a very able one, with high-vacuum experience, Harold Huggins. With Harold's expert help we soon not only outfitted the deposition system with the water-cooled rotating table (we purchased an ingenious *ferrofluidics* rotating high-vacuum feedthrough to achieve this), but also inserted into the deposition chamber a liquid nitrogen–cooled shroud, called *Meissner trap*, similar to the ones used in MBE systems. Liquid nitrogen shroud works most effectively when its surfaces are covered with Nb, which acts as an excellent getter for residual gases when it is cold. This significantly (by about a factor of 10) improved the base pressure (vacuum), dropping it to the low 10^{-8} Torr range, and in some instances even to the upper 10^{-9} Torr range. Of course films were deposited by sputtering, in the low-pressure atmosphere of research-purity Ar. This required a partial closing of the pump opening. So the real relevant vacuum was a bit higher: when the pump was partially closed (throttled), the pressure rose to the low 10^{-7} Torr range. Our Nb films had very decent T_C's over 9 K (sometimes T_C was even higher than the book value of 9.2 K, which indicates the presence of significant strain in a Nb film). As people in the field know, Nb film is a "canary in the coal mine," its T_C signaling the quality of the vacuum system. This is because Nb is sensitive to oxygen, exhibiting about 1 degree drop in T_C per 1 atomic percent of oxygen incorporated in a film. Of course, the amount of oxygen that will get into the film also depends on the deposition rate: the faster is the deposition, the cleaner is the film; our rates were close to 1 nm/s when a substrate was held stationary directly under the gun. This was only one of the options: having this versatile system, we had a variety of options as to how to make films and stacks of films, multilayers. The Nb film could be deposited with a substrate on a stationary table directly under the gun, or while the table was slowly rotating, passing the substrate under the Nb gun in each turn. This latter technique resulted in significantly lower average deposition rate, but it improved thickness uniformity, and allowed the substrate

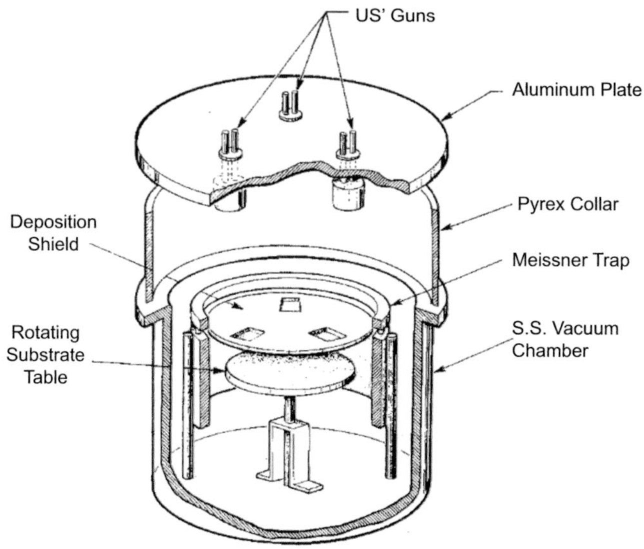

Figure 5.1 Magnetron deposition system, from Ref. 1. Meissner trap consists of two sections separated by a gap that admits a shutter (not shown). Provisions for water-cooling of the rotating table are also not shown.

to cool down while it was away from the gun, as substantial heating occurs at such high rates of deposition. To further decrease film's heating, we could thermally sink the substrate to the table while running cold water through the table. This proved particularly important later, when we deposited Al overlayers and second Nb electrodes: cold substrates prevented interdiffusion of Nb and Al and allowed for more controllable oxidation of the Al overlayer.

With three guns and the rotating table, the system was truly versatile, allowing for the simultaneous or sequential deposition of three different metals, or, if we wanted, three different alloys. We could also make *multilayers* of individual metals or even multilayers of alloys, if we wanted to. One system limitation was that having DC (rather than RF) sputter guns, we could not deposit dielectrics, but we did not plan to deposit them anyway, knowing that attempts to directly deposit tunneling oxide were futile, due to the extremely small oxide thickness of about 1 nm that was required in a tunnel junction. There was no way a deposited 1 nm oxide (or any other dielectric) could cover the base electrode uniformly and without pinholes. Thus, we were obtaining tunnel junctions by oxidation of a base electrode, as most people in the field. Oxidation is a self-limiting process (not always, but often, when an oxide is nice), with the resulting oxide thickness being in part determined by the same tunneling process that operates in a tunnel junction: when tunneling stops, oxide growth stops. This is as beautiful a mechanism for obtaining the right oxide thickness as one can imagine. In case of pure Nb, before Al overlayer came along, people resorted to RF-assisted growth of the tunneling oxide, mainly out of desperation, trying to fight Nb oxide problems (see below). With aluminum overlayer, there was absolutely no need for an RF-assisted oxide growth.

I started my work at Bell by learning how to make and measure semi-soft Nb–oxide–(Pb–Bi) junctions. To understand why these junctions were of interest, I have to say a few words about IBM's superconducting supercomputer project and the status of Josephson tunnel junctions in 1979–1980.

5.3 Superconducting Supercomputer Project at IBM and Its Extensions

In 1979, when I joined Bell Labs, a big Josephson supercomputer project was ongoing at IBM; it was called Josephson signal processor (JSP), and it was already in its 12th year, employing at that time well over 100 people.

IBM became interested in digital applications of Josephson junctions in 1964, only two years after Brian Josephson's famous theoretical prediction [2] and one year after its experimental verification at Bell Labs by J. M. Rowell and P. W. Anderson [3]. John Rowell in 1964 filed for a patent in which he suggested that switching of a Josephson junction between zero-voltage and finite-voltage states can be used in logic circuits [4], but Bell did not pursue the idea further. However, at IBM, a computer company, this was taken seriously. In 1966, at IBM, Juri Matisoo demonstrated subnanosecond switching in a Josephson junction [5]. Later it became clear that in low-capacitance junctions switching could occur on a picosecond scale, which was experimentally verified by the end of 1970s [6]. This was faster than any of the existing transistors at the time. In 1967 Matisoo proposed a flip-flop based on Josephson junctions [7] which could perform basic logic functions of AND and OR. The difference between Rowell's single junction switch and Matisoo's gate was that the latter was based on two junctions in a superconducting loop, an arrangement known as SQUID (superconducting quantum interference device). The close similarity of the behavior of a SQUID and of the two-slit light interferometer is the reason why SQUIDs are often called *interferometers*. Switching in a single junction and in a SQUID can be triggered by supercurrent exceeding its critical value called I_C. This can be achieved either by injecting additional current (a preferred scheme at the time), or by application of a magnetic field (the latter created by passing current through the nearby current line). Once the junction switches to the finite voltage state, it will stay there, unless it is *reset*; that is why this type of logic is called "latching logic." Resetting involves changing current direction. In the IBM project, the bunch of logic operations was performed during a cycle, and resetting of all of the gates was done at the end of a cycle with the use of an AC power supply.

Encouraged by fast switching, and facing serious power dissipation problems with semiconductor technology, in 1967 IBM established a program to study JJ technology with the aim of building a superfast computer (mainframe) based on it. The program grew steadily, and by 1979, they have been looking at all aspects of this technology, from materials to circuit design to packaging, and preliminary solutions (good or not so good) have been found for all these challenges. In the beginning of 1980, the status of the project was summarized in the March issue of the *IMB Journal of Research and Development* by its head, W. Anacker [8] and by the project's principal scientist, J. Matisoo [9], as well as by others. IBM circuits were made with (Pb–In–Au)–oxide–(Pb–Bi) junctions, i.e. both electrodes were soft, made of alloys of Pb; tunneling oxide was formed by a combination of thermal oxidation and RF-assisted oxidation [10]. We will discuss Josephson junctions and materials issues involved in their preparation in the next section.

By 1980, IBM circuits contained up to 10^2 gates (JJ interferometers); they were fabricated in 5 μm and, later, in 2.5 μm linewidth technology [11]. Memory cells were also based on superconducting interferometers, storing indefinitely (until they were read out) a single magnetic flux quantum Φ_0, persistent currents circulating in such a memory cell without dissipation [9]. While this was probably the most elegant feature of the JSP design when taken in isolation, construction of a working memory chip with a large number of identical memory cells (specifically, for *cache* memory) proved to be very difficult and was not achieved even by the end of the project [12]. While JSP, the way it was initially envisioned [8], was never built, a small prototype based on soft Pb alloy junctions was built in 1981–1982 (it was called cross-sectional model, or CSM) [12]. In that prototype, the cycle time was 3.7 ns, corresponding to clock rate of 270 MHz. Due to problems with cache memory, the memory in CSM was simulated.

Soft Pb alloy junctions were far from ideal; the main problem with them was in the high failure rate when they were subjected to repeated thermal cycling between 4.2 K and 300 K (see next section). Recognizing this and other problems, IBM researchers additionally developed more stable junctions with hard Nb base electrode and soft Pb alloy counterelectrode [13]. These were *edge* junctions, made at the edges of Nb films, for the reasons that will be made clear in the next section. These junctions were demonstrated in 1980, and this relatively late-stage development did not make it into the already quoted project summary published in March of 1980, nor were these semi-soft junctions used in the CSM prototype. They were more stable than their soft predecessors, but they carried with them old problems associated with Nb oxide tunnel barrier (see next section).

The effort at Bell was much smaller than at IBM, involving only half a dozen people. Bell's activity was established to "keep up with the Joneses" (or rather with the Watsons, in view of IBM's "Thomas Watson Research Center"). "Keeping up" meant having in-house expertise and access to the latest developments, to take off quickly, when the time came for this promising new technology to be commercialized (alas, this time never came). Of course, despite this small effort, Bell Labs would not be Bell Labs if people there would not be trying to do things better than in other places; in that instant, better than at IBM. At the end, what Bell indeed did better turned out to be our refractory junction technology. However, that development happened too late to be incorporated in the IBM project which closed down in September 1983. According to later accounts [12], the main reason for closing of the JSP project was insufficient speed advantage compared to rapidly developing silicon technology. Nonetheless, one cannot help but think that material problems with soft and semi-soft junctions also played a role. With respect to these material problems, our refractory JJ technology, which matured in 1983, the year when IBM effort was closed, was like a delivery of a life-saving antibiotic at the time of a funeral.

In 1979–1980, Japanese computer companies were worried about IBM gaining an early start on Josephson supercomputer. As a result, Japanese Ministry of International Trade and Industry (MITI) organized a large Japanese supercomputer project similar to IBM's that ran from 1981 to 1989. Several large companies participated: Fujitsu, Hitachi, Nippon Telephone and Telegraph (NTT), NEC, and Electro-Technical Lab (ETL). At the close of IBM project in 1983, the Japanese MITI-run project also came under scrutiny. Shinya Hasuo, the head of Fujitsu Laboratories JJ program, told me that what saved their project was our robust and reliable Nb/Al technology. After 1983, as IBM dropped out, for a while Japan became the main player in large-scale superconducting electronics. Nonetheless, many researchers in other countries, including USA, continued their work on a smaller scale, or even entered the field just at that time. For example, Sadeg Faris, IBM scientist who worked on the JSP project, decided to utilize our refractory junctions in his new company, Hypres, Inc., which he founded close to IBM Thomas Watson Center, in upstate New York, in 1983. His first product was to be ultrafast superconductivity-based sampling oscilloscope (which, as I recall, has been built but not sold in appreciable numbers). Hypres since then moved to many other specialized superconducting circuits, such as very fast A/D converters and the primary voltage standard, and today, in 2016, it still maintains excellent JJ fabrication facility (the foundry) and technical expertise in designing JJ circuits. Other chapters in this book will be the best guide to the present-day worldwide status of refractory JJ technology and its applications.

Why were IBM, and later Japanese companies, interested in a JJ computer? Short switching delay in a JJ was only part of the story. Equally important was the prospect of avoiding the problem faced by all large and densely packed semiconductor mainframes: namely, the problem with power dissipation. According to W. Anacker [8], a computer with parameters projected at the time for the Josephson machine (specifically: 1 ns processor, 2 ns cache, and 10 ns main memory, constructed in IBM 3033 system architecture), but implemented in a semiconductor version, would require removing 20 kW of power (!) from a volume of 640 cm^3, which was impossible, even with forced water and/or liquid nitrogen cooling. In contrast, Josephson mainframe with the same performance would dissipate less than 10 W of power [8], which could be comfortably handled by the liquid helium bath in which it would be immersed.

Indeed, superconductors and semiconductors operate at energy scales differing by a factor of about 10^2–10^3. A switching semiconductor passing current I (typically in the mA range) will dissipate power of the order of $P = E_g I/e$, where $E_g \approx 1$ eV is typical energy gap in a semiconductor (e.g., 1.1 eV in Si). Thus, power dissipated in a semiconductor gate is in milliwatts. In a superconductor such as Nb or Pb, the superconducting gap is $2\Delta \approx 3$ meV. Consequently, dissipation per gate in JSP project was between 3 and 5 μW [9, 11]. Furthermore, the advantage of small energy gap in superconducting logic can be realized in practice, because superconductors operate at 4.2 K where thermal energy $k_B T$ is about 100 times smaller than at room temperature, allowing for the reliable millivolt-scale operation, provided that the critical current I_C is not smaller than 0.1 mA [12].

Anacker's projections made in 1980 painted a rather bleak picture for fast semiconductor computers. It seemed that superconducting JJ technology might be the only way out. Yet, as we know, today in 2016, JJ supercomputers with superconducting memory still do not exist (although superconducting processors exist), while semiconductor computers routinely run with *clock rates* at several GHz. Why is the situation different today from what it was in 1980? I think the short answer is "because of the availability of CMOS gates with reduced power consumption." Silicon technology had a few tricks up its sleeve, tricks that were not yet known in 1980.[b]

Power dissipation prevents fast-switching transistors from being placed too close to each other, which implies larger processor size. The physical size of a processor, together with the transistor switching time, determines its ultimate speed. Even if switching delay were zero, there would be a propagation delay, which implies that a fast processor has to be small. Indeed, the time it takes a signal traveling in transmission lines with approximately 1/3 of the speed of light in vacuum to cross, for example, the distance from the processor to the memory and back, limits the rate at which computer can perform its operations, limits the *clock rate*. In 1979 it was believed that Si is limited to at best about 0.5 GHz clock rates. Of course, today, with computers routinely having 10 times higher clock rates, the arguments and estimates made back then seem less convincing. Nonetheless, in 1979, the appeal of superconducting supercomputer was strong, mainly because of fast switching and much lower power dissipation.

As we said, 270 MHz clock rate was demonstrated in the CSM prototype in 1981–1982. In the meantime, silicon technology was pushing its own clock rates, threatening to catch up with these numbers. What is more, there was a serious issue with respect to the *maximum* clock rate achievable with *latching logic.* It was known that high clock rates could produce uncontrollable switching known as *punchthrough*, first predicted by Bell Labs scientists Ted Fulton and Bob Dynes in 1971 [14]. In a Josephson computer, punchthrough implies a failure of logical operations. For 10 years since 1971, it was assumed that this detrimental effect has a sharp cutoff, that it does not take place at all at clock rates below a certain threshold. It was, however, realized in 1981 [15] that there is low but finite probability of uncontrollable switching at all clock rates, that there is a probability "tail" extending below that threshold. Therefore, the punchthrough could be detrimental for a large system, like a

[b]Martin Lepselter, at the time, Bell Labs director in charge of advanced chip development (essentially, all Si technology effort), once said to my friend Serge Luryi, in a context of comparing GaAs technology with Si technology: "Silicon is like a great white shark; it swims with its jaws wide open and just eats everything on its way; you cannot beat silicon!"

computer, where the large number of switching elements is making that small probability significant. In 1980 T. R. Gheewala estimated that punchthrough would become important in the IBM latching logic (specifically, in the AND gate) if it would switch with time delay of less than 250 ps, or with frequency above 4 GHz [11]. After publications [15], this estimate was revised, predicting trouble at about 1 GHz. Realization of this clock-rate limitation was one of the main reasons why IBM decided to end its JSP project [12].

In retrospect, it is clear that materials problems also contributed to the closing of the IBM project. Indeed, if the only problem were with the narrowing technical advantage of a JJ computer in view of a fast-developing silicon, then Japanese companies would not institute any meaningful continuation of the IBM project. However, they had such a continuation, and in fact, some of the goals of the IBM project were realized in Japan, using our Nb/Al–oxide–Nb technology. A 1.1 GHz clock 4 bit microprocessor was demonstrated in 1988 by Fujitsu [16], a 1 Kbit memory chip was demonstrated in 1987 by NEC [17], and a 4 Kbit memory in 1988 again by Fujitsu [18]. Shinya Hasuo, the head of Fujitsu Josephson effort, has been a Japanese champion of our technology from the very beginning.[c] In 1993, he wrote [19]:

> Although present Josephson microprocessors are not large, they can be operated over ten times faster than similar semiconductor microprocessors. This kind of performance is due largely to the reliability of Josephson junction technology. The niobium Josephson junction (Nb/AlOx/Nb) has served as the base upon which high-speed digital circuits have been developed because of its uniform, stable characteristics. Using niobium junctions, we can fabricate microprocessors having a few thousand gates and memory of a few kilobits.

It is interesting to note that, just as our breakthrough with junction materials appeared at the very end of the JSP project, a bit too late to make a difference, the punchthrough "speed limit" was

[c]I met Dr. Hasuo in 1983 at a conference where I reported on our new Josephson junctions. He then told me that the whole Japanese Josephson program was taking off because of our Nb/Al technology, and he said, "If we succeed, we will erect your statue made of niobium." Sorry for mentioning this flattery here, but I liked the way he put it.

also overcome at approximately the same time, at least conceptually. Ideas concerning the use of single flux quantum (SFQ) pulses (pulses which have an area equal to one flux quantum $\Phi_0 = 2$ mV ps) in logic operations date back to mid-1970s. K. K. Likharev, V. K. Semenov, and O. A. Mukhanov developed a complete logic family based on these ideas, rapid single flux quantum (RSFQ) logic, in the mid-1980s in Moscow State University [20]. In contrast to the latching logic, the 1's and 0's in RSFQ are not voltage states but instead the presence or the absence of an SFQ pulse in a given time interval. That new logic removed the suffocating clock-rate limit of the latching logic, opening entirely new horizons. Clock rates approaching hundreds of gigahertz became possible. For example, an astounding 770 GHz operation of a small logic circuit (T flip-flop) has been demonstrated at Stony Brook in 1999 [21]. This new logic, appearing in the mid-1980s, also came too late for the JSP project, but not too late for the future of JJ technology as a whole, which used that method for over 20 years, and now is using its extensions. Most SFQ logic circuits are fabricated in the same refractory Nb/Al–oxide–Nb technology that we are describing, the junctions being shunted by a resistor in parallel, to provide for the correct generation of Φ_0 pulses.

I am not qualified to review the present-day status of SFQ logic and the more recent impressive computational projects that have sprung up in various places. Today the world has moved beyond RSFQ, driven by the need to further improve energy efficiency. People are considering the *exascale* supercomputer, which will operate with 1000 petaflops, or with 10^{18} operations per second [22]. For such a machine, RSFQ's power dissipation, minute compared to that of CMOS, and even to latching Josephson logic, is still too high. The energy dissipated in each SFQ switching event (pulse generation) is fundamentally given by $\Phi_0 I_C$, which for typical $I_C = 50\ \mu$A constitutes 10^{-19} J [22]; with typical logic gate consisting of five JJs, the switching energy per gate is then 5×10^{-19} J. In comparison, at 4.2 K, $k_B T = 6 \times 10^{-23}$ J. Several orders of magnitude difference between these two numbers guarantees sufficiently small logic error rates. In RSFQ, this dynamic switching energy is a small part of the total energy budget. The static (standby) operation requires bias currents passing through the resistive bias network. As a result, the static bias power is about 0.8 μW per gate, which, at

20 GHz, is 60 times higher than the dynamic power associated with switching. New SFQ designs, such as energy-efficient rapid single-flux quantum (ERSFQ) or reciprocal quantum logic (RQL) reduce or completely remove that static power [22].

A superconducting processor has other important advantages beyond the low power dissipation: superconducting transmission lines are much better than normal-metal lines for transmitting high-frequency signals without distortions [20], and working at 4.2 K provides an advantage in terms of greatly reduced thermal noise compared both to room temperature and to liquid nitrogen boiling temperature of 77 K.[d]

So, in the late 1970s and early 1980s, it was assumed that superconducting interconnects and JJs would resolve problems faced by silicon technology; they were the way of the future, at least for the fastest supercomputers.

5.4 Josephson Junctions in 1980

Aluminum, which is quite hard (melting point of 660 C) is known to grow a beautiful self-limiting, pinhole-free tunneling oxide [23]. The Al–Al oxide–Pb junction was the first in which Ivar Giaever performed tunneling and discovered superconducting gap in lead in 1960 [24], the work for which he shared 1973 Nobel Prize with Esaki and Josephson. If Al had sufficiently high T_C, so that Al junctions could work in liquid helium, we would not be writing this book, as all junctions would be made of Al–oxide–Al, end of story. Alas, Al has transition temperature into superconducting state of 1.2 K (a little higher when it is disordered or amorphous). Using all-Al junctions requires reaching temperatures at least twice lower than T_C, which would require the use of special refrigeration techniques, making such junctions impractical. To operate in liquid He, at 4.2 K, only Pb and Nb among elemental superconductors have high-enough T_C's of 7.2 K and 9.2 K, respectively. Of these two, Pb is soft, with a melting

[d]I recall how at the end of the 1980s, when high-T_C superconductivity was the talk of the town, people discussed various digital logic schemes based on high-T_c junctions operating at 77 K. Listening to one of these discussions, Ted Fulton grumbled, "Wouldn't you want to have Nb junctions instead."

point of 327°C, while Nb is a strong, refractory metal with a melting point of 2468°C.

Pb–oxide–Pb junctions were made at IBM and elsewhere from the beginning of times. They exhibited decent $I-V$ curves, but were very fragile and plastic, a particularly troubling aspect being their inability to tolerate even minimal thermal cycling between room temperature and liquid Helium. In the course of addressing these problems at IBM, soft junctions were made somewhat stronger with additions of In, Au, and Bi. Much research went into these matters, and JSP project zeroed in on the base electrode made of Pb–In (12 wt%) and Au (4 wt%), and counterelectrode made of Pb (29 wt%) and Bi [10]. The oxide layer was formed by a combination of thermal and RF-assisted oxidation of the base electrode, and it turned out to consist mostly of In_2O_3, with a small admixture of PbO. Such junctions could withstand 100 cycles with about 5% failure rates [10]. While this was better than the behavior of pure lead junctions that would show close to 100% failure rates even in one cycle, clearly these alloyed junctions still had a long way to go before they would be acceptable for a serious processor technology. In addition, soft junctions could be easily destroyed, either by mechanical scratching during fabrication or in handling or by exposure to water, or even by exposure to humid air. Gerry Dolan once announced, with some glee, that he discovered that "lead dissolves in water": counterelectrodes in the junctions he made completely disappeared when immersed in water (lead oxides being water-soluble). Semi-soft Nb–oxide–Pb alloy junctions that were developed at IBM in 1980 [13] were much better in terms of cycling stability, but still suffered from the mechanical softness of the counterelectrode, as well as from the high dielectric constant of Nb oxide (see below).

A natural inclination then would be to explore Nb–oxide–Nb junctions. This proved to be difficult, as these junctions tend to come out as shorts: Nb, being a very reactive metal, interacts with its own oxide, devouring it as Saturn (Cronos) devoured his own children. With care, either with very long oxidation times, or with RF-assisted oxidation, it was possible to make non-shorted Nb–oxide–Nb junctions. However, all-Nb junctions, even if they could be made, were of inferior quality in terms of their $I-V$ characteristics.

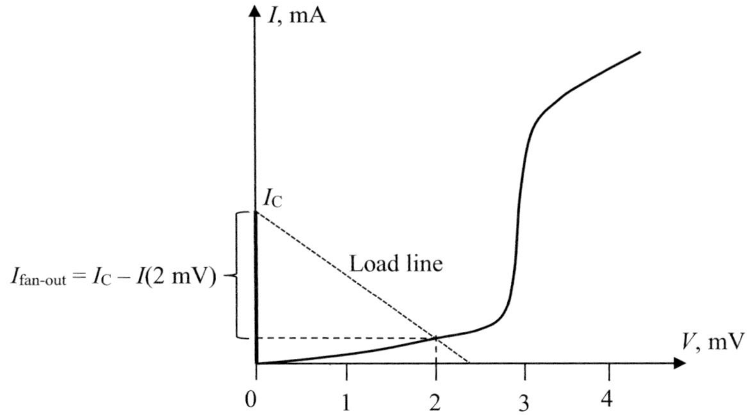

Figure 5.2 Schematics of Josephson zero-voltage and finite-voltage branches of a tunneling I–V. The load line is shown crossing I–V at the 2 mV point. V_m of this curve is approximately 12 mV. The fan-out current $I_{fan\text{-}out}$ deviating from the junction upon its switching is equal to I_C–I(2 mV).

The quality of Josephson junction I–V is often characterized by the figure of merit $V_m = I_c R$ (2 mV), where I_C is the critical Josephson current (maximum supercurrent) and R (2 mV) = 2 mV/I(2 mV) is the resistance at 2 mV, in the gap region, in ohms; here current is in mA. Thus we can rewrite that figure of merit as $V_m = 2I_C/I$(2 mV), where V_m is in mV. Why V_m characterizes the quality of a junction? In latching logic, the significance of V_m lies in its relation to the *fan-out* current, i.e., the current that will be deviated from the junction upon its switching. Indeed, a junction, when it switches from the zero-voltage state to the finite-voltage state, ends up at a point on the I–V curve determined by the *load line* $I = I_C - V/R_L$, where load resistance R_L is plugged in parallel to the junction. In latching logic, the point of intersection of the load line and the junction I–V is chosen in the gap region, just below the $(\Delta_1 + \Delta_2)/e$ current upturn, i.e., close to 2 mV. The junction switches from the top of the Josephson branch, at the point [0; I_C], to the point of its I–V characteristic at 2 mV; I(2 mV). The current that will be deviated away from the junction, the fan-out current, will be therefore equal to the difference between I_C and I(2 mV), as shown in Fig. 5.2.

The *relative* fan-out current (in relation to I_C), $I_{\text{fan-out}}/I_C = [I_C - I(2\text{ mV})]/I_C = 1 - I(2\text{ mV})/I_C = 1 - 2/V_m$. Now we understand why higher V_m is better: it increases the fan-out current.

At 4.2 K, theoretical BCS value of V_m in a junction with Nb gap reaches approximately 100 mV; at 2 K, it increases to thousands of millivolts [1], because $R(2\text{ mV})$ increases greatly at lower T. In contrast with the BCS value, in Nb–oxide–Nb junctions prepared at IBM, V_m was ranging between meager 1.5 mV and 8 mV [25], the latter figure obtained with considerable effort that included RF oxidation and two RF cleaning steps. Thus in the best IBM all-Nb junctions, the fan-out current would be equal to $1 - 2/8 = 3/4$ of I_C. Additionally, the I_C itself was reduced from the ideal BCS value by about a factor of 2 [25]. In terms of using such junctions in a latching logic, these aspects by themselves probably could be tolerated. The real problem (often unspoken) was with *reproducibility*: imperfect junctions (i.e., junctions with low values of V_m) are hard to make identical, hard to reproduce.

What causes low V_m in Nb junctions? Nb oxidation is notoriously complex: the layers adjacent to the tunneling oxide in both Nb electrodes are damaged by oxygen (bottom layer during oxidation, top layer due to Nb interacting with its own oxide), resulting in a gradient of properties. Detailed XPS and AES studies of oxidized Nb surfaces revealed a roughly layered system of oxides and sub-oxides, with Nb_2O_5 on top, NbO and, possibly, NbO_2 under it, and Nb with dissolved oxygen, NbO_x, $x = 0.02$, under NbO [26]. This picture provides the physical basis for the imperfect I–V and the low V_m. Indeed, a gradient of properties in the vicinity of the tunneling barrier, within the depth comparable to the *coherence length* of Nb (the latter being about 40 nm in bulk pure Nb and about 10 nm in small-grain Nb films), implies deviation of the density of quasiparticle states from the BCS shape. States appear in the superconducting gap [1, 27], which in the BCS picture supposed to be clean, free of quasiparticle states. These states in the gap will allow tunneling current to flow in the range of voltages where it could not have flown if the gap were clean. This is especially clear in Nb–oxide–Pb junctions, where the excess current shows up in the energy range $\Delta_{Pb} < eV < \Delta_{Pb} + \Delta_{Nb}$ [1]. If leaky oxide were responsible for the appearance of excess current, such current would exist at all bias voltages, from 0 to $(\Delta_{Pb} + \Delta_{Nb})/e$, not just above the Pb gap. The upturn of excess current at Δ_{Pb}/e is the telltale sign of the states in the gap effect. In Nb–oxide–Nb junctions, additionally the top electrode has its own imperfect layer adjacent to the tunnel barrier. The structure of that layer has been less studied than that of the bottom one, this top layer being buried under the top electrode, surface techniques thus being unable to access it directly. Nonetheless, it also contributes to excess current in tunneling. Thus in an all-Nb junction, with two unclean gaps, the excess current will be observed for all biases, from zero to the sum-gap. This has been often misinterpreted as the result of a leaky oxide, but in all likelihood, it is the result of the presence of quasiparticle states in both gaps, in the base electrode and in the counterelectrode.

It is hard to expect such complex phenomena to be reproducible; the exact form of the density of states and the resulting features of an I–V would be prone to variation from one batch of junctions to the other, even if they were nominally made in the same technology. Leo Tolstoy expressed the same thought in the famous opening line of *Anna Karenina*: "All happy families are alike; each unhappy family is unhappy in its own way." In other words, happiness is reproducible, while unhappiness is irreproducible.

Another serious problem with Nb junctions was the dielectric constant of native Nb oxide. The insulating (tunneling) oxide in Nb junctions is Nb_2O_5, niobium pentoxide. It has a high dielectric constant, $\varepsilon = 30$ [28]. This ε is about 3 times higher than the value found in the oxide grown on Pb–In–Au base electrode adopted in the IBM project: as we mentioned above, it was predominantly indium oxide In_2O_3 with $\varepsilon \approx 10$. Higher ε implies proportionally larger specific capacitance and thus three times slower logic—which was unacceptable for the JSP project. IBM's Nb oxide–Pb alloy junctions [13] were implemented as edge junctions to reduce the junction capacitance, in order to fight the high dielectric constant of Nb oxide; thus they were made on a Nb film edge, which has small area (small width in one direction).

Despite these problems, Nb was very appealing. Considerable body of work exists on semi-soft Nb–oxide–Pb junctions, which are much easier to make than Nb–oxide–Nb ones. I. Giaever and K. Megerle were the first to make tunnel junctions with Nb in

1961 [29]. Two years later, Giaever made Nb–oxide–Sn junctions on improved Nb films with near-bulk T_C [30]. The credit for making the first Nb–oxide–Pb Josephson junctions goes simultaneously to J. E. Nordman [31] and to L. O. Mullen and D. B. Sullivan [32], who published their papers side by side in 1969. All of them noted large excess current below the sum gap $\Delta_{Pb} + \Delta_{Nb}$ that was not present in all-Pb junctions. Nordman believed (incorrectly) that this excess current was the result of non-uniform junction surfaces. Nordman also noted the prominent "knee" structure above $\Delta_{Pb} + \Delta_{Nb}$. He studied junction stability in aging and found it to be improved over the stability of soft Pb junctions; yet parameters of his Nb–oxide–Pb junctions also changed with time. Mullen and Sullivan speculated that excess currents have something to do with the presence of NbO. That was generally correct, even though they offered no clear explanation as to how NbO produced that excess current; presumably they envisioned parallel tunneling paths into the superconducting Nb gap and into the normal metal NbO. Next K. Schwidtal published his impressive, detailed work on Nb–oxide–Pb Josephson junctions in 1972 [33]. He also noted characteristic excess current at $\Delta_{Pb} < eV < \Delta_{Pb} + \Delta_{Nb}$ and the "knee" structure just above the sum gap $\Delta_{Pb} + \Delta_{Nb}$. Schwidtal has shown that junctions were uniform, and correctly ascribed these features to the damaged near-barrier layer, interpreting them, for the first time, in a proximity-effect model. Schwidtal also noted great improvements in terms of storage and cycling stability of these semi-soft junctions compared with soft Pb–oxide–Pb junctions. R. F. Broom [34], his work becoming something of a classic in the field, made a careful study of temperature-dependent characteristics of Nb–oxide–Pb and Nb–oxide–Nb junctions in 1976.

The number of publications devoted to Nb–oxide–Pb (or Pb alloy) Josephson junctions is very large, and I will not attempt to list or review them here. Overall, these junctions had decent $I-V$ curves; when Josephson currents were present, reasonably high values of V_m in the range 10 mV–20 mV were generally found, and, owing to the strong Nb base electrode, such junctions exhibited decent cycling stability. Needless to say, the problem with high dielectric constant of Nb oxide remained in all of them. As we said above, IBM's technology of such junctions, developed in 1980, was based on edge junctions with small area and thus small capacitance [13].

This review of the junction technology in 1980 would be incomplete if I did not mention several attempts at improving Nb and some other junctions with the use of additional layers deposited over the base electrode, or over the tunnel barrier, or *instead* of an oxide tunnel barrier. As subsequent developments have shown, that was the promising direction. The intent was to improve tunneling spectroscopy in junctions of the Giaever type and, in some cases, to improve Josephson junctions. In the latter class of junctions, the class we are primarily interested in here, additional layers did not yet produce the desired improvements by 1980, and, as a result, these early attempts at resolving Nb problems in Josephson junctions were not highly publicized. In fact, I learned about them only when we did a literature search for our patent with J. M. Rowell towards the end of 1981, and when I prepared the bibliography for our article written with J. Kwo, published in 1984—Ref. 1 here.

Here is the brief discussion of overlayer studies before 1980 covering both types of junctions: Giaever junctions and Josephson junctions.

As far as I know, the earliest use of Al overlayer on top of a problematic base electrode was that of J. J. Hauser, D. D. Bacon, and W. H. Haemmerle in 1966 [35]; work performed at Bell Labs. The team addressed the problem of low gap values previously found in tunneling measurements on Nb_3Sn and V_3Si. It was suspected that the reason for reduced gaps was the damaged or off-stoichiometric near surface layer a few nanometers thick. With extremely short coherence length in A-15 superconductors, this presented a problem for tunneling which probes near-barrier layer to the depth of the order of coherence length. V_3Si films were made by sputtering, and covered with 8 nm of Al, which was then oxidized. Using Pb counterelectrode, I–Vs and their derivatives were measured. The correct V_3Si gap commensurate with the high T_C of this compound was determined. The authors of Ref. 35 concluded that superconducting gap of V_3Si was induced in a thin Al layer by proximity effect.

Years later, in 1978, at Stanford University, David Moore and coworkers also studied tunneling into A-15 compounds, including

V_3Si [36]. One of the coauthors of D. F. Moore was J. M. Rowell, who has spent some time in the Department of Applied Physics at Stanford. Ref. 36 describes Giaever junctions with 1.6–16 nm-thick Si overlayer on V_3Si. Resulting characteristics of V_3Si/Si-oxide–Pb junctions were improved compared with junctions without Si overlayer. The authors stated that they were unable to determine whether oxidized Si provided the whole barrier, or the barrier was formed by a combination of Si oxide and the native oxide of V_3Si (whatever that native oxide may have been—probably Si oxide as well?), the latter plugging the pinholes in a silicon layer. I give these details here to show that it was characteristic of the times to think that a 2 nm overlayer could not possibly be continuous. D. A. Rudman and M. R. Beasley additionally studied Nb, Nb–Sn alloy and several A-15 compounds covered with 2–3 nm of amorphous silicon [37]. They found reduced specific capacitances in these junctions. In [36, 37] silicon was used as an overlayer, to be oxidized for the formation of a tunnel barrier. In the work done at Sperry Research, Si was used as an artificial barrier. We will talk of this work a little later.

Over the years, after the work of Hauser et al. [35], Al overlayer technique was employed in a few tunneling gap measurements. For example, V gap was measured by covering V (which does not by itself produce a usable native oxide) with 5–7 nm of Al [38], where the Al overlayer was deposited at a lowered temperature, with the substrate cooled by liquid nitrogen.

The proximity tunneling study of Nb foils covered with 5–10 nm of Al is described in detailed experimental and theoretical papers published in 1979–1980 by E. L. Wolf, J. Zasadzinski, J. W. Osmun, and G. B. Arnold [39; see also chapters in this book written by E. L. Wolf, G. B. Arnold, and J. Zasadzinski]. The experimental study was performed on ultrahigh vacuum-cleaned, zone-refined, and electro-polished Nb foil covered in high vacuum with Al overlayer. Oxidized Nb/Al base electrode was then finished with indium counterelectrode. Measurements included direct I–V as well as $\mathrm{d}V/\mathrm{d}I$ vs. V and $\mathrm{d}^2V/\mathrm{d}I^2$ vs. V. The results were excellent, resolving problems previously encountered by others when tunneling into thermally oxidized Nb. Complete coverage of Nb by Al overlayer (*wetting*) was confirmed both by tunneling and by surface Auger analysis. The value of Nb gap found in that study,

$\Delta_{Nb} = 1.57$ mV, was probably the highest value yet reported. As described in the E. L. Wolf's chapter in this book, J. M. Rowell visited Ames Laboratory in the fall of 1979 and scrutinized the results.

We are now turning to Josephson junctions with Al overlayers. In 1972, R. B. Laibowitz and A. F. Mayadas of IBM published a paper [40] in which they described Josephson junctions with the structure Nb/Al–oxide–Nb, where Al overlayer thickness ranged from 40 nm to 160 nm. The junction quality was rather poor. We can estimate $V_m = I_C R(2\ \text{mV}) = 2\ I_C/I(2\ \text{mV}) \approx 3$ mV from Fig. 1 of Ref. 40 showing an I–V curve of a nominal Nb/Al(80 nm)–oxide–Nb junction. There is no doubt that this disappointing result stopped IBM team from pursuing this line of research further—a big mistake, in retrospect. The reason Nb/Al–oxide–Nb structure had such a disappointing debut was probably having to do with the incorrect way in which it was made into a Josephson junction. The authors of [40] commented themselves that the edges of Nb base electrode stripe were poorly, if at all, covered with Al overlayer, and as a result tunneling probably proceeded through the Nb oxide formed at the edges, apparently totally dominating the Al oxide tunneling. It is unclear why this obvious flaw was not corrected. In reading [40], one gets a feeling that, despite being based on the great concept, the work was performed poorly, and was unfinished and abandoned. There are features in that paper that do not correspond to what would be expected from having an Al overlayer 40 to 160 nm thick. Indeed, our experience with Al overlayers of such considerable thickness (in our case, Al layers of up to 100 nm were studied [1]) was that the shape of the tunneling I–V changes dramatically, reflecting proximity effect tunneling in SN–I–S or SN–I–NS structure. For one thing, one expects to find a greatly reduced sum gap $\Delta_1 + \Delta_2$ in structures with thick N layers. Yet nothing of the sort was observed in Ref. 40; the sum gap was 2.72 mV, which is close to the typical all-Nb junction sum gap at 4.2 K, supporting the view that what was really studied in [40] was a Nb–oxide–Nb edge junction, with Al overlayer playing no role whatsoever. The authors did not measure junction capacitance, which would tell them whether they are dealing with Nb or Al oxide.

There were two legal disclosures by Laibowitz and Mayadas, both in *IBM Technical Disclosure Bulletin.* That bulletin was a conduit for ideas IBM wanted to protect but did not feel like spending time and money patenting. The first disclosure [41] preceded the *Applied Physics Letters* [40] we just discussed by two years. Another [42] appeared in that bulletin one year after the APL, in 1973. These disclosures described similar junction structures: first Nb/Al-oxide–Nb, and then the structure with two Al layers, one covering Nb base electrode, and another deposited over the oxide formed on the first Al layer, forming a nominally symmetrical Nb/Al–oxide–Al/Nb structure. Such symmetrical junctions were *drawn* in [42], but they have not been actually *made*. These disclosures, serving only a legal purpose, contained no more technical information. Still, it is interesting to realize that R. B. Laibowitz and A. F. Mayadas envisioned the potential benefits of Nb/Al–oxide–Nb and Nb/Al–oxide–Al/Nb structures way back in 1970–1973. Should they have pursued the subject further and fabricated good refractory junctions in the early 1970s—which was well within their reach—the fate of the IBM supercomputer project could have been very different, at least in terms of materials (but that would imply—also in terms of on-chip uniformity and margins, which were so important for the latching logic based JJ computer). And should we at Bell Labs have known in 1980 about these disappointing 1972 results with Al overlayers, we would probably avoid tackling this structure altogether . . .

Al overlayers were also used in semi-soft Josephson junctions. For example, J. T. C. Yeh and C. C. Tsuei of IBM, in 1978, prepared Josephson junctions that had an alloy $Nb_{1-x}Al_x$ base electrode, compositions ranging from $Nb_{79}Al_{21}$ to $Nb_{65}Al_{35}$, and Pb counterelectrode [43]. Their hope was that Al would oxidize while Nb would not, thus forming a good tunnel barrier with low dielectric constant. Yet, for some reason, the dielectric constant was still high. This prompted Yeh and Tsuei to deposit a 2 nm Al overlayer over the base electrode, which was thermally oxidized, and then dielectric constant markedly decreased. The quality of the tunneling I–V also improved. Strangely, while $Nb_{1-x}Al_x$–oxide–Pb junctions showed decent Josephson current (Fig. 1 in Ref. 43), the I–V of the $Nb_{1-x}Al_x$/Al(2 nm)–oxide–Pb junction (Fig. 2 in Ref. 43) does not show any Josephson current. Overall impression from that short publication is somewhat similar to the one we get reading the above-mentioned publications of Laibowitz and Mayadas: great start, promising direction of research, golden opportunity within reach . . . and then the work appears unfinished and abandoned.

An example of an overlayer deposited over the barrier is provided in [44], where 0.8–2 nm of Cu was evaporated over the tunneling oxide, forming Nb–oxide–Cu/Nb junction. This prevented the formation of super-shorts, but the quality of the I–V was still very poor.

A. M. Goldman and coworkers used Al overlayers [45] on Nb_3Ge; they deposited 10 nm of Al, most of which was subsequently ion-milled away, and then oxidized what remained. In 1981, A. M. Goldman and coworkers [46] prepared semi-soft Nb, Nb_3Ge, $ErRh_4B_4$, and Au junctions with rare earth overlayers made of Er and Lu. This last publication followed our work on Nb/Al junctions, and so it cannot be considered part of the background.

In some cases, the tunneling oxide was eliminated altogether, whether native or that of the deposited overlayer, being replaced with artificial deposited barrier. In 1973, W. H. Keller and J. E. Nordman (that same Nordman who made some of the first Nb–oxide–Pb Josephson junctions [31]) made Nb Josephson junctions with Ge and InSb barriers and Pb counterelectrodes, obtaining good characteristics [47].

A considerable body of work describes Si artificial barriers. We have already referenced Stanford work on A-15s and on Nb that was published in 1979, the interest there being in measuring the gaps. Strictly speaking, those junctions should be characterized as having oxidized overlayers. At approximately the same time, an all-refractory Nb technology using Si artificial barriers was coming out of Sperry Research Center [48]. In that case, the barrier was formed from hydrogenated Si, which was additionally doped with n- and p-type impurities. Junction I–Vs in that early work coming out of Sperry were horrible; the V_m from Fig. 2 in [48] is approximately 2 mV, possibly the lowest I have ever seen in a publication. Then, in 1981, the Sperry group achieved a real breakthrough. They came up with what is known by the name "the whole wafer process" [49]. They deposited a blanket Nb–Si:H–Nb sandwich structure covering the whole wafer; here Si:H designates tunnel barrier made of hydrogenated silicon 7 nm thick. The deposition system was not opened until this structure was completed, which allowed for a degree of cleanliness and control that was unknown in the old junction fabrication techniques. All processing that defined tunnel junctions was done later, and it was done in an ingenious way, using what they called "selective niobium anodization process" (SNAP). Controllable anodization of Nb was used to the advantage, something that could not be done with Pb. The quality of the I–V curve of a resulting refractory junction was very decent, with $V_m \approx$ 12 mV when V_m was defined as $I_C R(1.7\ \text{mV})$ [49]; using our standard definition, with $V_m = I_C R(2\ \text{mV})$, the figure of merit would be about 9 mV. The authors of [48, 49] did not provide specific capacitance measurements, but it was clear that a thick Si barrier would have very low capacitance (and it did, as they showed later).

The idea of the clean whole-wafer deposition with subsequent patterning was a powerful one, a game changer for the refractory technology. The first Sperry publication describing this idea [49] more or less coincided with the start of our own work on Nb/Al-oxide–Nb system in 1981. We took that idea from the Sperry group and applied it, with some modifications, to our structure. This was perhaps the single most important outside influence that shaped our own project and made it attractive for technology. I will talk some more about whole-wafer processing when describing our work.

To summarize, there were three categories of problems encountered with Nb–oxide–Nb junctions:

(1) Problems that arise because of the messy and complex Nb oxidation of the base electrode, manifesting itself in the appearance of excess currents in the gap region and resulting in reduced quality I–V (low V_m)
(2) Problems that arise because Nb counterelectrode reacts with the tunneling oxide resulting in both reduced quality I–V and in the high likelihood of obtaining shorted junctions
(3) Problem with high dielectric constant of native Nb oxide, which was important for fast Josephson digital circuits

A solution suitable for Josephson integrated circuit technology would have to address all three problems at once. By 1980, such a satisfactory all-refractory solution has not been found (although

IBM's semi-soft Nb–oxide–Pb edge junction technology [13] was quite acceptable and could serve the JSP project if it continued beyond 1983). Given what we just described, the good solution may have been found in 1981, in the work of the Sperry group; it was also found, as will be described below, in our work on Nb/Al–oxide–Nb junctions, which have shown the first promising results in 1981.

5.5 Making Semi-soft Tunnel Junctions

I knew about problems with Nb, but I had to start somewhere, to learn how to make Josephson junctions and how to measure them. So in 1979 I started by making relatively large Nb–oxide–$Pb_{0.9}Bi_{0.1}$ junctions, with 0.2–0.5 mm^2 area, making them "by hand," with shadow masks, as opposed to making much smaller, micrometer-scale junctions using lithographic techniques and *clean room* processing. Making large junctions "by hand" is conducive to exploration of new materials and their combinations, compared with the more rigid and more disciplined lithographic definition of micrometer-scale junctions.

The junctions were made in stages. First, 2.5 cm long and 3 mm wide Nb base electrode stripe was deposited through the shadow mask in the magnetron deposition system I described in Section 5.2. Then Nb base electrode was exposed for hours or even days to air or to oxygen to form the tunneling oxide. The first mask was removed, and the edges of the base electrode stripe were painted with organic insulator (DUCO cement) to prevent tunneling into these edges, and finally the sample was covered with the second, orthogonal shadow mask for counterelectrode definition, and several $Pb_{0.9}Bi_{0.1}$ counterelectrodes (cross stripes) were evaporated through that second mask in a vacuum evaporator. We did not want to contaminate Nb system with soft metals, using a separate evaporator for $Pb_{0.9}Bi_{0.1}$. From the end of Nb deposition to the start of the pumping down in the second evaporator system, the Nb base was exposed to ambient atmospheres: first to dry oxygen that was admitted to the deposition chamber, but then, inevitably, to room air, which could be very humid, especially during the New

Jersey summer. I used to say that "tunneling season" in New Jersey lasts approximately from September to May. The quality of junctions noticeably decreased in the summer, when humidity was high.

The oxidized Si substrate containing several junctions was mounted on a low-temperature probe, which was pre-cooled in liquid nitrogen, and then lowered into a cryostat with liquid He at 4.2 K; one could also pump on He and get the temperature down to about 1.5 K, making tunneling characteristics a bit sharper than at 4.2 K.

In large junctions, the zero-voltage Josephson current is suppressed; it may be still visible, but it is smaller than the value predicted by the theory. However, even in large junctions one can still observe the detailed shape of the normal tunneling characteristic exhibiting the strong current rise at the sum of the gaps $\Delta_1 + \Delta_2$, and a smaller feature at the gap difference $\Delta_1 - \Delta_2$ (see Fig. 2.3 in E. L. Wolf's Chapter 2).

Like everyone who worked in superconductivity, from the beginning of my thesis at BNL, I was already "hooked" on observing the resistive phase transition, when resistance suddenly and mysteriously drops to zero at T_C. Now Josephson junctions added to it a new thrill of observing on the I–V the perfectly vertical (zero resistance) Josephson current at zero voltage, and, upon exceeding the critical current, seeing how the voltage jumps to a finite $\Delta_1 + \Delta_2$ value of about 3 mV, where Δ_1 and Δ_2 are the gaps of Nb and Pb-Bi. I knew that a Nobel Prize was awarded in 1973 to Brian Josephson for the prediction and the theory of that zero voltage super-current (see Chapter 1 written by the great man himself), and to Ivar Giaever, for observing the gap in tunneling. Beyond prizes, you knew that you are privileged to observe something incredibly subtle. That zero resistance occurs despite electron pairs passing through the layer of insulating oxide; that jump from zero-voltage to finite-voltage state occurs intrinsically in a time interval approaching 10^{-12} seconds; the "normal" tunneling characteristic appears despite superconductivity in both electrodes, reflecting the shape of the density of states and exhibiting the superconducting energy gap. Above all, it was exciting to remind yourself that what you are seeing is pure quantum mechanics in action, as there is no way to explain any of that classically.

Once I mastered semi-soft Nb–oxide–Pb alloy junctions, the time came to try something new. The first idea that came to mind was to replace Nb with another decent T_C superconductor that would be strong mechanically and would form a better oxide. It would have to be a compound or an alloy, as there is no elemental superconductor with sufficiently high T_C besides Pb and Nb. My collaborator on that work was Mark Levinson, like myself, newly hired young MTS. Our choice fell on Zr_2Rh, a material with $T_C = 11.3$ K, and, we hoped, the good and stable ZrO_2 oxide. Nobody had made films of Zr_2Rh before us, and nobody had tunneled into them, so it seemed interesting enough. Mark was making the films in his diode sputtering system, where he deposited Zr_2Rh onto a substrate held on a specially made heater, as cold films made in my system did not form the correct crystalline structure and were not superconducting. The best films were prepared at temperatures between 550 and 650°C (which was bad already—as this temperature is too high for applications). Then they were oxidized inside the deposition system in O_2 atmosphere, and post annealed, and eventually I finished the junctions with the Pb-Bi counterelectrode deposited in my evaporator. The best tunneling I–Vs were obtained when we capped the Zr_2Rh base electrode with Zr overlayer about 1 nm thick. I do not remember now if this was done before or after the first appearance of Al overlayers—probably after. Even the best I–V showed a significantly reduced gap in Zr_2Rh, and the estimated V_m was low, perhaps about 5 mV. So, overall it was not an exciting result; the material turned out to be capricious, and even with considerable care produced poor junctions. That is perhaps why, while we did most of this work in 1981, we published the results only in 1983 [50], even though by that time this line of research was made obsolete by the success of the Nb–Al system.

In 1979–1981, I also made and studied junctions with alloyed base electrodes hoping to substitute Nb oxide with a better oxide of the second alloy ingredient (an idea that apparently occurred approximately at the same time to C. C. Tsuei of IMB [43]). I made junctions with Nb_xTa_{1-x}, Nb_xZr_{1-x}, Nb_xAl_{1-x} (the same alloy as in Ref. 43), Nb_xHf_{1-x}, Ta_xZr_{1-x}, Nb_xHf_{1-x}, and $Nb_xTa_yZr_{1-x-y}$, finishing all of them with Pb–Bi counterelectrodes. It pains me to say that nothing came out of this work; the data, some of it quite decent and interesting, was not even published. The reason all of this work was forgotten and abandoned was the success of the Nb/Al system.

5.6 Metallic Superlattices and Tunneling into Nb/Al

In 1980 along came *metallic superlattices*. They were at first seemingly unrelated to refractory JJs, but they turned out to lead us right to them.

It was the time when several researchers started making stacks of metal layers, so-called metallic superlattices. After all, semiconductor superlattices such as GaAs/GaAlAs were the great success at the time (and still are), and "metal people" wanted their 15 minutes (or more) of fame too. One of the pioneers of this activity was Ivan Schuller, who made Nb/Cu multilayers [51]. Ivan was very proud of his multilayers, calling them LUCS, which stood for "layered ultrathin coherent structures" (Ivan thought that this abbreviation was close enough to the word "luck"). He and Charlie Falco studied these multilayers electrically, including superconductivity, and structurally, with X-rays [52], and found that individual layers were continuous even at the smallest thicknesses of the order of a few nanometers (hence the word "ultrathin" in LUCS). This was unexpected, and therefore interesting.

Indeed, in a couple of preceding decades, in the 1960s and 1970s, thin metal films were mostly deposited on insulating substrates, such as sapphire, or quartz, or silicon, or on semiconductors such as PbTe. The latter was done by my thesis advisor Myron Strongin, who researched ultrathin superconducting films while looking for *exitonic* superconductivity. It was found that inevitably, when the thickness dropped to a few nanometers (a few atomic layers), films became discontinuous; they formed "islands." So the fact that a metal can grow in a continuous manner in a stack, on top of another metal, was of interest. It was believed that metals growing on metals "wet" the surface, while metals growing on insulators and semiconductors do not. That was the general explanation for those ultrathin continuous layers. Note that metallic superlattices did not form true epitaxy, but showed instead some coherence, or partial crystallographic reference between the layers (hence the word "coherent" in LUCS).

So we started doing the same, but with a twist. We wanted to make superconducting superlattices, then to study their resistivity,[e] and to study the evolution of their T_C as a function of individual layer thickness, invoking proximity effect, superlattice effects, etc., and, finally, to tunnel into them, to perhaps find superlattice effects in tunneling.

People initially involved in that were John Rowell, Denis McWhan, and I. I was making multilayers, Denis provided expert and sophisticated X-ray analysis and structural modeling [53], and John was studying multilayers by tunneling. We tried different combinations of elements, starting with Nb/Cu, which had been already studied by I. Schuller [51]. We also made Mo/Al and Mo/Zr multilayers, but eventually our interest focused on Nb/Al, the two superconductors with very different T_C's and very different gap values, and one of them, Al, forming a good tunneling oxide.[f]

A little later John delegated tunneling measurements to Jochen Geerk, a brilliant experimentalist from the German government lab at Karlsruhe, who was visiting John's lab at Bell Labs for a few months. Jochen, like John, was an expert in tunneling. They both could use the celebrated W. L. McMillan's *inversion procedure*, which allowed one to interpret tunneling data in terms of Eliashberg function $\alpha^2 F(\omega)$, and to get from it the coupling parameter λ and the Coulomb pseudopotential μ^*. The inputs from tunneling into the McMillan–Rowell unfolding procedure [54] were the first and the second derivatives $\mathrm{d}V/\mathrm{d}I$ and $\mathrm{d}^2V/\mathrm{d}I^2$ vs. V. This is described in more detail in E. L. Wolf's chapter, in the context of doing this for Nb foils covered with Al.

So, I was making multilayers (superlattices) in my system, with varying thicknesses of individual layers, and then I was measuring

[e] I was always interested in resistivity of metals, and in 1986 I published a paper, *Phys. Rev.* **34**(3), 540, in which I deduced mean free paths in individual layers from the measurements of the total stack resistivity. I analyzed the data taken on our Nb/Al multilayers, and the data on Nb/Cu that was generously provided by Ivan Schuller.

[f] It is possible—although I do not recall such a discussion—that John Rowell was interested in Nb/Al system because he saw Nb/Al tunneling results in Edward Wolf's group, in Ames, in the fall of 1979, as I have said already.

their resistivity as a function of temperature between He and room temperature, and their resistive T_C, Denis was modeling their structure, and John and Jochen, and sometimes I, were tunneling into these superlattices. The junctions on multilayers were prepared in a manner I have described above for Nb–oxide–PbBi junctions, except that John and Jochen finished their samples with normal metal Ag counterelectrode, which was the standard practice when dealing with McMillan–Rowell inversion. While making samples, I was keeping record of how the stacks were made, and so we knew the exact number of layers and their order, including which layer was laid down last; not that we cared about this detail initially. This project started, I believe, around September or October of 1980. Looking at the old data, I see that I was measuring T_C's and even doing some tunneling into Nb/Al multilayers in October and November of 1980.

One day, either at the end of 1980, or perhaps in the first days of January of 1981, John Rowell showed up at the door of my lab and said to me something like this: "Jochen is tunneling into Nb/Al superlattices, and he noticed that when the top layer is Nb, tunneling is lousy as usual; but when the last layer is Al, it is beautiful. So why don't you make a thick film of Nb and cover it with a thin layer of Al, and we will tunnel into it." That was certainly an "eureka" moment, and the credit for it goes entirely to Jochen and to John, perhaps especially to John, for quickly making a logical leap from multilayers to a single Nb film. I did just that; on January 7, 1981, I prepared a sample covering thick Nb film (about 300 nm) with 5 nm of Al. When I made tunneling measurements, I saw the near-perfect tunneling characteristic, and the cleanest gap that I have ever seen. Soon afterwards, I made other Nb/Al(d_{Al})–oxide–PbBi samples, with d_{Al} ranging from a fraction of nanometer to a few nanometers. It is tempting to reproduce here a copy of one of those tunneling $I-V$ traces, which I still keep; in that case Al layer was only 0.72 nm thick; the sample was measured on January 19, 1981, data taken in pumped He, at $T = 1.7$ K.

I also supplied Nb(300 nm)/Al(d_{Al}) base electrodes to John and Jochen; after 16 hour room-air oxidations, their samples were finished with Ag counterelectrodes. Later, in the summer of 1980, I explored much thicker Al layers as well, being intrigued by the

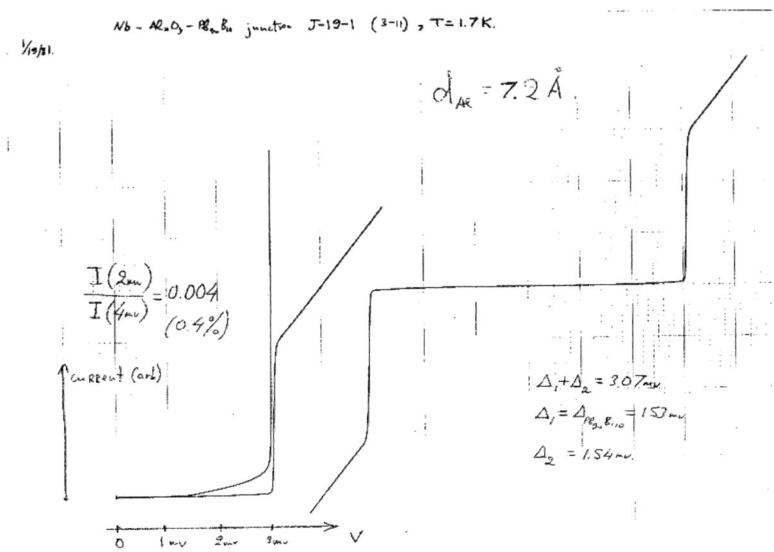

Figure 5.3 Original trace of an I–V of one of the first samples with thin Al overlayers. Here $d_{\mathrm{Al}} = 0.72$ nm, measurement taken at 1.7 K. Nb gap is 1.54 mV, and excess current at 2 mV is only 0.4% of the current at 4 mV.

changes in the tunneling I–V that were clearly associated with proximity effect.

For a pure Nb film, the oxidation rate was so slow that it took the whole day to obtain a tunnel junction with useful tunneling resistance. With Al overlayer, the required oxidation time was much shorter, about half an hour; the high bias (up to about 0.5 V) behavior of the dynamic resistance $\mathrm{d}V/\mathrm{d}I$ also changed markedly [55], exhibiting curvature previously found in pure Al junctions. This change of the tunneling behavior was already obvious at the remarkably small Al coverage of 0.18 nm; in other words, the marked changes were occurring with Al coverage corresponding to as little as one monolayer [55]. We speculated in [55] that at the smallest Al thicknesses, the tunneling oxide had to be a mixture of Al_2O_3 and Nb_2O_5. Later surface studies have shown that in fact Nb oxidation was completely suppressed even in such ultrathin layers (see Section 5.11).

We presented our tunneling studies of Nb/Al multilayers in the March Meeting of the American Physical Society [56] and submitted a paper describing tunneling into Nb/Al multilayers and into Nb with Al overlayers to the Proceedings of the 16th International Conference on Low Temperature Physics, LT-16, which took place in August 1981 in Los Angeles; the proceedings were published in 1982 [57]. Among other results, in that paper, John and Jochen described tunneling spectroscopy of Nb performed with d_{Al} up to 4 nm; they performed unfolding of the $\alpha^2F(\omega)$ and obtained $\lambda = 0.97$ and $\mu^* = 0.15$, all without a need to make corrections associated with proximity effect. In other words, at these Al thicknesses, none of the metallic unoxidized Al seemed to remain on the Nb surface, which was a little surprising (see Section 5.10).

Although I obtained first excellent tunneling characteristics of Nb/Al–oxide–Pb junctions, shown in Fig. 5.3, in January 1981, these I–V curves were not published until the end of that year, the reason being that John and I were required to fist file for a patent (see Section 5.7). Eventually they were published in the Proceedings of the International Electron Device Meeting in December 1981 [58]. In this paper we included a figure showing tunneling I–V curves of Nb/Al–oxide–$Pb_{0.9}Bi_{0.1}$ junctions with Al overlayer thicknesses, d_{Al}, ranging from zero (i.e., for the pure Nb base electrode) to 36.4 nm. I used the same figure in [1], which was published in 1984. It is rather informative, so I am also reproducing it here as Fig. 5.4.

As can be seen, there is a fairly large excess current in the pure Nb base electrode junction, this current starting at about 1.2 mV and increasing approximately linearly from that voltage to the sum-gap voltage of about 3.0 mV. It is important to note that there is no excess current at $V < 1.2$ mV. As we discussed in Section 5.3, and as is discussed in a little more detail in Ref. 1, this excess current should be therefore associated with imperfect layer of Nb adjacent to the tunnel barrier, rather than with a leaky oxide barrier. In junctions with Al overlayers, excess currents are significantly suppressed. The Nb gap is much cleaner under the Al overlayer. What is truly remarkable is that this happens even for d_{Al} as thin as 0.36 nm (3.6 Å), the point that was also stressed in Ref. 55 based on I–V curves at large bias voltages: a single monolayer of Al strongly modifies the tunnel barrier and cleans up the underlying Nb, suppressing its messy oxidation. The last two curves in Fig. 5.4 corresponding to d_{Al} of 11 nm and 36.4 nm show strong proximity

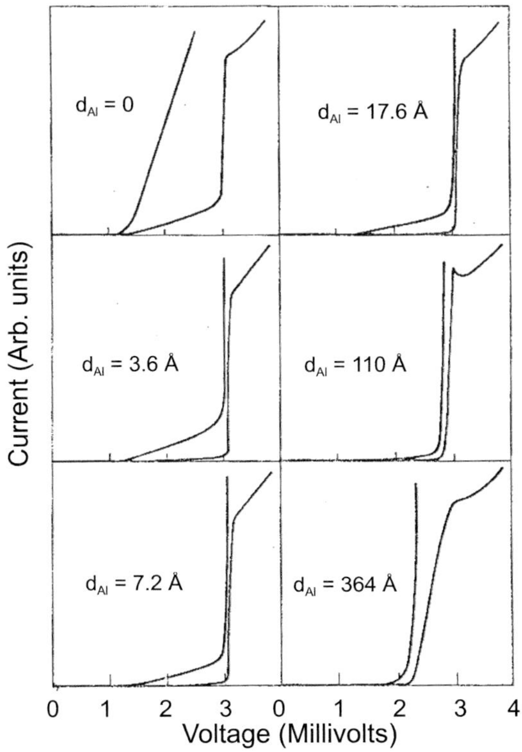

Figure 5.4 I–V characteristics of Nb/Al–oxide–$Pb_{0.9}Bi_{0.1}$ junctions measured at $T = 2$ K. Each curve shows $\times 10$ expansion of the current scale. The junction with $d_{Al} = 0$ (pure Nb) was oxidized for 4 days; the rest for 20 minutes. All junctions have areas of approximately 2×10^{-7} m^2 = 0.2 mm^2 (Scale: 1 Å = 0.1 nm).

effect features: the knee structure at $\Delta_{Nb} + \Delta_{PbBi}$, the sloping current upturn, and the reduced but very clean gap.[g]

[g]I wanted to make a detailed study of proximity effect tunneling in junctions with Al overlayers in a wide range of thicknesses, treating both quasiparticle proximity tunneling (shape of $I - V$) and much less studied Josephson proximity tunneling (reduction of $I_C R_n$). Vladimir Kresin, a theorist I knew well, in 1983 published a paper treating Josephson proximity tunneling, and I used some of his calculations. He predicted that $I_C R_n$ would depend on the coupling constant λ in the N layer, which made this study particularly interesting. I wanted also to interest Gerry Arnold in participating in this project. At the time, we had unique refractory junctions showing these effects very clearly (some of them described in Ref. 60). We published a short

5.7 The Sad Story of Our Patent Application

Prompted by the management (as I recall, it was my lab director who raised the alarm and prevented us from immediately publishing good tunneling curves of semi-soft junctions with Al layers), in the summer to fall months of 1981 John Rowell and I worked on a patent application. I had not yet made all-refractory junctions. At least that is how I remember it. We were assigned one of the patent attorneys from the patent office at Murray Hill; I forgot his name. We tried to think of all possible combinations of materials that would make Josephson junctions with easily oxidizable overlayers. Aluminum was, of course, at the center of our attention, but we also included Y, Zr, Ta, and various rare earths. We also legally covered additional aluminum (or other metal) layer above the tunneling oxide, thinking that perhaps it will be needed to protect the oxide from interacting with the hard counterelectrode.

In the process of writing a patent, the attorney performed the usual patent search, and this search exposed much earlier publications by Laibowitz and Mayadas [40–42] in which they described junctions with Nb/Al–oxide–(Al)/Nb structures. As I have said already, we did not know about that IBM work, and at first were somewhat taken aback. However, upon closer examination, we saw that Laibowitz and Mayadas used very thick Al layers, their minimum Al thickness being 40 nm. This of course contrasted with our thin overlayers. Secondly, we saw that they did not in reality see tunneling that was assisted by the Al overlayer and Al oxide, as was discussed in Section 5.3. That last aspect, whether the structure described in [40] actually worked or did not work, however, would not be relevant to the legal status of our patent claim. Therefore, we limited our patent to thin Al overlayers; I think we specified $d_{\mathrm{Al}} < 10$ nm, the argument being that thicker layers were unnecessary and even detrimental to junction quality because of the proximity effect.

---

paper (conference proceedings) on Josephson proximity tunneling with Huggins (I cannot find the reference now), but the big paper never materialized. I had to abandon this project, the reason being, in part, post-September 1983 pressure at Bell Labs to leave superconductivity altogether. It would have been a beautiful paper.

The fate of that patent was unusual. It was never properly filed in the US, or if it was filed, the patent attorney later abandoned it. He looked like an unhappy and nervous fellow; he left his position at Bell Labs abruptly, sometime in 1982, leaving our patent (and probably some other patents) unattended. Yet by some miracle it was filed in Japan; I still do not know how this happened. Some years later I actually saw our Japanese patent, translated into Japanese, so I am sure it really existed.

In 1990 or 1991, shortly after I moved from Bell to Stony Brook, I received a call from a Japanese man named Dr. Goto. He called from New York, and he wanted to visit me at Stony Brook University, to talk about my patent. I said yes, please come. So he came from New York City, and we sat down in my office. And Dr. Goto said, "I am working for a company called Seiko, the one that makes watches. Our company wants to enter the business of Josephson digital circuits, and so we would like to purchase rights to your Japanese patent on Nb/Al junctions." To which I answered, "Dr. Goto, you should not be talking about this to me, but to AT&T Bell Labs, because we, as employees of Bell Labs, signed all our legal rights to the company." That was the end of that very short conversation. I took Dr. Goto, who was a very nice man, to lunch, and we parted. When I described this story to some of my friends, they told me that I was an idiot. They said that Bell routinely allowed inventors to have some profit from their patents, and that all I had to do was to contact the intellectual property people at Murray Hill and to negotiate the terms. Unfortunately, it was too late. Sometimes I wonder if indeed John Rowell and I could have got some money out of this if I had acted more reasonably.

5.8 Nb/Al Refractory Junctions Are Emerging

It became pretty clear that there might be benefits for JJ technology coming out of these Al overlayers. However, it was not obvious that one can get rid of a soft counterelectrode. We counted on that, and in the patent, we of course included Nb counterelectrode, but there was no experimental proof of that, nor was it known how to approach the problem of geometrically defining such refractory junctions. So here came a moment when I could finally do what I was hired to do: I decided to replace the soft counterelectrode with Nb, to make a refractory Nb/Al–oxide–(Al)/Nb junction. While John Rowell, Ted Fulton, and others were supportive of this, it was no longer John's first priority, and all the subsequent work on refractory technology was done essentially without his participation (see also Section 5.12).

The first demonstration that this actually works, that the Nb counterelectrode no longer shorts the junction, I obtained with large "hand-made" JJs. I made these junctions sometime in the fall of 1981. To replace Pb–Bi with Nb counterelectrode, I had to develop a completely new procedure: indeed, DUCO cement insulation was no longer viable, as it would not survive the second Nb electrode deposition. Nevertheless, the junction area had to be defined away from the edges. We always covered the edges with insulators to avoid tunneling into potentially inferior material. This was especially important now, when a film of Nb was deposited and then covered with a thin layer of Al via sequential depositions through the same shadow mask. Clearly, the tapering edges of such a structure would likely have a problem with Al coverage. I do not recall now whether I knew then about the problems encountered by Laibowitz and Mayadas in Ref. 40—the problems which were the direct result of ignoring this edge exclusion rule. But how to define the junction area? I designed a system of three shadow masks shown in Fig. 2 of Ref. 1. The aim of these masks was to provide Ge insulation coverage over everything but the small square area in the middle of Nb/Al base electrode. Finally, the junction was finished with the deposition of Nb counterelectrode over that square. The results were most encouraging: junctions were not shorted, with very decent I–V curves (being large junctions, they did not show full Josephson current, so V_{m} could not be measured; assuming normal values of $I_{\mathrm{C}} R_n \approx 1.85$ mV, the V_{m} at 4.2 K could be estimated as $\sim$20 mV). One such I–V curve, measured at 1.55 K, is shown in Fig. 5.5.

The I–V shown here was measured in November 1981, yet I published it only in 1984, because at the end of 1981 we already made the first small junctions defined by lithography, and my "large" Ge-protected junctions, having served their role as a proof of principal, were put aside in favor of the technologically interesting

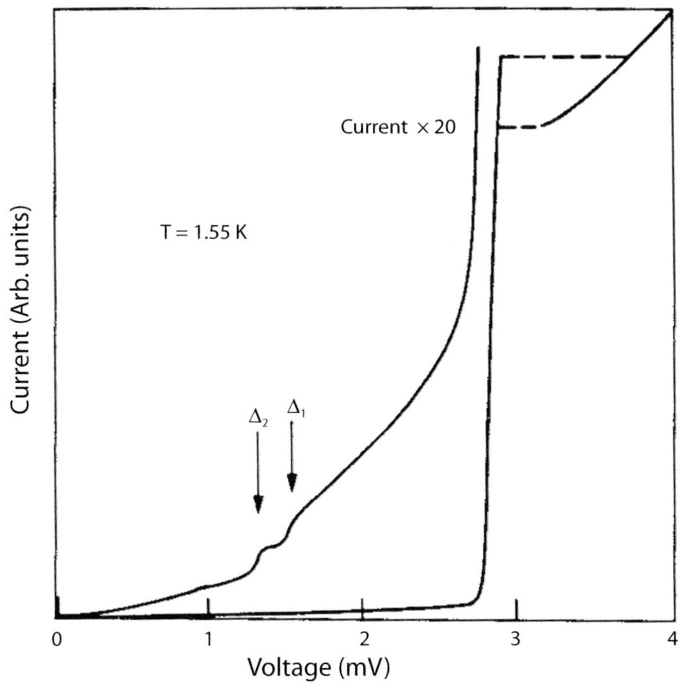

Figure 5.5 Tunneling I–V curve of one of the first all-refractory junctions with two Al overlayers, Nb/Al(2.5 nm)–oxide–Al(2.5 nm)/Nb, junction area defined with Ge insulator depositions through the system of shadow masks described in Ref. 1. The I–V was measured at 1.55 K; the base and counterelectrode gaps are $\Delta_1 = 1.53 \pm 0.03$ mV and $\Delta_2 = 1.32 \pm 0.03$ mV.

ones. The all-refractory curves published at the very end of 1981 [58] were obtained on $20 \times 20\ \mu m^2$ junctions.

5.9 The Whole-Wafer Process: SNEP-SNAP

Once I was convinced that Al oxide survives the second Nb deposition, and that I–V quality is reasonably good, the next step was to make it into a real technology, to define small junctions lithographically, and to develop a process for laying out junctions and their connections on a chip. My collaborators at that stage became my assistant Harold Huggins, who was very helpful in working out the process details, and Morris Washington, who was hired almost simultaneously with me to our department and was previously involved with making Josephson Pb alloy circuits. Morris was also helping to develop process details, and he was skilled in doing photolithography in a clean room. As I said at the end of Section 5.3, we were strongly influenced by the work of Harry Kroger and his group at Sperry Research. We were convinced that the way to go with our Nb/Al junctions was the whole wafer process, in part because of the cleanliness and control it offered, and in part because Nb could be very nicely anodized. Our initial whole-wafer activity was similar to Sperry: we defined junctions by placing photoresist pads over the areas where they will eventually appear, and anodized the unprotected areas of the Nb/Al–oxide–(Al)/Nb wafer through the Al oxide, but not all the way through the base electrode. Upon the removal of the photoresist this produced small islands (future junctions) sticking out of the Nb oxide sea produced during anodization. Then we cleaned the top surfaces of these junction islands by RF sputtering or ion milling, and deposited Nb connections to the individual junctions.

At this time—it must have been some time in 1982—we decided to modify the Sperry process, in order to move away from defining junctions with wet anodization, which in SNAP had to be done to a considerable depth. Thick anodic oxide was swelling in all directions, undercutting the junction area, making junction definition less precise. We wanted to define junctions by dry plasma etching instead.

The reader may remember how I described in Section 5.1 the ready availability at Bell Labs of great experts in every aspect of experiment or technology. This was one such instant. I asked my friend Nadia Lifshitz, and she immediately directed me to an expert in plasma etching who worked in her lab. To my shame, I cannot recall the name of the plasma-etching expert we consulted with. What is worse, I failed to thank him in our subsequent publications for his valuable advice. The advice was indeed valuable: we were told that Nb can be very nicely etched in the atmosphere of CF_4. What is more, when we explained that our structure contains an Al oxide layer, the expert said, "That is great, because Al oxide will act as a very effective etch stop; CF_4 does not etch Al_2O_3." We said, "But the

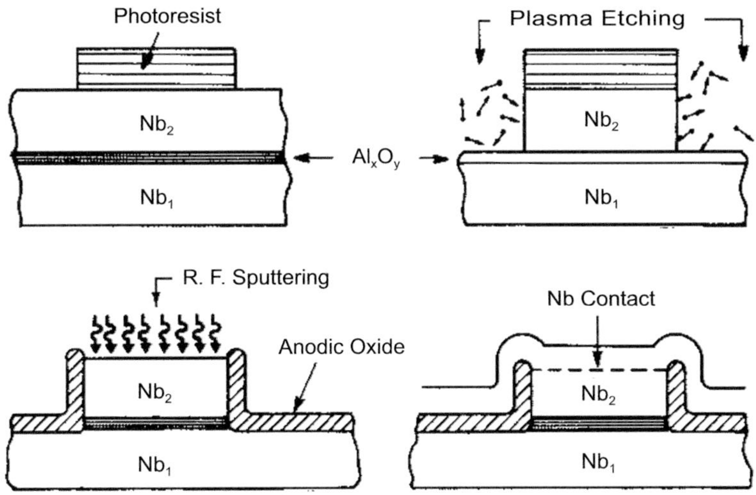

Figure 5.6 SNEP-SNAP process [1, 59]. R. F. sputtering operation is optional, as formation of a second junction at that surface is unlikely.

thickness of that Al_2O_3 is exceedingly small; it is a tunneling oxide only about 1 nm thick." No matter, said the expert, plasma etching in CF_4 will stop dead at that oxide layer. And it did! This was an unexpected gift, which made worrying about the right etching depth unnecessary, as stopping was automatic, just at the right depth, as soon as the top electrode was removed.

We still needed to perform an anodization, to create an insulating layer around the junction; however, this layer could be much thinner than the original anodization layer in SNAP; it presented no problems in terms of junction definition. We called the resulting selective niobium etching and anodization process SNEP-SNAP (I was of course thinking of a somewhat obscure expression "snip-snap," which means "quick, short, sharp, or smart" according to the dictionary). The SNEP-SNAP process is schematically shown in Fig. 5.6.

I want to show only one tunneling I–V of a typical good junction made by SNEP-SNAP; it has $V_m = 48$ mV at 4.2 K, which is very close to the theoretical V_m of an ideal BCS junction. After 1983, such I–Vs could be found in many papers, but to us, at the time (probably at the end of 1982, beginning of 1983) they still looked amazingly beautiful:

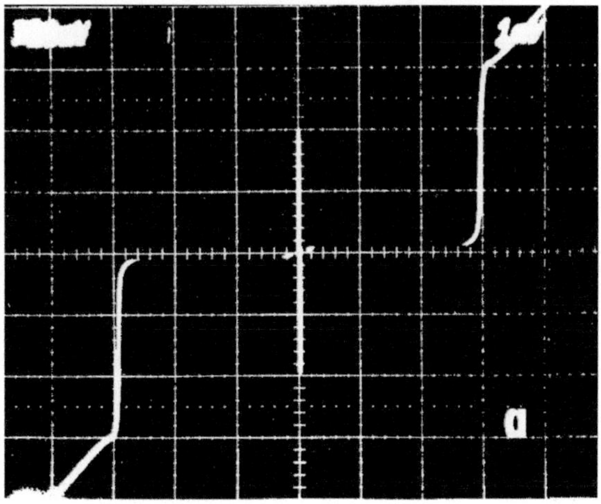

Figure 5.7 I–V measured at 4.2 K; junction size 10 μm × 10 μm; layer structure: Nb(300 nm)/Al(5 nm)–oxide–Nb(300 nm); critical current density $j_c = 480$ A/cm^2; $V_m = 48$ mV.

Christopher Marlowe's *Doctor Faustus*, when seeing (I believe, in Heaven) the Helen of Troy, says, "Was this the face that launched a thousand ships . . .?" Well, then, this is *the trace that launched a thousand chips.*

In September 1982, we submitted, with M. A. Washington and H. A. Huggins, the paper [59] that was published in March 1983 and was later quoted more than any other of our papers. In it we summarized the SNEP-SNAP process and described Nb/Al–oxide–Nb and Nb/Al–oxide–Al/Nb junctions, $d_{Al} = (5 \pm 0.5)$ nm, junction sizes 10 μm × 10 μm and 20 μm × 20 μm, critical current densities from 600 A/cm^2 to 1300 A/cm^2, and V_m from 15 mV to 35 mV (up to 260 mV at 2.0 K). Higher values of V_m were associated with symmetrical junction structures. The tunnel barrier was formed by in situ oxidation in the deposition system in 1 Torr of pure oxygen for 1–2 hours. We provided in [59] some of the details of SNEP, including our experience with Al oxide as a very effective etch stop (so that was the paper in which we should have acknowledged and thanked that plasma-etching expert). The base electrode gap ranged between 1.50 mV and 1.55 mV; that of the counterelectrode, between 1.25 mV and 1.45 mV. $I_C R_n$ product (the characteristic of the size of the critical current; see also E. L. Wolf's chapter) was found to be 1.71 ± 0.06 mV in junctions with single Al layer and 1.49 ± 0.06 mV in junctions with two Al layers, which was 20–25% lower than the theoretical weak-coupling value; most of this lowering could be ascribed to strong coupling in Nb. We stated in [59] that I_C vs. magnetic field **B** (the "diffraction pattern") had an ideal theoretical shape, indicating perfect uniformity of the critical current over the area of a junction. From that pattern we determined the value of the magnetic penetration length in Nb, $\lambda = (120 \pm 20)$ nm, which was slightly larger than the literature Nb value of (86 ± 5) nm. Measuring Fiske steps and using both these values of λ, we determined junction capacitance per unit area $C/A = (0.06 \pm 0.02)$ pF/μm^2, which should be compared to the pure-Nb value of 0.14 pF/μm^2 [13]. If we assumed the normal Al_2O_3 dielectric constant $\varepsilon = 8$, this gave us barrier oxide thickness in the range of 0.88 nm to 1.68 nm, which was most reasonable and consistent with independent estimates from tunneling [55] and from surface studies. We also commented on junction stability: storage of some of the junctions for a full year and several room temperature to 4.2 K cycles produced no detectable changes in the I–Vs, and, what was very important for applications, the structures survived baking of the photoresist at 150°C for 30 minutes. In other words, we showed that all the problems with Nb junctions that we cited in Section 5.3 were now resolved.

This APL [59] came out in March 1983, and five months later, in September, when I arrived at Fallen Leaf Lake (near Lake Tahoe in California), to the Josephson meeting organized, I believe, by IBM, I saw the "long faces" of my friends and colleagues and learned that IBM had just announced that they were closing the supercomputer project. I brought my beautiful new results to a funeral.

Over a year later, in July 1984, Harold Huggins and I submitted another good paper [60], which was also much quoted. Hinting at the recent closing of the IBM project, and in view of Japanese activity in the field, particularly using our new Nb/Al junctions, I opened that paper with the phrase, "Whether or not all-refractory

Josephson tunnel junctions will serve as switching elements in future (Japanese) supercomputers, they are of considerable interest . . .," etc. In [60] we reported the much-improved yield of good junctions and good junction parameters reproducibility from run to run. The single most important step leading to this improvement was heat sinking of the substrates to the water-cooled table in the deposition system—that is when my insistence of building a system with such a table paid off. Heat sinking to the table was done with thermal grease under the substrate and indium foil under the sample holder. I checked that small pieces of gallium, which melts at 26°C, did not melt when placed on the heat-sunk substrate. We even stopped using oxidized silicon chips for substrates, believing that stripping SiO_2 and using pure Si substrates further improved thermal contact between the growing structure and the table. Single-overlayer junctions were now as good as double-overlayer (symmetrical) ones; our V_m values ranged from 25 to 56 mV (!) at 4.2 K. In other words, the best junctions showed essentially ideal BCS characteristics. And, as all happy families, they were now alike. We also studied evolution of junction I–Vs and of critical currents in the range of Al thicknesses, d_{Al}, from 5 to 200 nm (that is when we obtained good proximity junctions that were mentioned above). Tunnel barrier was formed in situ in the deposition chamber, with table and the substrate held at approximately 15°C because of the water-cooling. The time of oxidation was always 10 minutes. The only parameter controlling the thickness of the tunneling oxide (and thus the critical current) was O_2 pressure, which was varied from 0.01 to 1 Torr, i.e., by a factor of 100. The corresponding critical current density varied from 4×10^4 to 10^3 A/cm^2, i.e., by a factor of 40. Note that way back in 1984 we were already making JJs with fairly high critical current densities of 40 kA/ cm^2. Later Alan Kleinsasser with his coworkers at IBM was able to push current densities in Nb/Al junctions to 400 kA/cm^2 [61].

While performing anodization of Nb, Harold Huggins monitored anodization voltage. One day he showed me a feature on the voltage vs. time trace that presumably indicated the instant when anodization profile was crossing the Al oxide and the remaining Al metal in the middle of the junction structure. We improved resolution of this feature by measuring dV/dt vs. t, and found that we can even clearly see the difference in the sharpness of that feature between the case of a substrate being heat-sunk or not being heat-sunk to the water-cooled table. We saw that we have a new spectroscopic tool, very simple and easy to use. We called it "anodization spectroscopy" (we even said in the paper "poor man's anodization spectroscopy") [60]. It was later explored by many, including measurements of anodization profiles of metallic superlattices; I cannot give detailed references on this interesting subject here.

5.10 Uniformity, Stability, and Cycling

At first we made chips with 8 junctions. On the basis of our experience with such samples, we mentioned preliminary results in terms of junction uniformity, reproducibility, and stability in our papers [59, 60]. However, serious testing of these properties requires making large arrays of identical junctions and their batch measurements. This we did with Morris Washington and John Gates, a young MTS from Area 52, who actually came to me and proposed that study. This was done in 1983, results appearing in print in 1984 [62]. The structure consisted of 15 series arrays of 50 junctions each; a single junction was included with each array for an individual measurement. The I–Vs were recorded using computerized test station developed by John Gates. The chips were subjected to 4900 thermal cycles between room temperature and 6 K, to storage at room temperature for two months, and finally to prolonged heating (annealing) performed in stages, with temperature held for 2 to 4 hours first at 100°C, then at 150°C, then at 200°C, then at 225°C, then at 250°C, and finally at 275°C. The overall results of this study were excellent. They were not only incomparably better than for Pb junctions, but also much better than the known results for Nb–oxide–Nb junctions [25], showing that the advantage, at least in part, resided in Al oxide rather than in the strength of the Nb electrode alone. It was interesting that junctions annealed at temperatures up to 250°C were stabilizing at a reduced value of the critical current, this reduction of I_C reaching –20% at 250°C. The amazing thing was that this decrease of the critical current was not accompanied by increased spread of I_C values among the junctions. In addition, despite this uniform decrease of critical current in all of the junctions subjected to annealing, the quality of I–Vs (V_m) was remarkably unchanging. This was so for temperatures up to 250°C. However, at 275°C, the critical current kept decreasing without showing a tendency to stabilize; this temperature was apparently over the threshold of a run-away behavior. We speculated that at this temperature oxygen was continually diffusing through the Nb towards the barrier. This indicated the limits for future thermal processing of such junctions. Our Nb/Al junctions were remarkably resilient.

5.11 Combination of Tunneling and Surface Studies; *Wetting* and Al Disappearance; Junctions with Y, Mg, and Er

Nb junction problems were evident in tunneling I–V curves, and surface studies of Nb oxidation [26] helped to understand the physical causes of these problems. Likewise, surface studies of Nb covered with Al overlayers contributed to a more detailed understanding of *how* these problems were resolved, illuminating some peculiarities that would remain a mystery if not for those surface studies.

Our success with Nb/Al junctions attracted the attention of Bell Labs experts in fields other than superconductivity and tunneling. Raynien (J.) Kwo, a young Bell Labs colleague, easily convinced me that we can benefit from X-ray photoemission spectroscopy (XPS) study of our Nb/Al structures, and she further engaged in that study an expert in XPS, G. K. Wertheim.

There were two puzzling aspects of tunneling into Nb/Al: one had to do with junction resistance, another with proximity effect.

If we fixed the way base electrode was oxidized (e.g., in air) and the time of oxidation (e.g., 30 minutes), and if we used the same counterelectrode material ($Pb_{0.9}Bi_{0.1}$), then the only remaining parameter that influenced junction resistance per unit area was the thickness of Al overlayer, d_{Al}. Theoretically, if Al indeed completely wetted the Nb surface and stayed at that surface, one would expect junction resistance vs. d_{Al} to quickly saturate and remain constant. The threshold value of d_{Al} above which junction resistance is not expected to change could be 1 nm, or perhaps 1.5 nm, but not much greater, this threshold corresponding to the thickness of Al overlayer such that all of it converted into Al_2O_3. Instead, junction resistance saturated only above $d_{Al} = 8$ nm [Ref. 1, Fig. 10; also see Ref. 55]. This was puzzling.

The second mystery had to do with expected I–V modifications at greater d_{Al}: we would expect to see familiar features associated with proximity effect with metallic Al once d_{Al} exceeded the said threshold value and some metallic Al have remained on the surface of Nb. And again, instead of 1–1.5 nm, the threshold was observed at about 10 nm; there was no visible proximity effect at lower Al thicknesses. This suggested that there was a lot less Al on the surface than what would be expected. But where did excess unoxidized metallic Al go?

These issues have been illuminated in XPS studies and accompanying near-surface structure modeling [1, 63], which indicated that for thicker nominal Al layers, substantial amount of Al was indeed missing from the surface. For example, in a sample with $d_{Al} = 11.2$ nm, about 9.6 nm of metallic Al was missing. The only reasonable assumption was that Al wetted *all* of the surfaces, not only the geometrical flat Nb film surface, but also internal surfaces of Nb grains. Our Nb films had $\sim$10 nm size grains; we assumed that Al diffused down these grain boundaries, soaking into the depth of the 300 nm thick Nb base electrode. Thus, Nb–Al geometry was a lot more complicated than the simple layered structure we had envisioned before the XPS study. Raynien Kwo found a nice justification for this behavior in the theory of surface segregation in binary alloys [64], which, when applied to the Nb–Al system, indeed predicted Al segregation to the surface, which we took as another way of stating that Al will wet all available Nb surfaces, including grain boundary surfaces. An elegant demonstration that this was indeed the case was obtained by preparing a Nb film at 650°C, this film having three times larger grains than the films prepared in my system at room temperature. In that case, when covered with Al, a lot less Al was missing from the geometrical surface [63]. In the study by E. L. Wolf et al. on recrystallized Nb foils [39], grains were so large (about 1 mm) that none of the Al was apparently missing.

This finding of the rapid grain boundary diffusion was making our system a bit less predictable, less well defined, at least in terms of the layer geometry. At the time, we span the story presenting it to be favorable to technology: we were saying that this effect relaxed requirements for the Al layer thickness control [63], which was true too. Later, when we realized the benefits of preparing junctions at a fixed cold water temperature of about 15°C [60], we revisited the surface study and found that grain boundary diffusion of Al was greatly reduced in such junctions [1]. The temperature during the Nb deposition onto freely placed substrates could rise to about 200°C, while in heat-sunk substrates it was close to 15°C. It was unlikely that Nb grain size would strongly depend on this modest temperature difference, suggesting that apparently the Al diffusion process itself was thermally activated: we speculated that Al diffusion down the grain boundaries was slow at 15°C, and much faster at about 200°C.

Later, in 1986, I teamed up with another group of surface scientists, this time using Auger electron spectroscopy, transmission electron spectroscopy, and scanning electron microscopy as tools. The paper that we have written [65] attempted to contrast the two pictures corresponding to "warm" and "cold" samples. The results contradicted some of the conclusions we reached earlier on the basis of XPS [1, 63]. Yet I do not have a complete trust in the conclusions of [65], as some of them appeared convoluted and not entirely self-consistent.

About a year later, the same team that studied XPS explored other reactive overlayers: Mg, Y, and Er [1, 66]. By that time, Alan Goldman and coworkers have also reported their study of erbium overlayers [46]. We found that Mg and Y were very similar to Al: they completely wetted Nb, and they also performed the "disappearing act," leaving little on the geometrical surface. Er was different: it allowed some Nb oxide to grow together with Er oxide; apparently it did not completely wet the Nb surface. The best junctions were with Y: they had the smallest excess current, marginally exceeding in $I-V$ quality the Al and Mg results, and they exhibited the largest Nb gap of 1.57 mV [66], the same gap value that has been seen by Wolf and coworkers [39]. I never tried to make hard junctions with Nb counterelectrode using Y overlayer, one reason being that yttrium oxide has high dielectric constant, and such junctions would be of less interest to Josephson technology than the Nb/Al ones. They may be of interest for other applications of Josephson junctions; I do not know if they were ever made.

5.12 Questions of Credit

As I said, John Rowell was quick to recognize that improved tunneling into the last Al layer in Nb/Al superlattice identified by Jochen Geerk implied new vistas for Nb tunnel junctions. Then he and I were working on a patent application, listing in it all kinds of possible counterelectrodes, including the hard ones, and including an obvious choice of a second Nb. But none of that has been yet shown to work; all-refractory junctions with Al overlayers have not been made yet. Towards the end of 1981, John Rowell, satisfied with the resolution of long-standing problems with Nb tunneling, was no longer participating in the development of all-refractory technology. As I said above, publication of excellent I–Vs of Nb/Al–oxide–$Pb_{0.9}Bi_{0.1}$ junctions was delayed due to a legal requirement of not putting the information into the public domain before the filing of Gurvitch–Rowell patent. The paper in which we were finally permitted to publish these I–Vs appeared at the very end of 1981 [58]. In that paper I also included the first I–V curves of Nb/Al–oxide–(Al)/Nb junctions that were prepared with the Nb anodization process (SNAP), as was described in Section 5.8. Even though John was not involved with all-refractory junction work, it was natural to include him as a coauthor in that paper because of the semi-soft junction data that was appearing there for the first time. In subsequent all-refractory papers [59, 60], which contained development of a SNEP-SNAP process and showed beautiful I–Vs of lithographically defined junctions of interest to technology, John did not participate at all, and his name was not in the lists of authors.[h]

[h]There was one more publication describing all-refractory technology, the paper published in *IEEE Transactions on Magnetics* MAG-19, No. 3, May 1983 (submitted

That is why I was surprised and saddened to read John's version of the story in *Supercurrents* [67], written by him 10 years later, in 1991, when I was already at Stony Brook and he also was no longer at Bell, having just left his high position at Bellcore to become a president of Conductus, Inc. In that article in *Supercurrents*, John describes his whole carrier in physics. Among his achievements he presents the story of the Nb/Al:Al_2O_3/Nb trilayers (on p. 13). He talks about Nb/Al multilayer tunneling (without mentioning my name), and how Jochen told him about good tunneling when Al was the last layer, and he immediately realized that this was the solution for the Nb technology. Then he mentions my name for the first time and he says (I quote): "Mike Gurvitch had an ideal deposition system for Nb/Al:Al_2O_3/Nb studies; he and Morris Washington fabricated fine Josephson junctions."

I think it should be clear from what I said above that my role was a bit more significant than to have an ideal deposition system for the job. I was told that this article in *Supercurrents* is not the only one in which John pointed to the *development of refractory Nb technology* as one of the highlights of his career, suggesting that it was essentially his achievement. John's reputation in superconductivity is rightfully very high. His contribution to the *inception* of Nb/Al technology with soft counterelectrodes is most significant and undisputed, as I think I described above honestly and precisely. Results obtained on semi-soft junctions with Al overlayers paved the way for the development of the refractory technology. However, John did not actually develop that *all-refractory* Nb/Al technology (the only one of interest to the industry, I might add). As was described above in some detail, I initiated this stage of the work, single-handedly made the first "large" refractory junctions, and then, with Washington and Huggins, developed SNEP-SNAP process for small, lithographically defined junctions. Here, I said it, and I do not wish to dwell on this any longer.

Nov. 30, 1982). That paper was essentially identical in its content to our ref. 59; it was written for the conference proceedings. In it, I included J. M. Rowell in the list of authors one more time, just out of respect. I remember very well how I brought an internal Bell Labs memo (which preceded all outside publications) to John's office for his signature, and how he said to me, signing the memo, "Mike, you ought to stop putting my name on these papers with which I had nothing to do."

5.13 Who Did What, Where, and When

The emergence of refractory Nb/Al JJ technology had many roots and involved, as we said above, many contributions from inside and outside of Bell Labs. Let me summarize these contributions, as I see them.

- General knowledge that oxidized Al forms good tunnel junctions. Here there were numerous contributions; by 1980 and for 20 years prior to that, it was common knowledge in the field. The first to make Al tunnel junction for tunneling into a superconductor was I. Giaever in 1960 [24].
- Quality Nb films made by sputtering; good, stable semi-soft junctions with Nb base electrode: common knowledge by 1980; I listed some of the contributions in Section 5.3.
- Early attempts at making junctions with Nb counterelectrode; work of Broom, Laibowitz, and their coworkers on Nb–oxide–Nb junctions at IBM [25]. Perhaps this activity can be qualified as defining the problem rather than resolving it; still, it formed an essential background. I should also mention here the early work of Laibowitz and Mayadas [40–42], which can be summarized as "bad junctions, right structures." As I said repeatedly, we did not know of that work until we wrote our patent with J. M. Rowell. Still, it preceded our work by 10 years and should not be forgotten.
- E. L. Wolf, J. Zasadzinski, J. W. Osmun, and G. B. Arnold: Nb tunneling spectroscopy [39] demonstrating good tunneling into Nb foils with thin Al overlayers, which were shown to completely wet Nb surface. This work could influence J. M. Rowell to pick the Nb/Al system for multilayered tunneling studies; if so, through him it influenced all of us working on refractory junctions at Bell.
- I. K. Schuller, C. M. Falco [51, 52], our own work [53]: metallic multilayers demonstrating ability of metals to cover other metals in very thin continuous layers.
- J. M. Rowell, J. Geerk, and M. Gurvitch: junctions with soft second electrode; very good tunneling results (I–Vs published later), barrier looking like Al oxide for very small thicknesses of Al overlayer [55]; resolution of problems with Nb tunneling

spectroscopy [56, 57], similar to what was found in [39], but with thinner Al overlayers.

- M. Gurvitch: first Nb/Al-oxide–(Al)/Nb "hand-made" all-refractory junctions with Ge insulation; first decent all-refractory I–Vs; existence proof for Nb/Al/Nb trilayers without which we would not venture into a lithographically defined junction process. Although done in 1981, before any SNAP work, this work was first described in print in 1984, in Ref. 1.
- Kroger and coworkers at Sperry Research [49]: all-wafer process for all-refractory junctions with amorphous Si barriers; use of Nb anodization (SNAP)—this work showed us the way of all-wafer processing for Nb/Al; in that it was very important.
- M. Gurvitch, H. A. Huggins, and M. A. Washington [59, 60]: modification of Sperry all-wafer process for our all-refractory junctions with Al; SNEP-SNAP; excellent quality junctions; gas phase CF_4 etching of Nb, Al oxide (tunnel barrier) as a stop etch; anodization spectroscopy.
- J. V. Gates, M. A. Washington, and M. Gurvitch [62]: early testing of junction arrays for uniformity, cycling stability, stability in aging, and stability at elevated temperatures.
- J. Kwo, M. Gurvitch, and others [1, 63, 65, 66]: XPS and other surface studies that helped to clarify some of the mysteries of Nb/Al structures. Demonstration that other reactive metals can also form useful overlayers on Nb.

Others did much since then: thousands of Josephson circuits; thousands of papers; diverse applications such as very fast A/D converters; the voltage standard based on Josephson junctions (over 10^5 Nb/Al/Nb junctions in a 10 V standard, see Chapter 10 written by J. Kohlmann); SQUIDs, including biomagnetic applications in heart and brain diagnostics and in disease localization; SIS mixers used in sub-millimeter wavelength radio astronomy; Josephson processors (currently up to 10^5 JJs); quantum computing. Much work went into process improvements: high current density junctions (we quoted [61]), low current density junctions, sub-micrometer junctions, excellent uniformity and reproducibility.

Sergey Tolpygo and coworkers at MIT Lincoln Lab developed a 10-layer process based on Nb/Al/Nb sandwich (JJ) structure uniformly covering 8″ (200 mm) Si wafers, with down to 200 nm circuit features, including junctions, circuits defined using 248 nm (deep UV) lithography [68]. According to [68], VLSI JJ densities with over 10^6 junctions per cm^2 are within reach. Impressive Nb/Al processes were also developed in other foundries in the USA (Hypres, TRW), in Japan (ISTEC), and in Europe (FLUXONICS foundry).

In the years following 1983, people in the field referenced our papers [mostly ref. 59 and 60], citation indexes running in the many hundreds. In more recent publications (e.g., in Ref. 68), our initial work is no longer remembered, the authors referencing more recent versions of the technological processes. Perhaps this is also a sign of recognition, and of maturity, inevitable in a field which is now 33 years old, and which is healthy and still developing.

I admire the work done by others on the Nb/Al system, and the work that led to using these junctions in various applications, but this chapter is already too long as it is. Activity outside of Bell was first directed at reproducing our results, then at incorporating them in large integrated circuits. It started shortly after the appearance of our first Nb/Al publications, and it seems to continue to this day. Wonderful work on the Nb/Al system has been done in the USA, in England, in France, in Germany, in Holland, in Sweden, in Ukraine, in Japan—probably especially in Japan. I am sure I missed some of the countries and some of the interesting applications. The reader will find many of the activities and applications having to do with refractory junction technology described in other chapters of this book.

References

1. Gurvitch, M., and Kwo, J. (1984). Tunneling and surface properties of oxidized metal overlayers on Nb, in *Advances in Cryogenic Engineering*, eds. Clarks, A. F., and Reed, R. P. (Plenum Publishing Corporation), **30**, pp. 509–533.
2. Josephson, B. D. (1962). Possible new effects in superconductive tunneling. *Phys. Lett.*, **1**, 251; also see Chapter 1 in this book.
3. Anderson, P. W., and Rowell, J. M. (1963). *Phys. Rev. Lett.*, **10**, 230; also Rowell, J. M. (1963). *Phys. Rev. Lett.*, **11**, 200.

4. Rowell, J. M. (1966). Supercurrent tunneling devices. U.S. Pat. No. 3,281,609 (filed in 1964 and granted in 1966).
5. Matisoo, J. (1966). Subnanosecond pair-tunneling to single-particle tunneling transitions in Josephson junctions. *Appl. Phys. Lett.*, **9**, 167.
6. Dhong, S. H., and Van Duzer, T. (1980). Minimum width control-current pulse for Josephson logic gates. *IEEE Trans. Electron Devices*, **27**(10), 1965; McDonald, D. G., Peterson, R. L., Hamilton, C. A., Harris, R. E., and Kautz, R. L. (1980). Picosecond applications of Josephson junctions. *IEEE Trans. Electron Devices*, **ED-27**, 1945.
7. Matisoo, J. (1967). *Proc. IEEE (Lett.)*, **55**, 2052; also, in *Proc. IEEE*, **55**, 172.
8. Anacker, W., (1980). Josephson computer technology: an IBM research project. *IBM J. Res. Dev.*, **24**(2), 107.
9. Matisoo, J. (1980). Overview of Josephson technology logic and memory. *IBM J. Res. Dev.*, **24**(2), 113; also, The superconducting computer. *Sci. Am.*, **242**(5), 50–65.
10. Ames, I. (1980). An overview of materials and process aspects of Josephson integrated circuit fabrication. *IBM J. Res. Dev.*, **24**(2), 188–194.
11. Gheewala, T. R. (1980). Design of 2.5-micrometer Josephson current injection logic. *IBM J. Res. Dev.*, **24**(2), 130.
12. Gallagher, W. J., Harris, E. P., and Ketchen, M. B. (2012). IEEE/CSC & ESAS European Superconductivity News Forum, No. 21.
13. Broom, R. F., Oosenbrug, A., and Walter, W. (1980). Josephson junctions of small area formed on the edges of niobium films. *Appl. Phys. Lett.*, **37**, 237; also Broom, R. F., Raider, S. I., Oosenbrug, A., Drake, R. E., and Walter, W. (1980). Niobium oxide-barrier tunnel junction. *IEEE Trans. Electron. Devices*, **ED-27**, 1998.
14. Fulton, T. A., and Dynes, R. C. (1971). Switching to zero voltage in Josephson tunnel junctions. *Solid-State Commun.*, **9**, 1069.
15. Jewett, R. E., and Van Duzer, T. (1981). Low-probability punchthrough in Josephson junctions. *IEEE Trans. Magn.*, **17**, 599; Harris, E. P., and Chang, W. H. (1981). Punchthrough in Josephson logic devices. *IEEE Trans. Magn.*, **17**, 603.
16. Kotani, S., Fujimaki, N., Imamura, T., and Hasuo, S. (1988). A Josephson 4b microprocessor. *Digest of Tech. Papers of 1988 Int. Solid-State Circuit Conf.*, 150–151; Kotani, S., Imamura, T., and Hasuo, S. (1990). A subnanosecond clock Josephson processor. *IEEE J. Solid State Circuits*, **25**, 117–124.

17. Wada, Y., Nagasawa, S., Ishida, I., et al. (1988). A 570 DS. 13 mW Josephson 1 Kbit RAM. *ISSCC Digest of Technical Papers,* 84; Nagasawa, S., Wada, Y., Hidaka, M., et al. (1989). 570-ps 13-mW Josephson 1-Kbit NDRO RAM. *IEEE J. Solid-State Circuits,* **24**, 1363–1371.
18. Suzuki, H., Fujimaki, N., Tamura, H., et al. (1989). A 4K Josephson memory. *IEEE Trans. Magn.,* **25**, 783–788.
19. Hasuo, S. (1993). Josephson microprocessors, in *New Superconducting Electronics,* eds. Weinstock, H., and Ralston, R. W., NATO ASI series, **251**, pp. 363–399.
20. Likharev, K. K., Mukhanov, O. A., and Semenov, V. K. (1985). Resistive single flux quantum logic for the Josephson-junction technology, in *SQUID-85,* de Gruyter, W., Berlin, pp. 1103–1108; also in Likharev, K. K., Semenov, V. K. (1991). RSFQ logic/memory family: a new Josephson-junction technology for sub-terahertz-clock-frequency digital systems. *IEEE Trans. Appl. Supercond.,* **1**(1), 3–28.
21. Chen, W., Rylyakov, A., Patel, V., Lukens, J., and Likharev, K. (1999). Rapid single flux quantum T-flip flop operating at 770 GHz. *IEEE Trans. Appl. Supercond.,* **9**(2), 3212–3215.
22. Guest Editor's column by MIT Lincoln Laboratory (2014). *The New Wave,* **20**(3), 1–11 (the NSA review of emerging technologies).
23. Mott, N. F. (1947). *Trans. Faraday Soc.,* **43**, 429; Knorr, K., and Leslie, J. D. (1973). *Solid State Commun.,* **12**, 615.
24. Giaever, I. (1960). Energy gap in superconductors measured by electron tunneling. *Phys. Rev. Lett.,* **5**, 147.
25. Broom, R. F., Laibowitz, R. B., Mohr, Th. O., and Walter, W. (1980). Fabrication and properties of niobium Josephson tunnel junctions. *IBM J. Res. Dev.,* **24**(2), 212–222. [Note that values of V_m listed on p. 217 should be in mV, not in V].
26. Grundner, M., and Halbritter, J. (1980). XPS and AES studies on oxide growth and oxide coatings on niobium. *J. Appl. Phys.,* **51**(1), 397–405; Lindau, I., and Spicer, W. E. (1974). *J. Appl. Phys.,* **45**, 3720.
27. Rowell, J. M., and Feldman, W. L. (1968). Excess currents in superconducting tunnel junctions. *Phys. Rev.,* **172**, 393.
28. Basavaiah, S., and Park, K. C. (1975). *IEEE Trans. Electron Devices,* **MAG-11**, 774; Basavaiah, S., and Greiner, J. H. (1976). *J. Appl. Phys.,* **47**(9), 4201–4202.
29. Giaever, I., and Megerle, K. (1961). Study of superconductors by electron tunneling. *Phys. Rev.,* **122**, 1101.

30. Neugebauer, C. A., and Ekvall, R. A. (1964). Vapor-deposited superconductive films of Nb, Ta, and V. *J. Appl. Phys.*, **35**(3), 547; the authors thank I. Giaever for performing tunneling measurements, I-V-s shown in the paper.
31. Nordman, J. E. (1969). Thin-film Josephson junctions using getter-sputtered niobium. *J. Appl. Phys.*, **40**(5), 2111–2115.
32. Mullen, L. O., and Sullivan, D. B. (1969). Fabrication of tunnel junctions on niobium films. *J. Appl. Phys.*, **40**(5), 2115–2117.
33. Schwidtal, K. (1972). DC and AC Josephson effect in sputtered Nb–NbO_x–Pb junctions. *J. Appl. Phys.*, **43**, 202.
34. Broom, R. F. (1976). Some temperature-dependent properties of niobium tunnel junctions. *J. Appl. Phys.*, **47**, 5432.
35. Hauser, J. J., Bacon, D. D., and Haemmerle, W. H. (1966). Energy gap of V_3Si. *Phys. Rev.*, **151**, 296.
36. Moore, D. F., Zubeck, R. B., Rowell, J. M., and Beasley, M. R. (1979). Energy gaps of the A-15 superconductors Nb_3Sn, V_3Si, and Nb_3Ge measured by tunneling. *Phys. Rev. B*, **20**(7), 2721–2738.
37. Rudman, D. A., and Beasley, M. R. (1980). Oxidized amorphous-silicon superconducting tunnel junction barriers. *Appl. Phys. Lett.*, **36**(12), 1010–1013.
38. Rybal'chenko, L. F. (1979). *Sov. J. Low Temp. Phys.*, **5**(11), 610–615.
39. Wolf, E. L., Zasadzinski, J., Osmun, J. W., and Arnold, G. B. (1979). *Solid State Commun.*, **31**, 321; also by the same authors (1980). *J. Low Temp. Phys.* **40**(1/2), 19–49.
40. Laibowitz, R. B., and Mayadas, A. F. (1972). Josephson junctions with Nb/Al composite electrodes. *Appl. Phys. Lett.*, **20**(7), 254–256.
41. Laibowitz, R. B., and Mayadas, A. F. (1970). *IBM Technical Disclosure Bulletin*, **13**(5), 1197–1198.
42. Laibowitz, R. B., and Mayadas, A. F. (1973). Multilayer Josephson memory device. *IBM Technical Disclosure Bulletin*, **15**(11), 3316–3317.
43. Yeh, J. T., and Tsuei, C. C. (1979). Tunneling studies on thin film Nb-Al alloys. *IEEE Trans. Magn.*, **MAG-15**, 1.
44. Hawkins, G., and Clarke, J. (1976). *J. Appl. Phys.*, **47**, 1616.
45. Buitrago, R. H., Goldman, A. M., Toth, L. E., and Cantor, R. (1979). *IEEE Trans. Magn.*, **MAG-15**, 589.
46. Umbach, C. P., Goldman, A. M., and Toth, L. E. (1982). *Appl. Phys. Lett.*, **40**(1), 81.

47. Keller, W. H., and Nordman, J. E. (1973). Nb thin-film Josephson junctions using a semiconductor barrier. *J. Appl. Phys.*, **44**(10), 4732–4738.
48. Kroger, H., Potter, C. N., and Jillie, D. W. (1979). *IEEE Trans. Magn.*, **MAG-15**, 488.
49. Kroger, H., Smith, L. N., and Jillie, D. W. (1981). *Appl. Phys. Lett.*, **39**, 280; Smith, L. N., Kroger, H., and Jillie, D. W. (1983). *IEEE Trans. Magn.*, **MAG-19**, 787.
50. Levinson, M., and Gurvitch, M. (1983). *J. Appl. Phys.*, **54**(8), 4683–4685.
51. Schuller, I. K. (1980). *Phys. Rev. Lett.*, **44**, 1597.
52. Schuller, I. K., and Falco, C. M. (1982). *Microstructure Science and Engineering/VLSI*, ed. Einspruch, N. G. (Academic Press, New York), p. 183.
53. McWhan, D. B., Gurvitch, M., Rowell, J. M., and Walker, L. R. (1983). *J. Appl. Phys.*, **54**(7), 3886–3891.
54. McMillan, W. L., and Rowell, J. M. (1969). *Superconductivity*, ed. Parks, R. D. (Marcell Dekker, New York), Chapter 11.
55. Rowell, J. M., Gurvitch, M., and Geerk, J. (1981). Modification of tunneling barriers on Nb by a few monolayers of Al. *Phys. Rev. B*, **24**(4), 2278–2281.
56. Geerk, J., Gurvitch, M., McWhan, D. B., and Rowell, J. M. (1981). *Bull. Am. Phys. Soc.*, **26**, 441.
57. Geerk, J., Gurvitch, M., McWhan, D. B., and Rowell, J. M. (1982). Proc. of the 16th International Conference on Low Temperature Physsics, LT-16, Physica, **109–110B**, 1775.
58. Gurvitch, M., Rowell, J. M., Huggins, H. A., Washington, M. A., and Fulton, T. A. (1981). Nb Josephson tunnel junctions with thin layers of Al near the barrier. **IEDM-81**, 115–117.
59. Gurvitch, M., Washington, M. A., and Huggins, H. A. (1983). High quality refractory Josephson tunnel junctions utilizing thin aluminum layers. *Appl. Phys. Lett.*, **42**(5), 472.
60. Huggins, H. A., and Gurvitch, M. (1985). Preparation and characteristics of Nb/Al-oxide-Nb tunnel junctions. *J. Appl. Phys.*, **57**(6), 2103.
61. Miller, R. E., Mallison, W. H., Kleinsasser, A. W., et al. (1993). Niobium trilayer Josephson tunnel junctions with ultrahigh critical current densities. *Appl. Phys. Lett.*, **63**, 1423.
62. Gates, J. V., Washington, M. A., and Gurvitch, M. (1984). Critical current uniformity and stability of Nb/Al–oxide–Nb Josephson junctions. *J. Appl. Phys.*, **55**(5), 1419–1421.

63. Kwo, J., Wertheim, G. K., Gurvitch, M., and Buchanan, D. N. E. (1982). X-ray photoemission spectroscopy study of surface oxidation of Nb/Al overlayer structures. *Appl. Phys. Lett.*, **40**(8), 675.
64. Miedema, A. R., and den Broeder, F. J. A. (1979). On the interfacial energy in solid-liquid and solid-solid metal combinations. *Z. Metallkunde*, **70**, 14.
65. Chang, C. C., Gurvitch, M., Hwang, D. M., and Blonder, C. W. (1987). Auger electron spectroscopy, transmission electron microscopy, and scanning electron microscopy studies of Nb/Al/Nb Josephson junction structures. *J. Appl. Phys.*, **61**(11), 5089–5097.
66. Kwo, J., Wertheim, G. K., Gurvitch, M., and Buchanan, D. N. E. (1983). XPS and tunneling study of air-oxidized overlayer structures of Nb with thin Mg, Y and Er. *IEEE Trans. Magn.*, **MAG-19**(3), 795–798.
67. Rowell, J. M. (1991). Do the experiment anyway. *Supercurrents*, 13.
68. Tolpygo, S., Bolkhovsky, V., Weir, T., Johnson, L., Gouker, M., Oliver, W. (2015). Fabrication process and properties of fully planarized deep-submicron Nb/Al–AlO_x/Nb Josephson junctions for VLSI circuits. *IEEE Trans. Appl. Supercond.*, **25**(3), 1101312.

Chapter 6

Refractory Niobium Nitride NbN Josephson Junctions and Applications

Jean-Claude Villegier*
INAC CEA-Grenoble, Grenoble Alpes University,
1 Allee des Chaumes, 38640 CLAIX, France
jc.villegier@hotmail.fr

The refractory nitride Josephson tunnel junctions with NbN superconducting thin-film base electrode ($T_c \approx 16.5$ K) were studied after 1982 using thermal or plasma oxide tunnel barriers, showing attractive I–V nonlinear characteristics only when a superconducting soft metal (Pb–In) was used as counterelectrode. A few years later, introduction of in situ deposited tri-layers including MgO or AlN tunnel barrier permitted a large improvement in the junction quality, yield, and reproducibility, opening new application fields. Thereafter, NbN (or one of its parent compounds such as $Nb_{1-x}Ti_xN$) was established as a possible alternative to refractory Nb Josephson devices (operating at or below 4.2 K) for building superconductive electronic circuits achieving higher frequencies, up to the THz range for SIS mixers or oscillators as well as operating temperature up to 10 K. "Self-shunted" nitride SNS and SS'S junctions, such as NbN–TaN_x–NbN JJs, are taking advantage

*Former Senior Adviser.

Josephson Junctions: History, Devices, and Applications
Edited by Edward Wolf, Gerald Arnold, Michael Gurvitch, and John Zasadzinski

ISBN 978-981-4745-47-5 (Hardcover), 978-1-315-36452-0 (eBook)
www.panstanford.com

of large $R_N I_C$ products induced by efficient quasi-particle diffusion and good nitride barrier matching at interfaces with nitride electrodes, which solve frequency limitation problems induced by the capacitance of SIS junctions. RF front ends, MUX, ADC, and digital RSFQ nitrides circuits have been demonstrated, some of them optimized to operate near 10 K inside closed-cycle refrigerators for minimizing energy dissipation in space satellite applications. But NbN circuits are today still less mature than the established Nb–Al–AlO_x–Nb JJ state-of-art technology (~20k JJs with $J_J \approx$ 8 kA/cm^2, ~1 μm^2) suitable to achieve prototypes of low dissipation integrated circuit SFQ processors described in [1] *and the references inside*. However, important properties of NbN technology seem more economically relevant for the future: conventional Si–CMOS foundry factories should be used to establish a NbN–SiO_2 stack-layered process, stable up to ~350°C, not sensitive to hydride species diffusion, while present Nb trilayers are degraded above 180°C and hydrogen diffusion is detrimental to junction yield. A recent NbN IC processing demonstration (0.5 μm linewidth) has been done on 8-inch silicon wafers at the CEA-LETI CMOS facility, making it possible to integrate further NbN ADC and processors chips reliably designed and fabricated with high yield at reasonable cost in any CMOS foundry. Another advantage of NbN thin and very thin (few nanometers thick) films and nanowires with very short electron–phonon relaxation time leads to fast, light, or current-sensitive self-resetting disjunctions switches with jitter of a few picoseconds and fast heat dissipation in the substrate. On-chip integration of nitride front-end circuits such as resonators, microwave filters, registers, highly sensitive and fast photosensors (KIDs, SNSPD, HEB, etc.), and superconducting–semiconducting interfaces circuits widen the field of superconductive electronics applications from mK up to 10 K.

Moreover, recently, deep submicrometer size (~0.01 μm^2), very low capacitance NbN–MgO–NbN SIS and SQUID circuits were achieved using a self-aligned process. Such junctions, beside their attractive properties for nitride IC, are able to emit or detect in a controllable way single microwave photons and could be a new building block of quantum information circuits.

6.1 Early Niobium Nitride Devices

6.1.1 Applied SIS Josephson Tunnel Junctions Until 1983

Following Chapter 2 by E. L. Wolf on the refractory Josephson junction theory and further chapters on circuits developments oriented towards superconducting niobium, we introduce a short and not exhaustive development on the specific interest found to apply the B1-cubic refractory phase of niobium nitride (NbN) electrodes films whose critical temperature (T_c) is close to 17 K and others parent nitride compounds such as $Nb_{1-x}Ti_xN$ ($T_c \approx 15.5$ K) for achieving alternate superconducting electronic and optoelectronic devices.

In most cases, the circuit integration success relies on the development of reliable and reproducible Josephson junctions of superconductor–insulator–superconductor (SIS), superconductor–normal *metal*–superconductor (SNS), or superconductor–*lower* T_c superconductor–superconductor (SS'S) types described in Chapter 2.

Experimentally the insulating tunnel barrier of a SIS junction could be a native oxide formed thermally or in an oxidizing plasma on the base electrode or an oxide layer formed on a very thin "overlayer" deposited on the base electrode such as Nb–Al–AlO_x–Nb trilayer junctions widely applied today, described by M. A. Gurvitch in Chapter 5, whose theory is developed in Chapter 4 by G. B. Arnold. The good Al (6–8 nm) surface coverage is due to the affinity of Al for Nb.

Alternately, an in situ deposited dielectric or semiconductor nanolayer barrier should be a solution for SIS junctions, requiring a very uniform conformal coating of the base electrode and the built-up of an uniform barrier potential interfaced with both refractory Nb or NbN electrodes. Several barrier materials have been studied without complete success, as shown by S. T. Ruggiero, in the period 1970–1990 [2]. For example, in the early 1970s, Cardinne et al. [3] at Air Liquide, took barrier modeling arguments for using low barrier potential materials such as amorphous semiconductor films to improve the barrier transmission and the high-frequency coupling to Josephson junctions (see Fig. 6.1). They were fabricating Nb–aGe–Nb SIS junctions where the $\sim$8 nm thick amorphous germanium

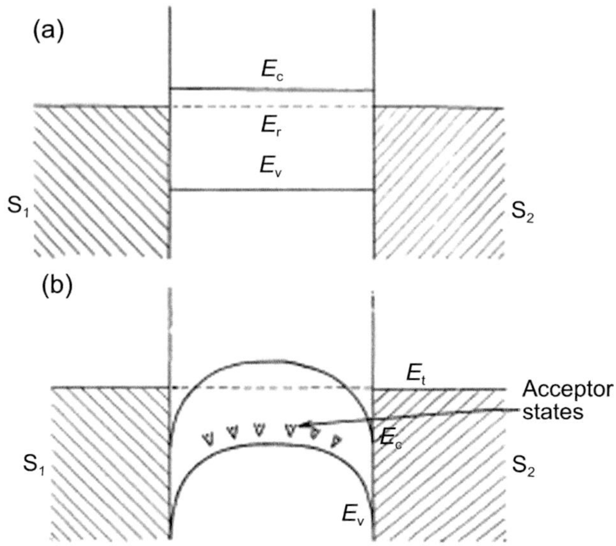

Figure 6.1 Energy barrier models in the case of a low-gap semiconductor (E_V and E_C are the valence and conduction band energies, respectively) or insulator. (a) Ideal barrier: S_1 and S_2 are superconductive electrodes. (b) Cardinne et al. model of an amorphous semiconductor without surface states [3].

barrier should have Fermi and barrier potentials pinned by the large amount of defects ($\sim10^{21}$ cm^{-1}) in the bulk but also by the Nb–aGe–Nb interface states. However, pursuing their study in the same deposition equipment, we found by transmission electron microscopy (TEM), that the Ge barrier layer sputtered on Nb in the range of 20–200°C substrate temperature was amorphous but not uniform in thickness and present pinholes making most of the junction short-circuited [4]. Even oxidizing the aGe barrier to fill pinholes with NbO_x was not successful due to generation of stress and formation of poor Ge oxides.

So before 1983, applications-driven studies were mostly done on SIS junctions and SQUIDs circuits where at least the counter-electrode was made of a soft metal such as lead alloys (Pb–In, Pb–Bi, etc.). Pb(In)–PbO_x–Pb *alloy* and Nb–NbO_x–Pb(In) SIS Josephson tunnel junctions and circuits were improved step by step, still facing major issues both in storage, thermal cycling, poor reliability, as well problems of logic gates operation, memory cell integration, and so on.

These problems unsolved by large research teams had led to the closing of major Josephson projects in 1983–1985 (IBM, Bell Labs, CEA-LETI, etc.) [5].

However, in 1980, G. Dousselin and G. Rosenblatt showed that growing the oxide barrier on the edge of a superconducting Nb or NbN base electrode (~0.3 nm thick) rather than on its flat surface could lead to very-small-area Josephson-"edge" (or "ramp-edge") junctions and greatly improve the circuit downsize scaling and integration [6]. This technology was used later for achieving three-terminal devices as other superconducting nanocircuits [7] and, more recently, for achieving ramp-edge-type HTS YBCO Josephson circuits with the junction of both electrodes oriented toward a–b crystal planes across native or others types of barriers [8].

6.1.2 Success and Limitations of Refractory Trilayer Processes

Then came the breakthroughs of introducing the refractory Nb trilayer junction processing based on the selective niobium anodization process (SNAP-1981) of H. Kroger et al., reactivating the amorphous a-Si(H) barrier junctions (Nb–aSi–Nb) [9, 10] and on the selective niobium etching process (SNEP-1983) of M. Gurvich et al. for Nb–Al–AlO_x–Nb SIS junctions [11].

SNEP, closely associated with the use of a thin Al overlayer oxidized, is still today successfully applied to most Josephson circuit fabrications described in Chapter 5 by M. Gurvich. One year later, we introduced the selective niobium overlap process (SNOP) during the 1984 Applied Superconductivity Conference in San Diego [12]. The SNOP was applied to NbN trilayer junctions and recently again for achieving very-small-area NbN–MgO–NbN SIS junctions and SQUIDs circuits described in paragraph 6.4.5 [13].

In situ elaboration of a SIS trilayers in a sputtering or other physical vacuum deposition (PVD) equipment without breaking the vacuum is a big advantage to achieve uniform and reproducible junction barrier nanolayers interfacing with electrodes films. Such a SIS (or SNS) multilayer coating is collectively patterned over

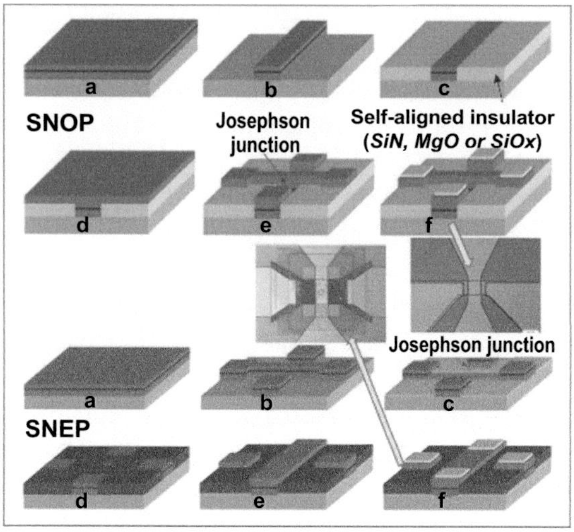

Figure 6.2 Comparison of selective niobium overlap process (SNOP) using three photomasks (or e-beam levels) and SNEP trilayer processing (4 photomasks) applied to NbN–MgO–NbN Josephson tunnel junctions [15].

a wafer containing several chips. Each chip may be transformed in a more or less large network of interconnected Josephson junctions (together with inductors, capacitors, resistors elements). In such superconducting integrated circuits (SICs), a low dispersion (σ value) of functional elements parameters is mandatory. The junction areas (about 10 μm^2 in the 1990s, $\sim$1 μm^2 in 2015, and estimated to be $\sim$0.1 μm^2 near 2025) and linewidth should be defined accurately by lithography [14]. Junction area is determined by a window patterned in a SiO_2 or others dielectric layers in SNEP, by anodic Nb_2O_5 in SNAP or by direct electrodes overlap in SNOP. Processing of SNEP and SNOP applied to NbN–MgO–NbN junctions is described on Fig. 6.2 [15]. More recently, the planarized all-refractory technology for superconductors (PARTS) was introduced at IBM [16]. It permits the use of lithographic tools and techniques of semiconductor microelectronics to scale down junction size and linewidths uniformly on large Si (or glass, sapphire, etc.) wafers. Such kind of planarized processing by chemical–mechanical polishing (CMP) is presented in Fig. 6.3 when applied to a NbN stack processed on 8-inch Si wafers in the CEA-Grenoble LETI

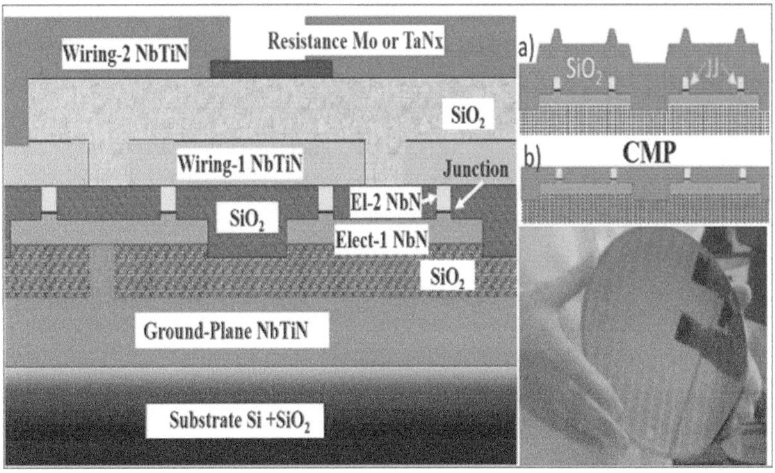

Figure 6.3 Cross section of the stack of planarized layers applied to the realization of NbN–NbTiN RSFQ circuits on 8-inch Si (+SiO_2) wafers. *Right*: Detail of the chemical–mechanical polishing (CMP) process applied here to a self-shunted (SS'S) Josephson junction with TaN_x barrier and whole-wafer photo. Reproduced with permission from Villegier [17].

C-MOS platform [17]. Earlier, such planarized circuit fabrication has been applied to the fabrication of most of the Nb ICs, but necessitates specialized foundries because of the constraints in temperature and specificities of niobium–aluminum junction processing [18, 19].

6.2 Niobium Nitride Tunnel Josephson Junctions

SIC circuit fabrication requires to build a multilayer stack of patterned dielectric, resistive, and superconducting elements such as a ground-plane, DC and HF interconnecting wirings, junction and resistor levels. Processing SIC with a high enough yield becomes a critical issue which requires specific foundries or finding common grounds with semiconductor circuit foundries' capabilities [17].

One of the drawbacks of Nb–Al–AlO_x–Nb refractory junctions shown by Pavolotski et al. [20] and other authors [19, 21] is poor junctions' aging properties during storage at room temperature, leading to a change in the junction normal resistance Rn combined with a change of sub-gap resistance R_j and in a degradation of the junction quality factor R_J/R_N. This is particularly due to the large diffusion channels inside the Nb BCC crystal structure, leading to inclusion of hydrogen (or oxygen) atoms at barrier interface [21]. So, the progress of the Nb IC integration and of the SFQ circuit fabrication, according to microelectronics facilities' rules and capabilities, suggests strong modifications in the choice of the Nb junction barrier (such as introduction of self-shunted barriers [22] and/or increase in SIS barrier transparency). The removal of aluminum overlayers and of Nb anodic oxide layers, the introduction of better dielectric quality insulator layers and of nitride thin layers to prevent hydrogen diffusion inside the layer stack [19], and the improvement of dry etching processing conditions should be mandatory in the future Nb SIC. Such Nb junction and circuit technology issues are described in more detail in Chapters 4, 5, 6, 7, and 11. Here we introduce the niobium nitride (NbN) technology bringing alternate fabrication solutions and higher-performance devices for some of the SIC applications.

6.2.1 Introduction of NbN Film Textures

The superconductive properties of cubic phase of niobium nitride were observed with a T_c of 16 K in 1941. It is interesting to note that a superconducting columbium nitride ribbon (*columbium was the old name of niobium*) was demonstrated in 1946 by Robert M. Milton from John Hopkins University as a sensitive and fast (*transition edge*) bolometer device operating in infrared detection [23]. So, before the discovery of the BCS theory, before the Josephson effects, and two years before the invention of the "transistor" or any integrated circuit (IC), a tiny ribbon of NbN was proved to be "a sensitive high speed bolometer operating at about 16 K."

Among different superconducting refractory nitrides compounds which can be obtained by different technics in thin films, NbN is a type II superconductor in the dirty limit which presents the highest T_c in the FCC cubic-B1 phase with pmm4 isotropic properties. NbN nitrogen-deficient cubic structure films are often granular with large grain boundaries resistances. They show a single energy gap with strong electron–phonon coupling ($2\Delta/k_B T_c \approx 4.2$), a ratio

$\rho_{300K}/\rho_{20K} < 1.3$ conduction of poor metallic type, a low value of the product Fermi surface wave vector times elastic electron mean free path ($k_F.l$), and a large value of London penetration depth [2, 24–26].

Cathodic RF sputtering and chemical vapor deposition (CVD) were used to deposit among the first NbN superconducting thin films in the 1970s with controlled polycrystalline columnar mostly (111) texture oriented and good T_c values [27–29]. However, at that time the NbN deposition conditions under RF sputtering or CVD, the difficult choice of the substrate materials, the large substrate heating during deposition, the defects due to ion bombardment in RF sputtering, the uncontrolled carbide precipitates inclusions, and so on, did not favor the feasibility of practicable Josephson tunnel junctions or of other applied NbN thin films devices.

The study of the role of sputtering conditions and substrate material on NbN cubic film texture and epitaxy at low enough substrate temperature was a clue in the 1980s [24, 30]. The crystalline matching of NbN with hexagonal sapphire orientations was studied by Noskov et al. [31] and a documented study of NbN epitaxy or textured films on different substrates was reported by A. I. Braginski and J. Talvacchio [24]. As shown by a TEM picture in Fig. 6.4 (left), very thin (from 2 to 20 nm thick) NbN epitaxial layers

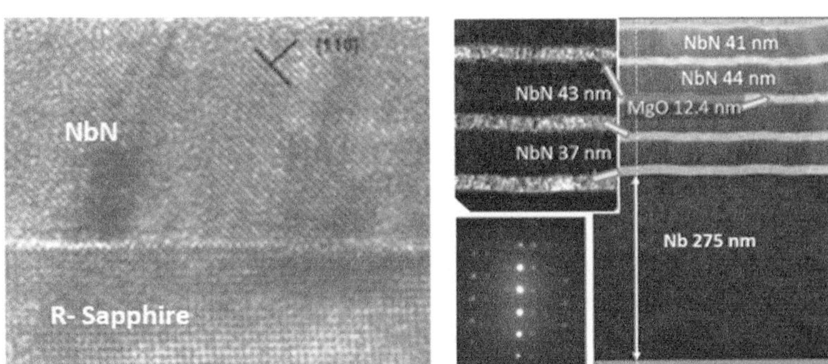

Figure 6.4 *Left*: TEM cross section of single-crystal NbN film sputtered at 600°C on R-plane sapphire substrate [32, 33]. *Right*: TEM cross section and diffraction of a [MgO–NbN](×4) multilayer sputtered on an underlayer Nb film (275 nm) [33].

can be grown uniformly by DC magnetron sputtering at ~600°C on noncubic substrates such as R-, A- or M-plane orientations of sapphire wafers. We show that thin epitaxial NbN films are (135) oriented on R-plane Al_2O_3 and paved by twin domains [32, 33]. On the contrary, NbN on M-Al_2O_3 is untwined, leading to a lower resistivity, an increased critical current density at 4.2 K, and a higher critical temperature ($T_c \approx 13$ K for a very thin 4 nm film). Fortunately, the cubic B1 structure of NbN is more tolerant to defects than the close hcp NbN phase, which presents a much lower critical temperature. However, it is difficult to optimize the grain texture and the resistive and superconducting properties when the NbN films are sputtered on unmatched substrates like silicon or amorphous SiO_2 more common in microelectronics. The Bruce van Dover group at Bell labs sputtered the first good-quality NbN films at low temperature by DC magnetron sputtering, a PVD method which is easy to manage in thin film stacks [34].

6.2.2 Use of Templates in NbN Heterostructures

Using moderate substrate temperatures, Kosaka and Onodera in 1974 showed that it is possible to epitaxially grow cubic B1 NbN on cubic MgO (100) single-crystal suitably polished and surface cleaned [35]. Taking into account this possibility, Yamashita et al. [36] demonstrated the epitaxial growth of a NbN film on top of a cleaned silicon wafer covered with a MgO buffer layer or with an Al_2O_3 buffer. More recently, TiN [37] and Nb_5N_6 [38] were also used as buffer layers on Si. For example, a better (100)-oriented NbN texture observed by X-ray diffraction (XRD) (Table 6.1) shows the improvements in the resistivity, in the surface resistance at 10 GHz, and in the gap frequency of a 250 nm DC magnetron-sputtered NbN film which is epitaxial on MgO (100) single crystal, (100) polycrystalline textured on a RF-sputtered MgO (20 nm thick) layer deposited on Si (100) wafer and with a columnar (111) self-texture when deposited on a (100) oriented cleaned silicon wafer [39, 40].

Three directions of successful research were issued from such results:

Table 6.1 Crystal texture and DC and RF resistance of 250 nm NbN films deposited at 20°C on MgO (100) and Si (100) wafers covered with or without an RF-sputtered 20 nm MgO buffer [40]

Layers	NbN/MgO (100)	NbN/MgO/Si(100)	NbN/Si (100)
XRD:I(200)/I(111)	>100	8	0.07
ρ_{300K}(μΩcm)	60	100	140
$R_{S\text{-}4K}$ (at 10 GHz)	7 μΩ	10 μΩ	10 μΩ
$\nu(\Delta_0)$ (THz)	1.4	1.3	0.9

(a) The possibility of growing hetero-epitaxial rock-salt structure stacks based on NbN ($a = 0.439$ nm) on cleaned Si (100) wafers [$a_{45°} = 0.384$ nm $= 0.543$ nm$/2^{1/2}$], using buffer layers, such as MgO ($a = 0.419$ nm) [36] or TiN ($a = 0.425$ nm) [37]. This leads to epitaxial superconducting NbN junction circuits integrating low RF resistance wirings and resonators with low specific inductances. This makes it possible to integrate detectors, microwave filters, and readout circuits (see Fig. 6.11).
(b) The achievement of all-epitaxial SIS NbN–MgO–NbN mixers and circuits on MgO (100), using also a thin epitaxial MgO barrier [41].
(c) The use of a 8–20 nm thick (100) oriented self-textured MgO film RF-sputtered on any amorphous substrate or on Si wafers covered by silica (SiO_2) was applied by several laboratories to improve the (100) texture of NbN films, which are smoothers with a lower resistivity than the (111) self-textured NbN layer wiring and junction electrodes [2, 24, 42].

6.2.3 NbNO$_x$ Barriers in NbN SIS Junctions

For Josephson IC applications one of the main challenges in the 1980s was in the fabrication of reproducible micrometer or submicrometer SIS Josephson tunnel junctions, inductively and capacitively matched with transmission lines and resistors integrated in a reliable stack of layers. Besides the development of lead alloys junctions followed by the Nb–NbO$_x$ ones, the H. Hayakawa ETL team reported on the completion of high-quality

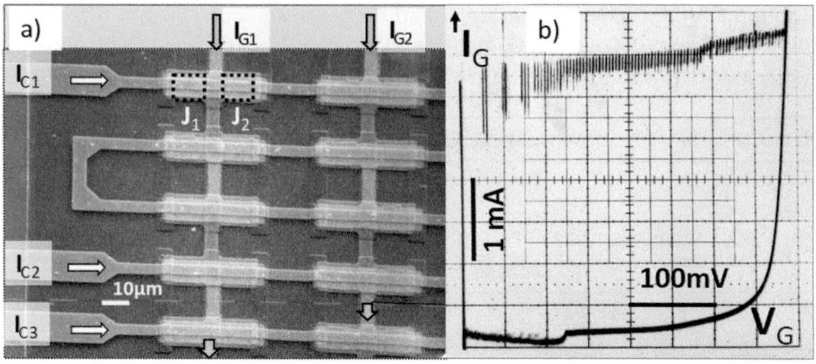

Figure 6.5 SEM picture of addressable 100 DC SQUIDs memory with 6 µm control lines and 10×10 µm^2 junctions. Current–voltage characteristics at 4. 2 K of $100 \times$ NbN–NbNO$_x$–Pb(In) junctions SQUIDs ($<J_C> = 1.74$ kA/cm^2) in series. Reproduced with permission from Villegier [12].

NbN–NbNO$_x$–Pb SIS junctions and DC SQUIDs. They deposited a ZnO film mask before NbN RF-sputtering deposition at 500°C, followed by the sputter-etching cleaning of NbN base-electrode and thermal oxidation [43, 44].

As observed by X-ray photoemission (XPS) [45] and junction conductance [12], the $(NbN)_2O_{5-X}$ thermally or RF-plasma [46]-formed barriers at room temperature in an oxygen gas partial pressure were similar to the Nb_2O_{5-X} barrier of the Nb–NbO$_x$–Pb junctions, showing an effective barrier potential of about 0.3 eV [12]. As shown in Fig. 6.5, during that period of time, NbN–oxide–Pb(In) and NbN–oxide–NbN junction arrays and DC SQUID memory arrays with control lines (4 or 6 µm linewidths) were also developed at CEA LETI with a dispersion on Josephson currents and on tunnel resistances of a few percent [12], mainly due to lithographic dispersion and to flux trapping. NbN–oxide–NbN junction and SQUIDs arrays were also fabricated [12], but the ratio of the sub-gap resistance R_{SG} measured at 3 mV to normal resistance R_N (R_{SG}/R_N) was decreased in comparison to those of NbN–oxide–Pb(In) junctions. Prof. John Bardeen, during his visits to CNRS-Grenoble in 1983–84 and to CEA-LETI, was curious about the attractive developments of NbN (instead of Nb) Josephson tunnel junction applications to voltage standard, SIS mixers, SQUID magnetometers, and logic gates. During the same period, others groups were developing high-quality NbN–oxide–PbIn junctions [47, 48]. Even ETL achieved integrated NbN–oxide–NbN Josephson logic circuits (i.e., 8-bit ripple carry adder) with Nb wirings [49] and Hitachi high-quality NbN–oxide–NbN junctions and a 3 kBit logic gate array, comprising 22,000 junctions [50].

In parallel, stimulated by H. Kroger [51] and by others [3], several groups were studying amorphous semiconductor barriers. NbN–aSi–NbN and NbN–aSi–Nb junctions were obtained with quality factors of the same order to those of NbN–oxide–NbN SIS [52–55], often better than in Nb–NbO_x–Nb SIS.

For most applications, it is important to deposit in a controllable and reproducible way, thin and smooth polycrystalline NbN or $Nb_{1-x}Ti_xN$ [56] junction electrodes and wiring lines in the thickness range 200–350 nm with low enough resistivity, low kinetic inductance, and large critical temperature. Good dielectric SiO_2 or Si_3N_4 layers deposited by reactive sputtering or plasma-enhanced chemical vapor deposition (PE-CVD) at low temperature must separate electrodes and lines [17].

6.3 NbN Junctions for IC Applications

6.3.1 From $NbNO_x$ to MgO and AlN Barriers in NbN SIS Junctions

As explained above, it is attractive to use high-frequency quality (low loss tangent at low temperature in the microwave frequency range) single-crystal substrates such as MgO (100) or sapphire, matching more or less with cubic B1 nitride or to RF-sputter (or deposit by pulse laser [PLD]) a MgO buffer layer on Si+SiO_2 wafers to improve NbN or NbTiN film texture and quality. The study of in situ MgO artificial tunnel barrier deposition, then, comes naturally with the interest to achieve NbN–MgO–NbN SIS "trilayers" grown epitaxially or at least textured [2].

Several laboratories were developing such NbN–MgO–NbN junctions since 1983 that were fabricated with SNEP, SNOP, or other processes with Josephson current densities J_c ranging from 1 to

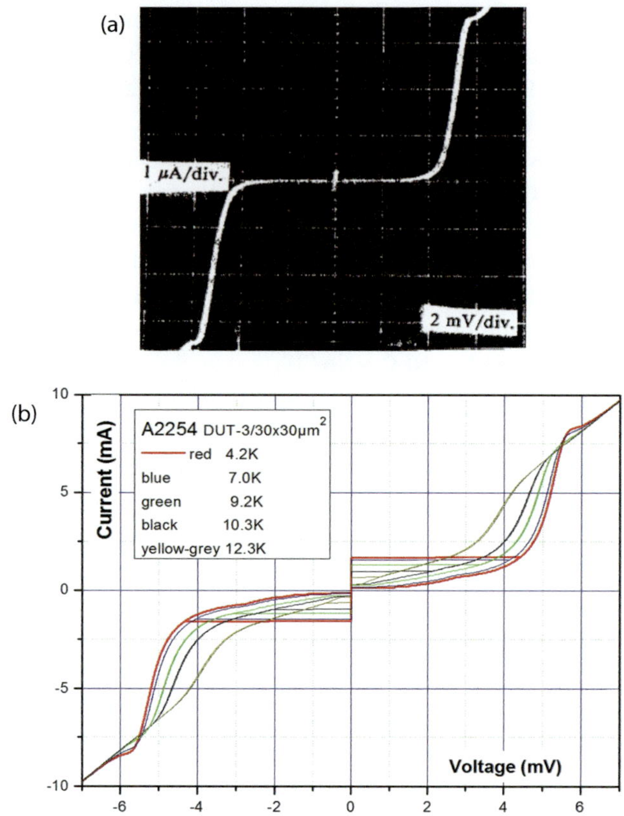

Figure 6.6 NbN–MgO–NbN junction characteristics on MgO/SiO_2–Si substrate. (a) Current–voltage characteristics of a 0.5 μm^2 SIS at 4 K. (b) A 30×30 μm^2 SIS at different temperatures ranging from 4.2 to 12.3 K [39, 60].

40 kA/cm^2 [39–41, 56–64]. Typical current–voltage characteristics of SNOP NbN-MgO-NbN SIS with a ~200 nm thick MgO or SiO_2 self-aligned insulator film, RF-sputtered, are presented in Fig. 6.6.

Experimental conditions for optimizing the junction energy gap (i.e., $\Delta_{BE} + \Delta_{CE}$), the sub-gap leakage current, and the Josephson current density lead to different quantitative results depending on the junction trilayer deposition conditions and on post-deposition temperature during processing [2, 60].

From these experimental studies, one should emphasize on different factors influencing the MgO barrier properties and consequently the junction properties:

(a) The substrate choice (MgO template) and NbN base-electrode deposition conditions (temperature, Ar-N2 plasma type, etc.) [2, 60].
(b) The MgO barrier conductance is often dominated by high-transparency channels, associated with excess Josephson and sub-gap currents signed by a subharmonic gap structure, as shown in Fig. 6.6(b) [60]. Dieleman et al. [64] and other groups demonstrate that it is due to multiple Andreev reflection (MAR) mechanisms. Contrary to Nb–Al–AlO_x–Nb junctions, a post-annealing of the NbN–MgO–NbN patterned junction (or unpatterned trilayer) a few hours in the temperature range of 200–300°C has the benefic effect of removing those conductance extra-channels, decreasing apparent Josephson current, but more efficiently sub-gap conductance and improving the overall current–voltage characteristics [39, 60]. The value of the energy gap is not affected by the junction annealing up to 350°C (when current density J_c is not affected by self-heating). Figure 6.7 shows that the effective MgO barrier height observed at about 1.3 eV is increased inside the trilayer after annealing for a few hours and tends to reach the expected 2.8 eV MgO bulk barrier height [60]. One also can observe similar MAR origin behavior with other artificial barriers deposited on NbN or MgB_2 electrodes [65–67].
(c) We observed like in NbN–oxide–NbN junctions [12, 66] that the values of the energy gaps of the NbN base electrode and of the NbN counterelectrode are different [60]. The base-electrode gap $\Delta_{BE} \approx 3$ meV can be sharp according to $T_c \approx 16$ K, but the polycrystalline granular nature of the NbN counterelectrode (with low electronic mean free path) grown in contact with the barrier leads to a depressed superconductivity close to the barrier interface associated with in-plane dispersion of grains gaps $\Delta_{CE} \approx 1.5$–2.5 meV, later observed by scanning tunneling microscopy (STM) [68, 69]. The challenge is to improve the growth conditions and (100) texture of NbN CE.

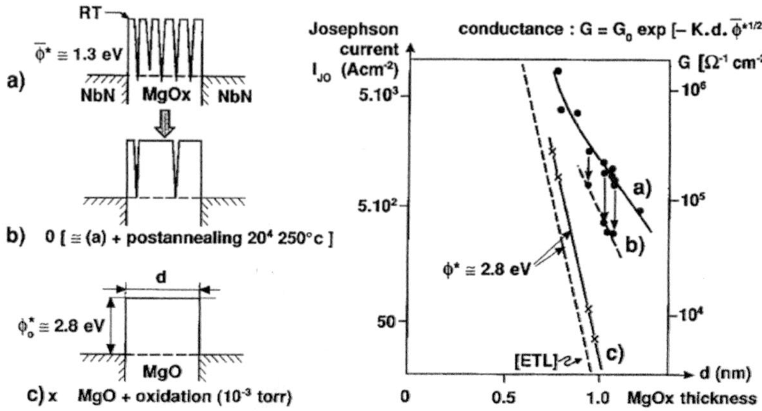

Figure 6.7 Josephson current density I_{J0} (A/cm^2) or junction conductance G (S/cm^2) of NbN–MgO–NbN tunnel junctions vs. MgO-deposited barrier thickness, before and after wafer post-annealing at 250°C for 2 hours [39, 60].

$Nb_{1-x}Ti_xN$ [for $x = 0.4$, $T_c \approx 15.4$ K] films presents a slightly lower T_c than NbN films (~16 K), but they are attractive for their lower RF losses and surface resistance up to 1 THz in the submillimeter wave receivers applications and good-quality NbTiN–MgO–NbTiN tunnel junctions were fabricated [56].

High-quality NbN tunnel junctions were also fabricated using AlN_x as barrier [70–72]. Radical nitridation of an Al layer deposited on a base NbN layer is the most successful to form the barrier [70] with low sub-gap leakage currents, and high critical current density (J_c) of up to 15.6 kA/cm^2. The maximum-to-minimum spread in I_c was ±1.5% for a series array of 200 junctions with a J_c of 4.4 kA/cm^2 [70].

It was also observed that MgO/AlN/MgO heterostructure barrier can be tuned to high current density and RSJ characteristics between 9 and 14 K [17].

6.3.2 NbN and NbTiN SNS and SS'S Junctions

As described by E. L. Wolf in Chapter 2, SNS junctions whose barrier is either a normal diffusive metal or a superconductor or a constriction could deliver Josephson current with interesting properties with regards to applications. They rely on a different mechanism named Andreev reflection, described first by Andreev [73] and in the same period by De Gennes et al. [74].

When the barrier is a dirty superconducting material with a smaller T_c' than the electrodes, the induced coherence length $\xi_N(T)$ in the barrier of diffusion constant D should be in the form (for $T < T_c'$):

$$\xi_N(T) = \sqrt{\frac{\hbar D}{2\pi k_B T}} \sqrt{1 + \frac{2}{\ln \frac{T}{T_c'}}}$$

and Josephson critical current:

$$I_0(T) = \frac{\pi \Delta_0^2}{4eR_N k_B T_c} \left(1 - \frac{T}{T_j}\right)^2 \frac{L/\xi_N}{\sinh L/\xi_N}$$

Likharev [75, 76] gave developments related to the different parameter configurations of Josephson weak-links. The approximate relation between junction current density, energy gap, junction length, and coherence length is

$$J_c(T) \propto \Delta(T)^2 \sqrt{T} e^{(L/\xi_n(T_c))\sqrt{T/T_c}}.$$

Various barrier materials associated with NbN electrodes have been studied with limited applied interest [2]. A breakthrough was achieved in 2000 by Kaul et al. [77], obtaining internally shunted NbN Josephson junctions with a sputtered TaN_x barrier. The TaN_x film resistance has been tuned to obtain a large value of $R_N I_C$ (>0.5 mV at 4.2 K) and substitution of NbN to NbTiN electrodes more suitable for nonlatching logic applications at 4.2 K [78, 79].

Such NbN–TaN_x–NbN "self-shunted" nitride SNS or SS'S junctions, depending on the resistivity of TaN_x and barrier thickness, have been studied by Setzu in comparison to NbN–MgO–NbN junctions [15].

The resistivity of DC magnetron sputtered TaN_x films developed at CEA-INAC could be varied from a few hundred micro-ohms to a few hundred milliohms by increasing the N_2 pressure during reactive sputtering in Ar–N_2 gas mixture, from a pure Ta target [15, 17, 80]. By adjusting the TaN_x barrier thickness, nonhysteretic I–V characteristics of such NbN–TaN_x–NbN junctions are presented in Fig. 6.8, showing a parabolic shape with up-turn curvature at 4.2 K

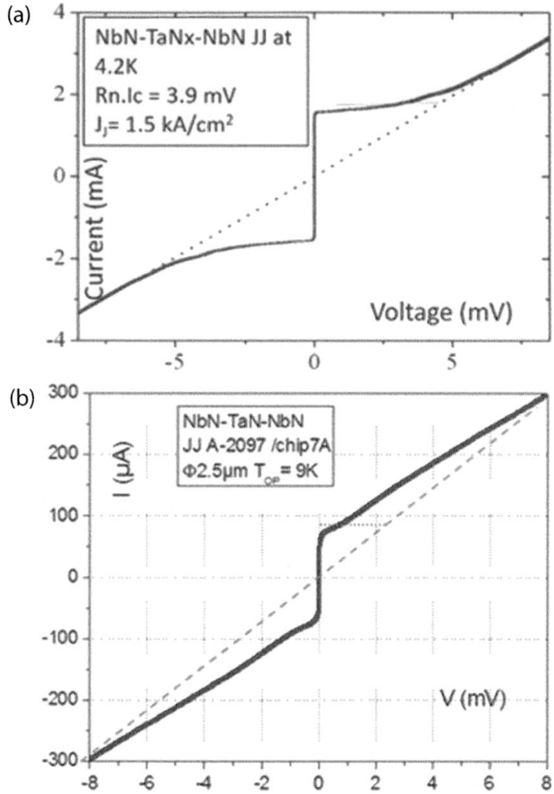

Figure 6.8 Current–voltage characteristics at 4 K (a) and 9 K (b) of refractory Josephson junctions of the form NbN–TaN_x–NbN showing large $R_N I_C$ values of about 3.9 mV at 4.2 K and 2 mV at 9 K from Setzu [15] and Villegier et al. [17].

[78, 79], as expected for the RSJ model (see Chapter 2). As shown in Fig. 6.8, $R_N I_C$ values of 3.9 mV were observed at 4.2 K and of 2 mV at 9 K, with J_C and R_N of magnitudes that are suitable for single-flux quantum digital circuits [15, 17].

Large $R_N I_C$ products are induced by efficient quasi-particles diffusion and good crystalline matching of nitride barriers (TaN_x or TiN_x [81]) at interfaces with NbN (or NbTiN) electrodes in the same cubic B1 crystalline phase, which solve frequency limitations induced by external shunt of the large capacitance of SIS junctions.

The temperature dependence of $R_N I_C$ of the junctions with ~10 mΩcm barrier resistivity was measured for various barrier thicknesses and the coherence length of the barrier was determined to be about 5 nm [15].

The high resistivity superconducting TaN_x barrier and the dielectric MgO tunnel barriers are compared in Fig. 6.9: for example, to obtain a Josephson critical current density of $J_c \approx 8$ kA/cm^2, one has to deposit ~0.8 nm of MgO or about 12 nm of TaN_x. The TaN_x barrier thickness is easy to control accurately and the slope $\delta J_C/\delta d$ is much lower than for the MgO tunnel barrier. The reproducibility of TaN_x resistivity and of barrier conductance is, however, more critical, but across a 2×2 cm^2 chip all the junctions are internally damped with a large $R_N I_C$ and a small spread is observed when a CMOS foundry process is applied with a good control of nitride films uniformity and lithography [17].

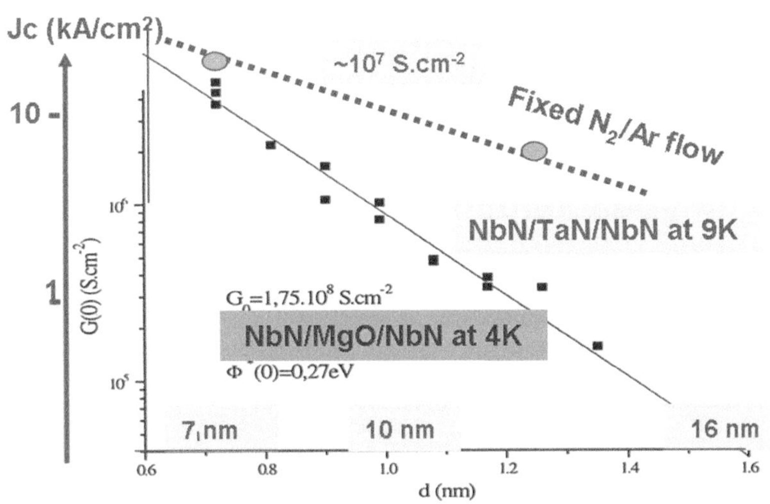

Figure 6.9 Comparison of TaN and MgO barrier transmission given as Josephson current density J_c (kA/cm^2) or conductance G (S/cm^2) versus barrier thickness (the TaN scale is [10×] the MgO barrier scale) of self-shunted NbN–TaN–NbN and NbN–MgO–NbN junctions from Setzu [15] and Villegier [42].

6.4 NbN Digital Circuits and Other Applications

6.4.1 First Digital Circuits Based on NbN–Oxide–NbN Junctions

The first applications of NbN (substituted to Nb) Josephson tunnel junctions integrate lead alloys as junction counterelectrode and wiring levels, as described in the first paragraph. So applications were limited due to the drawback of the mechanical properties of lead alloys generating hillocks during storage and thermal cycling [5]. Research studies were limited to Josephson integrated circuit demonstrators [30] and to HF circuits: voltage standard and SIS heterodyne mixers (see Chapters 7 and 10) [39].

A more promising solution comes when a reliable NbN "trilayer dry process" could be achieved with NbN–oxide–NbN junctions, even if the energy gap was decreased below theoretical expectations and sub-gap current could not be reduced to a low value [12]. In Japan, remarkable demonstrations were first accomplished using thermal or plasma oxidized NbN junction base-electrodes and dry-etching processing [49, 50]:

- An 8-bit logic adder (four junction logic [4 JL], 2.5 μm linewidth) of 2800 NbN–oxide–NbN junctions with Nb wiring levels was successfully achieved in 1984 at Electrotechnical Laboratory [49].
- Two years later, Hitachi was developing 1.5 μm square NbN–oxide–NbN junctions and fabricating 3K-logic-gate array using 23,000 junctions demonstrating successful operation [50].

But these very promising demonstrations occurred, unfortunately, at the time when mostly all research efforts and teams were switched towards the newly discovered HTS materials and they focused their studies on the difficult task of getting uniform YBaCuO Josephson junctions (see IEEE papers' content of Applied Superconductivity conferences of 1986, 1988, and 1990).

6.4.2 HF Applications of NbN–MgO (or AlN)–NbN Junctions at 2 K and 10 K

Since 1986, the analog NbN (or NbTiN) SIS junction circuits have been benefitting immensely from the introduction of MgO (or thin AlN) barriers instead of $NbNO_x$.

Several laboratories have developed NbN/MgO/NbN SIS junction mixers for radio-astronomy, operating in a larger bandwidth and at higher frequencies than Nb SIS mixers mostly limited to the Nb gap frequency near 650 GHz [60, 61].

Typically, NbTiN/MgO/NbTiN SIS junction mixers have been achieved at JPL-CALTEC with a receiver noise temperature of 250 K DSB near 1 THz [56], even better than the results reported at that time with NbN devices at similar frequencies. The production of a large number of optimized NbTiN (and NbN) SIS receivers (tuned in the band 750–950 GHz) is still an important task for the Atacama Large Millimeter/submillimeter Array (ALMA) of telescopes [82]. An overview and a description of the NbN, NbTiN SIS receivers and also figures of merit of hot-electron bolometer (HEB) THz receivers based on thin NbN or other nitride microbridges are given in Chapter 7 by Valery Koshelets et al.

It should be emphasized that the domain of metrology described in Chapter 10 by Johannes Kohlmann also benefits from NbN junctions, thanks to the uniformity of NbN/TiN_x/NbN junctions and to the possibility of stacking NbN junctions in a mixed 2D–3D configuration well coupled to a microwave structure. A 10 V programmable Josephson voltage standard circuit, made of 327 680 NbN/TiN_x/NbN/TiN_x/NbN junctions uniform in double-junction stacks, in collaboration with ETL (now AIST) and NIST, has been successfully demonstrated [81].

Before the year 2000, the most original and promising NbN IC technology focused on space applications was in TRW [59, 83, 84].

NbN–MgO–NbN Josephson junction and circuits were developed in parallel to the TRW Nb multilevel technology [84]. TRW was able to demonstrate a 60 GHz NbN single-flux quantum (SFQ) counter circuit in 1991 [85] and later developed 12 mask-level NbN circuits [84].

One of the specific characteristics of NbN circuits was to operate at 10 K in RF front-ends packaged in a space satellite or rocket, inside a cryogenic remote system, with lower energy and lower weight levels (a factor 4 to 10) than expected for Nb RF front-ends operating at 4 K.

The successively published NbN operating IC devices were as follows:

- The design in collaboration with JPL of a 10 K NbN A/D converter for readout of large Si:As BIB, 128×128 IR focal plane array sensors [86].
- The design and fabrication of the NbN ADC was later improved to 12-bit SFQ counter operating at 10 Msps and reading IR images at 100 frames/s at 9 K and dissipating 0.3 mW [87].
- A further improvement in the system bandwidth concerns the cryopackaging using a dedicated superconducting multichip module (MCM) and the implementation of a NbN digital signal processor (DSP) located behind the ADC, all circuits operating at 10 K [88].
- Finally a 16-bit, 10 K SFQ ADC counter was successfully optimized associated with several NbN DSP modules including first-in-first-out (FIFO) memory registers (~2000 JJ each) and published in 2001 [89], after the TRW Redondo Beach Superconducting Laboratory was fused with Northrop Grumman Laboratories (NGST).

6.4.3 Internally Damped NbN Junctions Applied to RSFQ Technologies

In order to produce efficient rapid single-flux quantum logic circuits (RSFQs), following a logic concept of quantified Josephson voltage (or current) impulsion carrying a single-flux quantum ($\Phi_0 = 2.07$ mV ps) proposed by K. K. Likharev et al. [1] (see references inside the paper), a reliable and reproducible process based on refractory superconductors is mandatory. It should be compatible with very large scale integrated (VLSI) multilevel circuit fabrication rules [84]. Details on RSFQ circuit schemes and functions can be

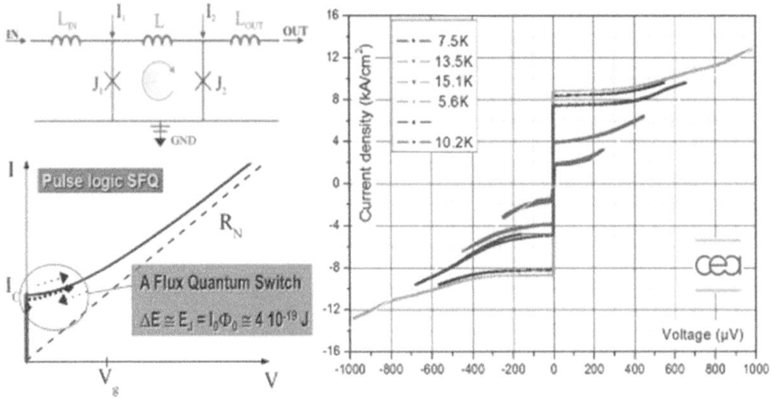

Figure 6.10 Basic SQUID gate in rapid single flux quantum logic (*left*) and current density–voltage characteristics at different temperatures (from 5.6 to 15.1 K) of an internally damped NbN–TaN_x–NbN Josephson tunnel junction suitable for RSFQ applications.

found in [90, 91], together with recent examples of state-of-the-art circuit fabrication made of Nb junctions [92] and IC realization [93].

A typical SFQ SQUID gate should only be made of critically damped RSJ Josephson junctions with a large enough $R_N I_C$ (i.e., cut-off frequency) value, acting as nonlinear inductors, interconnected by simple inductors, as shown in Fig. 6.10. Internally damped junctions. such as NbN–TaN_x–NbN [17, 77] or NbTiN–TaN_x–NbTiN [78, 94] junctions presented above, are especially suitable for making compact and scalable to µm^2-scale SFQ logic gates.

The development of useful nitride RSFQ circuits integrating internally damped junction was limited until now to a few demonstrations, such as MUX and DEMUX, new HF SFQ circuits [78], which should operate at 4.2 K, using either an Nb junction technology upgraded to higher $R_N I_C$ values or the "penta-level" Nb/NbTiN/TaN_x/NbTiN/Nb junction and Nb wirings proposed by van Duzer et al. [78].

Recently, an analytical head containing a sigma–delta NbN ADC was designed at Savoy University and implemented at CEA for the French Centre d'Etudes Spatiales (CNES) to operate at 9–10 K in a closed-cycle refrigerator devoted to Space Telecoms satellites [17, 95]. The circuit, fabricated on 8-inch Si wafers at CEA (LETI-CMOS

Table 6.2 Stack layers of a NbN IC circuit fabrication (from CEA [17])

Layer	Thickness (nm)	ρ_{20K} (μΩ.cm)	Deposition
NbTiN	650	250	DC-mag
Mo	45	~200	DC-mag
SiO_2	300	Insulator	PE-CVD
NbTiN	350	250	DC-mag
SiO_2	300	Insulator	PE-CVD
NbTiN–NbN	150–150	250	DC-mag
TaN_x	8	10^4	DC-mag
NbN–NbTiN	150–150	250	DC-mag
SiO_2	300	Insulator	GE-CVD
NbTiN (GP)	550	250	DC-mag

and INAC) platforms, integrates internally damped submicrometer NbN–TaN_x–NbN planarized junctions and 0.7 μm linewidth wirings. The layer stack is shown in Fig. 6.3 and Table 6.2. A penta-layer NbTiN–NbN–TaN_x–NbN–NbTiN is in situ DC magnetron sputtered (DC-mag) in order to efficiently couple NbTiN wirings (with lower inductivity than the NbN lines) to the junction NbTiN–NbN bilayer electrodes. Figures 6.8 and 6.10 show typical I–V curves at 9–10 K and Fig. 6.12 (*left*) a toggle flip-flop (TFF) SFQ gate.

As explained above, internally damped NbN, NbTiN Josephson junctions (without adding external damping resistor) are attractive for VLSI digital or mixed analog/digital SFQ circuits. Internally damped junctions and issued logic gates lead to a natural down-size scalability and to larger cut-off frequency due to large $R_N I_C$ values mostly independent of critical current density J_c, in comparison to the externally damped junctions by resistors whose cut-off frequencies are strongly reduced by the resistor, as discussed in [96]. The eventually found drawback of too large specific inductance of NbN films due to large penetration depth and kinetic inductance can be reduced by introducing Nb layers or NbTiN extralayers inside the stack.

The use of a semiconductor foundry to fabricate NbN superconductive electronics circuits (which is not possible for Nb IC without strong processing conditions and tool modifications) ensure the reliability of submicrometer lithography and cost-efficient techniques such as statistical process control. This could be used

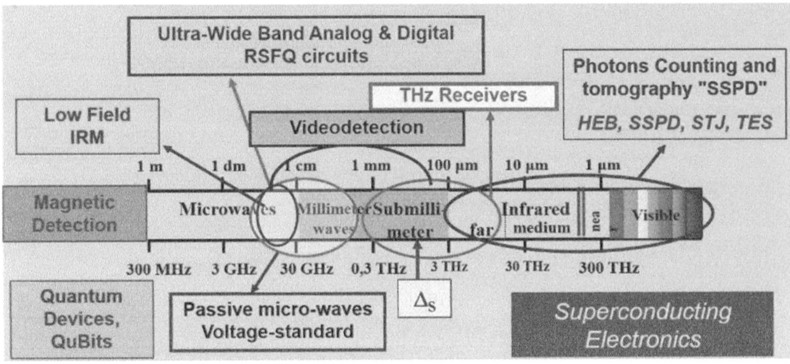

Figure 6.11 Main classes of applications of Nb or NbN superconducting electronic devices and integrated circuits (SIC). Because of a shift of the gap Δ_S toward higher frequency, NbN analog and digital devices bring a possible shift to a higher bandwidth, frequency, and operating temperature in comparison to Nb. Moreover, some of the detector devices based on fast disjunctions (e.g. SSPD) are more specific to NbN [42].

to optimize fabrication costs, to track derives and defects in the process, and to ensure the long-term availability of superconductive electronics chips.

6.4.4 NbN Devices Offer Wider Applications Than Nb Ones

Forecasting superconductive electronics technology is not an easy task, but it is again, as in 1980 with IBM project, and in 1987 with new high-T_c superconductivity, an actual subject in 2015 [97], where refractory NbN or other nitride junctions and emerging nanostructures should have specific advantages for different devices and applications tentatively listed in a frequency (or bandwidth) scale in Fig. 6.11 [42]. What should be the benefits of substituting nitrides with Nb-based junctions or to introduce new nitride active or passive devices?

(a) Recent logic and memory programs are focused on large-scale integrated circuits for developing energy-efficient green supercomputers and data centers [98]. Greatly improved energy efficiency gates, such as reciprocal quantum logic (RQL),

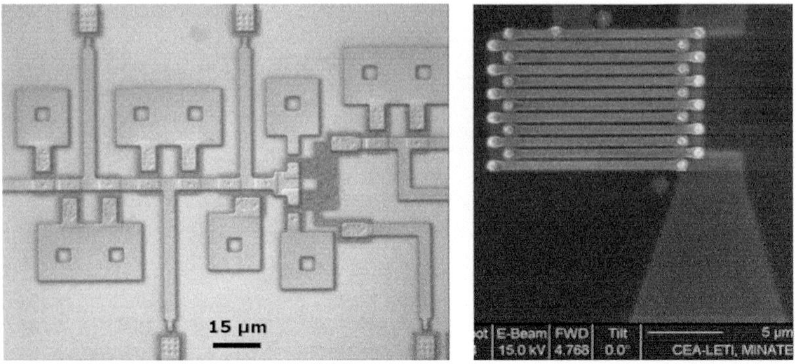

Figure 6.12 *Left*: Toggle flip-flop (TFF). *Right*: Top view of an SNSPD pixel (0.35 μm linewidth) interconnected on chip by an array of 24 NbN–TaN_x–NbN Josephson tunnel junctions (0.5 μm diameter) plus an injector gate, above a ground plane.

efficient SFQ (ERSFQ, eSFQ), LR-RSFQ, and low-voltage RSFQ, have been designed and are now developed [99]. SFQ ICs and HF front-ends should have the benefit of higher areal density, higher frequency bandwidth, and lower jitter of internally damped NbN junctions.

(b) Nanowires and nanobridges, patterned in a few nanometers thick NbN or NbTiN, epitaxially grown films, operate under superconducting DC current bias as fast controllable (~100 ps) disjunctions, sensitive to a single photon, a current pulse, or an HF wave modulation [68]. NbN advantage resides in its very short electron–phonon time of a few picoseconds with low jitter [68, 100]. NbN superconducting nanowire single photon detectors (SNSPDs) are widely developed and can be integrated on chip (see Fig. 6.12, left) and read by SFQ circuits [101]. Low-noise THz hot-electron bolometer mixers (HEBs), pulse discriminators [102], and new "nanocryotron" (nTron) made of superconducting NbN nanowire as a three-terminal electrothermal device [103] are presently developed. The domain of quantum computation should also benefit from NbN qubits [104].

6.4.5 Scaling of NbN Josephson Junction Size

One example of successfully down-scaling NbN–MgO–NbN SIS junction size was obtained to achieve single-photon emission [13]. A simple DC voltage bias (V) on a small-capacitance Josephson junction leads to emission of microwave radiation via inelastic Cooper-pair tunneling. In this process a tunneling Cooper pair emits one or several microwave photons with a total energy of 2 eV [105]. The observed average photon emission rate is well explained within the so-called P(E) theory [106], but this theory does not predict the statistics of the emitted photons. Alexander Grimm at CEA INAC achieved very-small-area, good-quality NbN–MgO–NbN junctions and DC SQUIDs using an e-beam patterning variant of the SNOP process [12, 13] (Figs. 6.13 and 6.14) and was able to

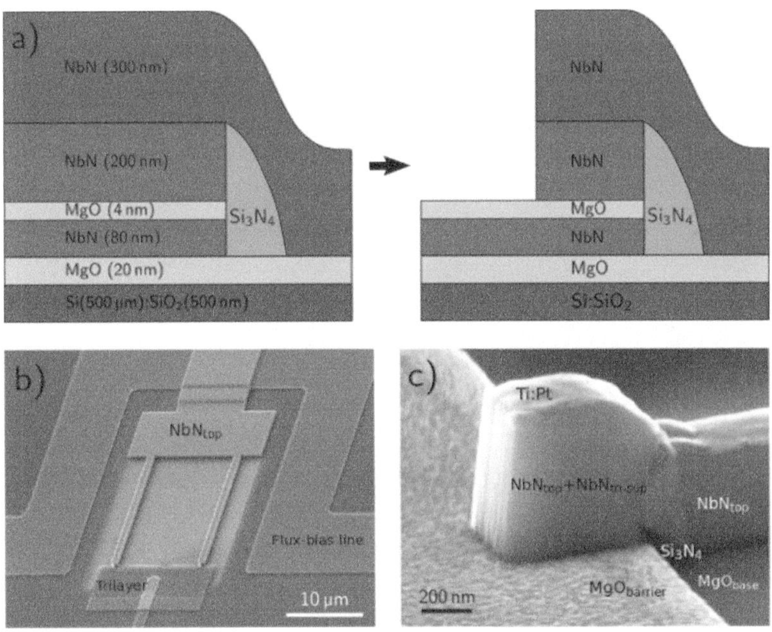

Figure 6.13 (a) Step in the trilayer with self-aligned spacer covered by an NbN layer forming the top electrode before (*left*) and after (*right*) etching. (b) SEM view of a SQUID. (c) SEM micrograph of a Josephson junction. The Si_3N_4 spacer extends well beyond the lower NbN of the trilayer, insulating it from the NbN top electrode, reproduced with permission from Grimm [13].

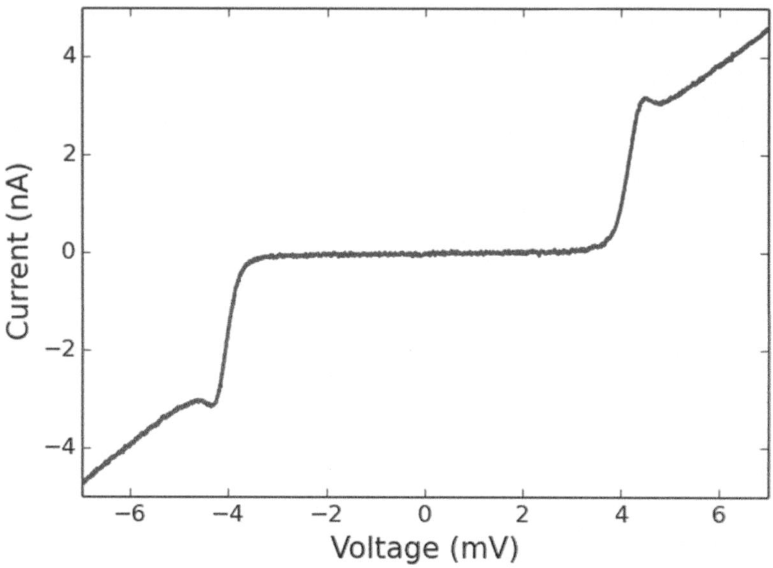

Figure 6.14 Typical current voltage characteristic of a SQUID consisting of two 150 nm×150 nm NbN-MgO-NbN Josephson junctions in parallel, reproduced with permission from Grimm [13] (DC Josephson current has been suppressed by applying a magnetic field through the SQUID loop).

emit single photons, detect them, and study the statistics of their quantum emission.

Future applications of such single microwave photon emission sources up to the THz range to quantum information are straightforward.

In conclusion, taking advantage of recent demonstrations of processes, such as at CEA, it is possible to forecast NbN Superconducting Processors integrating other on-chip HF devices fabricated onto 8-inch Si wafers in any medium-size classical Si-CMOS facility. Future implementation of ULSI refractory nitride processors ($N_{JJ} \approx 10^9/\text{cm}^2$) supposes further down-scaling of NbN junction size ($\sim$0.1 μm^2) with a large cut-off frequency ($R_N I_C >$1 meV) and low spread ($\sigma \approx 1\%$). This looks achievable in a CMOS foundry as soon as compact, energy-efficient logic circuits, dense memory gates, and fast three-terminal amplifiers are designed.

Acknowledgments

The author thanks numerous former collaborators from CEA-Grenoble, including M. Aurino, E. Baggetta, S. Bouat, P. Cavalier, R. Chicault, R. Espiau, N. Hadacek, V. Larrey, M. Levis, L. Maingault, M. Reymermier, R. Setzu, and L. Vieux-Rochaz. The author also thanks several CEA, university, and CNRS colleagues, especially C. Antoine, A. Grimm, M. Hofheinz, and P. Febvre, as well as Fluxonics colleagues.

References

1. Likharev, K. K. (2012). Superconductor digital electronics, *Physica C*, **482**, 6–18.
2. Ruggiero, S. T. (1990). Artificial tunnel barriers, in *Superconducting Devices*, eds. Ruggiero, S. T., and Rudman, D. A. (Academic Press, Boston), Chapter 10, pp. 373–390.
3. Cardinne, P., Nordman, J., and Renard, M. (1974). On the use of low barrier potential materials to improve high frequency coupling to Josephson tunnel junctions. *Rev. Phys. Appl.*, **9**, 167–171.
4. CEA LETI Internal Report, DRET Contract (1974).
5. Robinson, A. L. (1983). IBM drops superconducting computer project. *Science*, **222**, 492–494.
6. Dousselin, G., and Rosenblatt, J. (1980). Behavior of edge-grown low-capacitance superconducting tunnel junctions. *J. Appl. Phys.*, **51**, 802–804.
7. Buhrman, R. A. (1984). Superconductor microstructures. *Physica*, **126B**, 62–69.
8. Fujimaki, A., Kawai, K., Hayashi, N., Horibe, M., Maruyama, M., and Hayakawa, H. (1999). Preparation of ramp-edge Josephson junctions with natural barriers. *IEEE Trans. Appl. Supercond.*, **9**(2), 3436–3439.
9. Kroger, H., Smith, L. N., and Jillie, D. W. (1981). Selective niobium anodization process for fabricating Josephson tunnel junctions. *Appl. Phys. Lett.*, **39**, 280.
10. Kroger, H., Potter, C. M., and Jillie, D. W. (1978). Niobium Josephson junctions with doped amorphous silicon barriers. *IEEE Trans. Magn.*, **MAG-15**, 488–489.
11. Gurvitch, M., Washington, M. A., and Huggins, H. A. (1983). High quality refractory Josephson tunnel junctions utilizing thin aluminum layers.

Appl. Phys. Lett., **42**, 472–474; Gurvitch, M., Washington, M. A., Huggins, H. A., and Rowell, T. M. (1983). Preparation and properties of Nb Josephson junctions with thin Al layers. *IEEE Trans. Magn.*, **MAG-19**, 791–794.

12. Villegier, J.-C., Vieux-Rochaz, L., Goniche, M., Renard, P., Vabre, M. (1985). NbN tunnel junctions. *IEEE Trans. Magn.*, **MAG-21**, 498–504.
13. Grimm, A. (2015). Josephson photonics: Statistics of photons emitted by inelastic Cooper pair tunneling. PhD thesis, Grenoble Alpes University.
14. Anders, S., et al. (2010). European roadmap on superconductor electronics - status and perspectives. *Physica C*, **470**(23–24), 2079–2126.
15. Setzu, R. (2007). Etude et réalisation de jonctions Josephson en NbN à barrière semi-métallique en TaxN, PhD thesis, J. Fourier Grenoble University.
16. Ketchen, M. B., Pearson, D., Kleinsasser, A., et al. (1991). Sub-μm, planarized, Nb-AlOx-Nb Josephson process for 125 mm wafers developed in partnership with Si technology. *Appl. Phys. Lett.*, **59**, 2609–2611.
17. Villegier, J.-C., Bouat, S., Aurino, M., Socquet-Clerc, C., and Renaud, D. (2011). Integration of planarized internally-shunted submicron NbN junctions. *IEEE Trans. Appl. Supercond.*, **21**, 102–106.
18. Hinode, K., Nagasawa, S., Sugita, M., Satoh, T., Akaike, H., Kitagawa, Y., and Hidaka, M. (2003). Pattern-size-free planarization for multilayered large-scale SFQ circuits. *IEICE Trans. Electron.*, **E86-C**(12), 2511–2513.
19. Tolpygo, S. K., Amparo, D., Yohannes, D. T., Meckbach, M., and Kirichenko, A. F. (2009). Process-induced variability of Nb/Al-AlOx/Nb junctions in superconductor integrated circuits and protection against it. *IEEE Trans. Appl. Supercond.*, **19**(3), 135–139.
20. Pavolotsky, A., Dochev, D., and Belitsky, V. (2011). Nb/Al-AlOx/Nb junction properties variations due to storage and mounting. Proc. of the 22nd Int. Symposium on Space Terahertz Technology, Tucson, USA, pp. 1–5.
21. Hinode, K., Satoh, T., Nagasawa, S., and Hidaka, M. (2008). Hydrogen-inclusion-induced variation of critical current in Nb/AlOx/Nb Josephson junctions. *J. Appl. Phys.*, **104**, 023909-1–023909-6.
22. Olaya, D., Dresselhaus, P. D., and Benz, S. P. (2010). Niobium-silicide junction technology for superconducting digital electronics. *IEICE Trans. Electron.*, **E93C**, 463–467.

23. Milton, R. M. (1946). A superconducting bolometer for infrared measurements. *Chem. Rev.*, **39**, 419–433.

24. Braginski, A. I., and Talvacchio, J. (1990). MBE growth of superconducting materials, in *Superconducting Devices*, eds. Ruggiero, S. T., and Rudman, D. A. (Academic Press, Boston), Chapter 8, pp. 273–324.

25. Gurvitch, M., Remeika, J. P., Rowell, J. M., Geerk, J., and Lowe, W. P. (1985). Tunneling, resistive and structural study of NbN and other superconducting nitrides. *IEEE Trans. Magn.*, **MAG-21**(2), 509–513.

26. Lamura, G., Villegier, J.-C., Gauzzi, A., Le Cochec, J., Laval, J. Y., Placais, B., Hadacek, N., and Bok, J. (2002). Granularity-induced gapless superconductivity in NbN films: evidence of thermal phase fluctuations. *Phys. Rev. B*, **65**, 104507 (1–4).

27. Keskar, K. S., Yamashita, T., and Onodera, Y. (1971). Superconducting transition temperatures of RF sputtered NbN films. *Jpn. J. Appl. Phys.*, **10**(3), 70; Oya, G., and Onodera, Y. (1974). Transition temperatures and crystal structures of single-crystal and polycrystalline NbN films. *J. Appl. Phys.*, **45**, 1389–1397.

28. Shy, Y. M., Toth, L. E., and Somasundaram, R. (1973). Superconducting properties, electrical resistivities and structure of NbN thin films. *J. Appl. Phys.*, **44**, 5539.

29. Aubert, A., and Spitz, J. (1975). Structure and composition of thin films of niobium nitride and carbone nitride obtained from reactive cathodic sputtering. *Le Vide*, **30**(175), 1–7 (*In French*).

30. Villegier, J.-C., and Veler, J-C. (1983). Investigation of high rate magnetron sputtering of niobium films for Josephson integrated circuits. *IEEE Trans. Magn.*, **19**(3), 946–950.

31. Noskov, V. L., Titenko, Y. V., Korzhinskii, F. I., Zelenkevich, R. L., and Komashko, V. A. (1980). Heteroepitaxial layers of niobium nitride on sapphire. *Sov. Phys. Crystallogr.*, **25**, 504–508.

32. Espiau de Lamaëstre, R., Odier, Ph., and Villégier, J.-C. (2007). Microstructure of NbN epitaxial ultrathin films grown on A-, M-, and R-plane sapphire. *Appl. Phys. Lett.*, **91**, 232501-1–232501-3.

33. Villegier, J.-C., Bouat, S., Cavalier, P., Setzu, R., Espiau de Lamaëstre, R., Jorel, C., Odier, P., Guillet, B., Mechin, L., Chauvat, M. P., and Ruterana, P. (2009). Epitaxial growth of sputtered ultra-thin NbN layers and junctions on sapphire. *IEEE Trans. Appl. Supercond.*, **19**(3), 3375–3378.

34. Bacon, D. D., English, A. T., Nakahara, S., Peters, F. G., Schreiber, H., Sinclair, W. R., and van Dover, R. B. (1983). Low temperature deposition

and properties of superconducting NbN by reactive DC magnetron sputtering. *J. Vac. Sci. Technol. A*, **1**, 365.

35. Kosaka, S., and Onodera, Y. (1974). Epitaxial deposition of niobium nitride by sputtering. *Jpn. J. Appl. Phys.*, **13**(Suppl 1), 613.
36. Yamashita, T., Hamasaki, K., Kodaira, Y., and Komata, T. (1985). Nanometer bridge with epitaxially deposited NbN on MgO film. *IEEE Trans. Magn.*, **21**(2), 932–934.
37. Zhang, J. J., Su, X., Zhang, L., Zheng, L., Wang, X. F., and You, L. (2013). Improvement of the superconducting properties of NbN thin film on single-crystal silicon substrate by using a TiN buffer layer. *Supercond. Sci. Technol.*, **26**, 045010 (1–6).
38. Jia, X. Q., Kang, L., Gu, M., Yang, X. Z., Chen, C., Tu, X. C., Jin, B. B., Xu, W. W., Chen, J., and Wu, P. H. (2015). Fabrication of a strain-induced high performance NbN ultrathin film by a Nb5N6 buffer layer on Si substrate. *Supercond. Sci. Technol.*, **27**, 035010 (1–6).
39. Villegier, J.-C., Radparvar, M., Yu, L. S., and Faris, S. M. (1989). RF-sputter-deposited magnesium oxide films as high quality adjustable tunnel barriers. *IEEE Trans. Magn.*, **25**(2), 1227–1230.
40. Villegier, J.-C., Delaet, B., Larrey, V., Febvre, P., Tao, J. W., Angenieux, G. (1999). Extraction of material parameters in NbN multilayer technology for RSFQ circuits. *Physica C*, **326–327**, 133–143.
41. Kawakami, A., Wang, Z., and Miki, S., (2001). Low-loss epitaxial NbN/MgO/NbN trilayers for THz applications. *IEEE Trans. Appl. Supercond.*, **11**, 80–83.
42. Villegier, J.-C. (2011). Tutorial: microelectronic issues & prospects in disruptive superconducting integrated circuits. Colloquium Microelectronics August 30th 2011; Joao Pessoa Brazil.
43. Shinoki, F., Takada, S., Kosaka, S., and Hayakawa, H. (1980). Fabrication of high quality NbN/Pb Josephson junction. *Jpn. J. Appl. Phys.*, **19**(1), 591–594.
44. Kosaka, S., Shinoki, F., Takada, S., and Hayakawa, H. (1981). Fabrication of NbN/Pb Josephson tunnel junctions with a novel integration method. *IEEE Trans. Magn.*, **MAG-17**, 314.
45. Darlinski, A., and Halbritter, J. (1987). Angle-resolved XPS studies of oxides at NbN, NbC, and Nb surfaces. *Surf. Interface Anal.*, **10**, 223–237.
46. Broom, R. F., Raider, S., Oosenbrug, A., Drake, R. E., and Walter, W. (1980). Niobium oxide-barrier tunnel junction. *IEEE Trans. Electron Devices*, **27**(10), 1998–2008.

47. Hikita, M., Takei, K., and Igarashi, M. (1983). Ar ion bombardment effects of NbN/Pb Josephson junctions with plasma oxidized barriers. *J. Appl. Phys.*, **54**, 7066.
48. van Dover, R. B., Bacon, D. D. (1983). Properties of NbN/Pb Josephson tunnel junctions. *IEEE Trans. Magn.,* **19**(3), 951–953.
49. Kosaka, S., Shoji, L., Aoyagi, M., Shinoki, F., Tahara, S., Ohigashi, H., Nakagawa, H., Takada, S., and Hayakawa, H. (1985). An integration of all refractory Josephson logic LSI circuit. *IEEE Trans. Magn.*, **21**, 102–109.
50. Yano, S., Tarutani, Y., Mori, H., Yamada, H., Hirano, M., and Kawabe, U. (1987). Fabrication and characteristics of NbN-based Josephson junctions for logic LSI circuits. *IEEE Trans. Magn.*, **23**(2), 1472–1475.
51. Kroger, H. (1980). Josephson devices coupled by semiconductor links. *IEEE Trans. Electron Devices*, **27**(10), 2016–2026.
52. Shoji, A., Shinoki, F., Kosaka, S., and Hayakawa, H. (1981). Tunneling characteristics of NbN/NbN Josephson junctions with glow-discharge-produced amorphous silicon barriers. *Jpn. J. Appl. Phys.*, **20**, L587.
53. Shinoki, F., Shoji, A., Kosaka, S., Takada, S., and Hayakawa, H. (1981). Niobium nitride Josephson tunnel junctions with oxidized amorphous silicon barriers. *Appl. Phys. Lett.*, **38**, 285–286.
54. Jillie, D. W., Kroger, H., Smith, L. N., Cukauskas, E. J., and Nisenoff, M. (1982). Niobium nitride-niobium Josephson tunnel junctions with sputtered amorphous silicon barriers. *Appl. Phys. Lett.*, **40**, 747–749.
55. Kuriki, S., and Matsuda, M. (1987). DC SQUIDs made of NbN/a-Si/NbN tunnel Junctions. *IEEE Trans. Magn.*, **MAG-23**(2), 1064–1067.
56. Kooi, J. W., Stern, J. A., Chattopadhyay, G., Le Duc, H. G., Bumble, B., and Zmuidzinas, J. (1998). Low-loss NbTiN films for THz SIS mixer tuning circuits. *Int. J. Infrared Millimeter Waves*, **19**(3), 373–383.
57. Shoji, A., Masahiro, A., Kosaka, S., Shinoki, F., and Hayakawa, H. (1985). Niobium nitride Josephson tunnel junctions with magnesium oxide barriers. *Appl. Phys. Lett.*, **46**, 1098–1100.
58. Le Duc, H. G., Khanna, S. K., and Stern, J. A. (1987). All NbN tunnel junction fabrication. *Int. J. Infrared Millimeter Waves*, **8**(10), 1243–1248.
59. Kerber, G. L., Cooper, J. E., Morris, R. S., Spargo, J. W., and Toth, A. G. (1989). NbN/MgO/NbN Josephson tunnel junctions fabricated on thin underlayers of MgO. *IEEE Trans. Magn.*, **25**(2), 1294–1297.

60. Larrey, V., Villegier, J.-C., Salez, M., Miletto-Granozio, F., and Karpov, A. (1999). Processing and characterization of high Jc NbN superconducting tunnel junctions for THz analog circuits and RSFQ. *IEEE Trans. Appl. Supercond.*, **9**(2), 3216–3219.
61. Kawakami, A., Uzawa, Y., and Wang, Z. (2003). Development of epitaxial NbN/MgO/NbN-superconductor-insulator-superconductor mixers for operations over the Nb gap frequency. *Appl. Phys. Lett.*, **83**(19), 3954–3956.
62. Senapati, K., and Barber, Z. H. (2009). Sidewall shunted overdamped NbN-MgO-NbN Josephson junctions. *Appl. Phys. Lett.*, **94**, 173511-1–173511-3.
63. Thakoor, S., et al. (1987). Insulator interface effects in sputter deposited NbN/MgO/NbN tunnel junctions. *J. Vac. Sci. Technol. A*, **5**, 1721.
64. Dieleman, P., Bukkems, H. G., Klapwijk, T. M., Schieke, M., and Gundlach, K. H. (1997). Observation of Andreev reflection enhanced shot noise. *Phys. Rev. Lett.*, **79**(18), 3486–3489.
65. Chen, K., Cunnane, D., Shen, Y., Xi, X. X., Kleinsasser, A. W., and Rowell, J. M. (2012). Multiple Andreev reflection in MgB2/MgO/MgB2 Josephson junctions. *Appl. Phys. Lett.*, **100**, 122601 (1–3).
66. Chicault, R., and Villegier, J.-C. (1987). New phonon structures observed by tunneling in granular NbN films. *Phys. Rev. B*, **36**(10), 5215–5224.
67. Blaugher, R. D., Przybysz, J. X., Talvacchio, J., and Buttyan, J. (1987). Processing of All-NbN tunnel junction series arrays. *IEEE Trans. Magn.*, **23**(2), 673–675.
68. Romestain, R., Delaet, B., Renaud-Goud, P., Wang, I., Jorel, C., Villegier, J.-C., and Poizat, J.-Ph. (2004). Fabrication of a superconducting niobium nitride hot electron bolometer for single-photon counting. *New J. Phys.*, **6**, 129 (1–15).
69. Escoffier, W., Chapelier, C., Hadacek, N., and Villégier, J.-C. (2004). Anomalous proximity effect in an inhomogeneous disordered superconductor. *Phys. Rev. Lett.*, **93**, 217005 (1–4).
70. Akaike, H., Funai, T., Naito, N., and Fujimaki, A. (2013). Characterization of NbN tunnel junctions with radical-nitrided AlNx barriers. *IEEE Trans. Appl. Supercond.*, **23**(3), 1101306 (1–6).
71. Makise, K., Terai, H., and Wang, Z. (2012). Resistively shunted NbN/AlN/NbN tunnel junctions for single flux quantum circuits. *Phys. Proc.*, **36**, 116–120.

72. Makise, K., Terai, H., and Uzawa, Y. (2016). NbN/AlN/NbN/TiN tunnel junctions on Si (100) substrate for superconducting devices. *IEEE Trans. Appl. Supercond.*, 26(3), 1–3.
73. Andreev, A. F. (1964). Thermal conductivity of the intermediate state of superconductors. *Sov. Phys.-JETP*, **19**(5), 1228–1231.
74. De Gennes, P. G., and Guyon, E. (1963). *Phys. Lett.*, **3**, 168; De Gennes, P. G. (1964). *Rev. Mod. Phys.*, **36**, 225.
75. Likharev, K. K. (1979). Superconducting weak links. *Rev. Mod. Phys.*, **51**, 101–159.
76. Golubov, A. A., Kupriyanov, M. Yu, and Il'ichev, E. (2004). The current-phase relation in Josephson junctions. *Rev. Mod. Phys.*, **76**(2), 411–469.
77. Kaul, A. B., Whiteley, S. R., Van Duzer, Th., Lei, Yu, Newman, N., and Rowell, J. M. (2001). Internally shunted sputtered NbN Josephson junctions with a TaNx barrier for nonlatching logic applications. *Appl. Phys. Lett.*, **78**, 99.
78. van Duzer, T., Zheng, L., Meng, X., Loyo, C., Whiteley, S. R., Yu, L., Newman, N., Rowell, J. M., and Yoshikawa, N. (2002). Engineering issues in high-frequency RSFQ circuits. *Physica C*, **372–376**, 1–6.
79. Yu, L., Newman, N., Rowell, J. M., and Van Duzer, T. (2005). Incorporation of a frequency-dependent dielectric response for the barrier material in the Josephson junction circuit model. *IEEE Trans. Appl. Supercond.*, **15**(3), 3886–3900.
80. Setzu, R., Baggetta, E., and Villegier, J.-C. (2008). Study of NbN Josephson junctions with a tantalum nitride barrier tuned to the metal-insulator transition. *J. Phys. Conf. Ser.*, **97**, 012077 (1–6).
81. Yamamori, H., Ishizaki, M., Shoji, A., Dresselhaus, P. D., and Benz, S. P. (2006). 10 V programmable Josephson voltage standard circuits using NbN/TiNx/NbN/TiNx/NbN double-junction stacks. *Appl. Phys. Lett.*, **88**, 042503.
82. Uzawa, Y., Fujii, Y., Gonzalez, A., Kaneko, K., Kroug, M., Kojima, T., Miyachi, A., Makise, K., Saito, S., Terai, H., and Wang, Z. (2015). Tuning circuit material for mass-produced terahertz SIS receivers. *IEEE Trans. Appl. Supercond.*, **25**(3), 2401005 (1–5).
83. Kerber, G. L., Abelson, L. A., Elmadjian, R. N., and Ladizinsky, E. G. (1999). Characteristics of junctions and resistors fabricated using an all-NbN superconductor integrated circuit foundry process. *IEEE Trans. Appl. Supercond.*, **9**(2), 3267–3270.
84. Abelson, L. A., and Kerber, G. L. (2004). Superconductor integrated circuit fabrication technology. *Proc. IEEE*, **92**(10), 1517–1533.

85. Spargo, J. W., Cooper, J. E., Kerber, G. L., King, G. R., Morris, R. S., and Toth, A. G. (1991). A 60-GHz NbN single-flux quantum counter circuit. *IEEE J. Solid-State Circuits*, **26**(6), 884–886.

86. Eaton, L. R., Durand, D. J., Sandell, R., and Spargo, J. W. (1995). Design of a 10K NbN A/D converter for IR focal plane array sensors. *IEEE Trans. Appl. Supercond.*, **5**(2), 2457–2460.

87. Durand, D., Dalrymple, B., Eaton, L., Spargo, J., Wire, M., Dowdy, M., and Ressler, M. (1998). 10K NbN ADC for IR sensor applications. *Appl. Supercond.*, **6**(10–12), 741–750.

88. Johnson, M. W., Durand, D., Eaton, L., Leung, M., Spooner, A., and Tighe, T. (1999). NbN circuits and packaging for 10 Kelvin IR focal plane array sensor signal processing. *IEEE Trans. Appl. Supercond.*, **9**(2), 4357–1360.

89. Sun, A. G., Dalrymple, B., Durand, D. J., Herr, Q. P., Johnson, M. W., Luine, J. A., and Spooner, A. (2001). 10K NbN DSP module for IR sensor applications. *IEEE Trans. Appl. Supercond.*, **11**(1), 312–317.

90. van Duzer, T., and Turner, C. W. (1998). *Principles of Superconductive Devices and Circuits*, 2nd ed. (Prentice Hall, Upper Saddle River, NJ).

91. Gross, R., and Marx, A. (2005). Digital electronics, in *Applied Superconductivity: Josephson Effect and Superconducting Electronics* (Walther-Meißner-Institut, Garching), Chapter 5, pp. 215–268.

92. Tolpygo, S. K., Bolkhovsky, V., Weir, T. J., Oliver, W. D., and Gouker, M. A. (2015). Fabrication process and properties of fully-planarized deep-submicron Nb/Al-AlOx/Nb Josephson junctions for VLSI circuits. *IEEE Trans. Appl. Supercond.*, **25**, 1101312.

93. Herr, A. Y., Herr, Q. P., Oberg, O. T., Naaman, O., Przybysz, J. X., Borodulin, P., and Shauck, S. B. (2013). An 8-bit carry look-ahead adder with 150 ps latency and sub-microwatt power dissipation at 10 GHz. *J. Appl. Phys.*, **113**(3), 033911 (1–6).

94. Yu, L., Gandikota, R., Singh, R. K., Gu, L., Smith, D. J., Meng, X., Zeng, X., Van Duzer, T., Rowell, J. M., and Newman, N. (2006). Internally shunted Josephson junctions with barriers tuned near the metal–insulator transition for RSFQ logic applications. *Supercond. Sci. Technol.*, **19**, 719–731.

95. Villegier, J.-C. (2008). Report on fast NbN SFQ analytical head fabrication. R&T. CNES-Toulouse Fr.

96. Hadacek N. (2002). Etude de Films Minces et de Jonctions Josephson en Nitrures Supraconducteurs (TiN et NbN). Application à la Logique RSFQ. PhD thesis, Joseph Fourier Grenoble University.

97. Massachusetts Institute of Technology Lincoln Laboratory. (2014). Forecasting superconductive electronics technology. *The Next Wave*, **20**(3), 1–10.

98. Scott Holmes, D., Ripple, A. L., and Manheimer, M. A. (2013). Energy-efficient superconducting computing - power budgets and requirements. *IEEE Trans. Appl. Supercond.*, **23**, 1701610 (1–10).

99. Herr, Q. P., Herr, A. Y., Oberg, O. T., and Ioannidis, A. G. (2011). Ultra-low-power superconductor logic. *J. Appl. Phys.*, **109**(10), 103903 (1–8).

100. Delacour, C., Pannetier, B., Villegier, J.-C., and Bouchiat, V. (2012). Quantum and thermal phase slips in superconducting niobium nitride (NbN) ultrathin crystalline nanowire: application to single photon detection. *Nano Lett.,* **12**, 3501–3506.

101. Miki, S., Yamashita, T., Wang, Z., and Terai, H. (2014). A 64-pixel NbTiN superconducting nanowire single-photon detector array for spatially resolved photon detection. *Opt. Express*, **22**, 7811.

102. Ejrnaes, M., et al. (2011). Characterization of superconducting pulse discriminators based on parallel NbN nanostriplines. *Supercond. Sci. Technol.*, **24**, 035018.

103. McCaughan, A. N., and Berggren, K. K. (2014). A superconducting-nanowire three-terminal electrothermal device. *Nano Lett.*, **14**, 5748–5753.

104. van Woerkom, D. J., Geresdi, A., and Kouwenhoven, L. P. (2015). 1 minute parity lifetime of a NbTiN Cooper-pair transistor. *Nat. Phys.*, **11**, 547–550.

105. Hofheinz, M., et al. (2011). Bright side of the Coulomb blockade. *Phys. Rev. Lett.*, **106**, 217005.

106. Ingold, G.-L., and Nazarov, Y. V. (1992). In *Single Charge Tunneling*, eds. Grabert, H., and Devoret, M. H. (Plenum, New York).

Chapter 7

Applications in Superconducting SIS Mixers and Oscillators: Toward Integrated Receivers

P. N. Dmitriev, L. V. Filippenko, and V. P. Koshelets
Kotelnikov Institute of Radio Engineering and Electronics (IREE), Mokhovaya Street 11, Building 7, 125009 Moscow, Russia
pavel@hitech.cplire.ru, lyudmila@hitech.cplire.ru, valery@hitech.cplire.ru

Currently, Nb-based tunnel junctions are basic elements of most low-T_c superconducting electronic devices and circuits. In particular, the superconductor–insulator–superconductor (SIS) mixers that employ high-quality Nb-based tunnel junctions have the noise temperature limited only by the fundamental quantum value $hf/2k_B$ [1–13], where h is Planck's constant, f is frequency, and k_B is Boltzmann's constant. The SIS receivers are presently used in both ground-based and space millimeter and submillimeter radio telescopes. At higher frequencies, the lack of compact and easily tunable local oscillators (LO) motivates the direct integration of a superconducting local oscillator with the SIS mixer. A superconducting integrated receiver (SIR) [14–17] comprises on one chip all the elements needed for heterodyne detection. Being lightweight and having low power consumption, in combination with nearly quantum-limited sensitivity and a wide tuning range

Josephson Junctions: History, Devices, and Applications
Edited by Edward Wolf, Gerald Arnold, Michael Gurvitch, and John Zasadzinski

ISBN 978-981-4745-47-5 (Hardcover), 978-1-315-36452-0 (eBook)
www.panstanford.com

of the superconducting local oscillator, the SIR becomes a perfect candidate for many practical applications. In this chapter, an overview of the Nb-based tunnel junction's technology at Kotelnikov IREE in Moscow (as well as the SIR and the superconducting LO developments) is presented. By improving on the fully Nb-based SIR, we have developed and studied Nb–AlN–NbN circuits that exhibit an extended operation frequency range. Continuous frequency tuning for the phase-locked superconducting LO has been experimentally demonstrated in the range of 350–750 GHz. These achievements enabled the development of a 480–650 GHz integrated receiver for the atmospheric-research instrument TELIS (TErahertz and submillimeter LImb Sounder) [18–21]. Further developments of the Nb-based SIS junction technology, along with examples of its laboratory applications, will be also presented.

7.1 Nb-Based Tunnel Junctions for Low-Noise SIS Receivers and Superconducting Oscillators

To realize a quantum-limited performance, the SIS tunnel junctions with a high current density, high energy gap, and extremely small leakage currents are required. In this section, the technological aspects for fabrication of high-quality Nb-based tunnel junctions with parameters required for low-noise SIS are discussed. The fabrication technology of the Nb–AlO_x–Nb tunnel junctions is based on the fact that a very thin Al layer can completely cover the base Nb electrode [22–24], somehow "planarizing" the column-like structure of the Nb film. This Al layer is subsequently oxidized and the top Nb electrode is deposited on the oxidized layer to form a so-called trilayer structure. The SIS tunnel junctions were fabricated from the trilayer structure by using the selective niobium anodization process (SNAP) [25] or the selective niobium etching and anodization process (SNEAP) [23, 26]. The stress, surface morphology, superconducting characteristics, and crystal structure of sputtered Nb films, as well as the surface coverage of thin Al layers deposited on Nb films with different sputtering parameters, were evaluated in order to judge their applicability for fabrication of high-quality Nb-AlO_x–Nb junctions [27–30].

7.1.1 Niobium Tunnel Junctions with an AlO_x Barrier

For low-noise operation, the SIS tunnel junctions with a small leakage current $I_l(V)$ under the gap voltage and minimal energy gap spreading δV_g are required. This is especially important for relatively low-frequency devices ($f \approx$ 100–300 GHz), since δV_g has to be much smaller than the frequency quantum hf/e and the leakage current at a bias voltage of about V_g—$hf/2e$ determines the noise of the mixer. Any additional structure on the IVC of the junction considerably decreases the operation range of the mixer. According to the procedure described above, a thin Al layer covers the Nb base electrode [22, 23]. As a consequence, a residual Al layer appears between the Nb and the isolator barrier, and the tunnel structure is Nb/Al/AlO_x/Nb. It results in suppression of the Nb gap and the appearance of the so-called knee structure due to the proximity effect. A study of the knee's dependence on the thickness of the base Nb electrode and the additional Al layer was performed [31]; the experimental results were compared with numerical calculations based on the microscopic theory of the proximity effect.

The SIS tunnel junctions were fabricated [31] by using SNEAP on the crystalline Si substrates covered by a buffer layer of Al_2O_3 (d = 80 nm). A trilayer structure, Nb–Al/Al_2O_3–Nb, was deposited in a single vacuum run by using DC magnetron sputtering for both Nb and Al films (P_{Ar} = $1{\cdot}10^{-2}$ and $5{\cdot}10^{-3}$ mbar; the deposition rate was approximately 2 and 0.2 nm/s for Nb and Al, respectively) [32]. The substrates were thermally attached to the holder under temperature control. Pure oxygen at an appropriate pressure was used for the formation of the tunnel barrier (oxidation temperature 300 K, time = 20 min). The SIS junction area was defined by RIE followed by anodization; the thermally deposited SiO layer of about 270 nm was used as an insulator.

According to the Werthamer tunnel theory, the IVC of the Nb–Al/Al_2O_3–Nb (S-S'-I-S) tunnel junction depend on the quasiparticle density of states (DOS) in the S' layer (Al). The DOS can be calculated on the basis of the microscopic proximity effect model for S-S' bilayers [33]. The model assumes a short electron mean free path (dirty limit conditions) in both S (Nb) and S' (Al) materials, resulting in the following parameters:

$$\gamma = \frac{\rho_S \xi_S}{\rho_{S'} \xi^*_{S'}} = \sqrt{\frac{D_{S'}}{D_S} \frac{N_{S'}(0)}{N_S(0)}}, \; \gamma_B = \frac{R_B}{\rho_{S'} \xi^*_{S'}}.$$

Here, $\xi_S = \sqrt{D_S/2\pi T_{cs}}$, $\xi^*_{S'} = \sqrt{D_{S'}/2\pi T_{cs}}$, $D_{s,s'}$, $\rho_{s,s'}$, and $N_{s,s'}(0)$ are the coherence lengths, the diffusion coefficients, the normal state resistivities, and the electronic densities of states in the normal states of S and S' metals, T_{cs} is the critical temperature of S metal, and R_B is the product of the resistance of the S-S' boundary and its area. The best fit with the experiment gave [31] the following set of parameters: $\xi_{Nb} = 15$ nm, $\xi_{Al} = 40$ nm, $\gamma = 0.3$, and $\gamma_B = 1$. Calculations in the model above show that the DOS in the S' layer has an energy gap of $\Delta_g < \Delta_{Nb}^{bulk}$ with a large weight of filled subgap states within the energy range $\Delta_g < E < \Delta_{Nb}^{bulk}$. That leads to the appearance of the knee structure on the IVC.

The definition of the knee value is illustrated in Fig. 7.1a [31]. The knee current I_k is defined as the point of maximum deflection of the IVC from the R_n line. I_k is normalized to the quasiparticle current jump I_g at the gap voltage V_g. The value of I_g is evaluated as a current at V_g between the lines of R_n and leakage resistance R_j. The value of V_g is determined at the crossing of the bisector between R_n and R_j with the measured IVC (see Fig. 7.1a).

The experimentally measured IVCs at different thicknesses of the base Nb electrode for $d_{Al} = 9$ nm are presented in Fig. 7.1b, and the currents are normalized to $I(4\text{ mV})$. The values of the knee determined from both theoretical and experimental curves, as well as the measured values of V_g, are listed in Table 7.1 [31].

Table 7.1 Parameters of Nb–AlO$_x$–Nb junctions (area $A = 7200\ \mu^2$) for $d_{Al} = 9$ nm [31]

d_{Nb}, nm	I^*_k/I^*_g (theory for $d_{Al} = 8$ nm)	I_k/I_g	V_g, mV
35	0.04	0.055	2.75
50	0.85	0.075	2.77
75	0.145	0.105	2.79
100	0.185	0.17	2.82
150	0.245	0.195	2.84
200	0.285	0.245	2.86
350	0.325	0.21	2.86
500	0.325	0.225	2.86

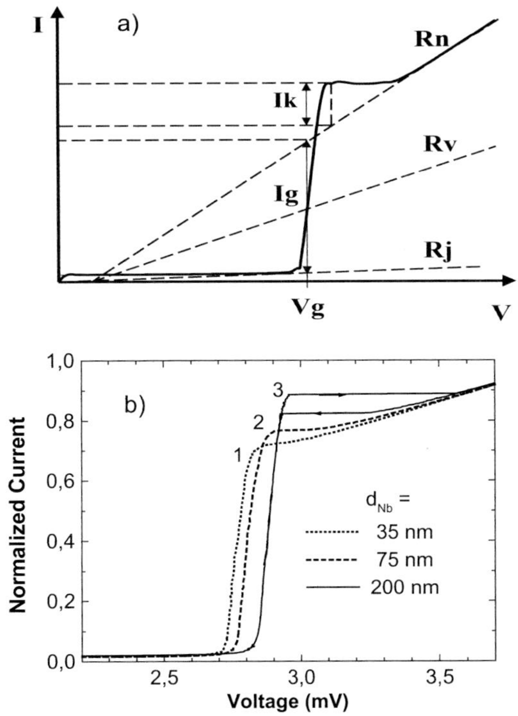

Figure 7.1 (a) Definition of the main parameters for the model SIS IVC. (b) Experimentally measured IVCs at $d_{\text{Al}} = 9$ nm for three different thicknesses of the base electrode. Reproduced with permission from Dmitriev et al. [31].

The normalized knee value $K^{\text{Nb}} = I_{\text{k}}(d_{\text{Nb}})/I_{\text{k}}$ (200 nm) is shown in Fig. 7.2. One can see that the experimental dependence coincides well with the theory up to $d_{\text{Nb}} = 200$ nm. At further increase of the Nb thickness, the surface morphology of the sputtered Nb films changes considerably [29]. As a result, the Al layer is not uniform and the measured knee (averaged over the junction area) is lower than the calculated one.

To avoid the morphology effect, a thin Nb base electrode ($d_{\text{Nb}} = 50$ nm) was used to study the knee's dependencies on the Al thickness. The experimental IVCs for different Al thicknesses are shown in Fig. 7.3 [31]. It should be noted that the Al thickness decreased with oxidation; therefore, 1 nm was subtracted from the

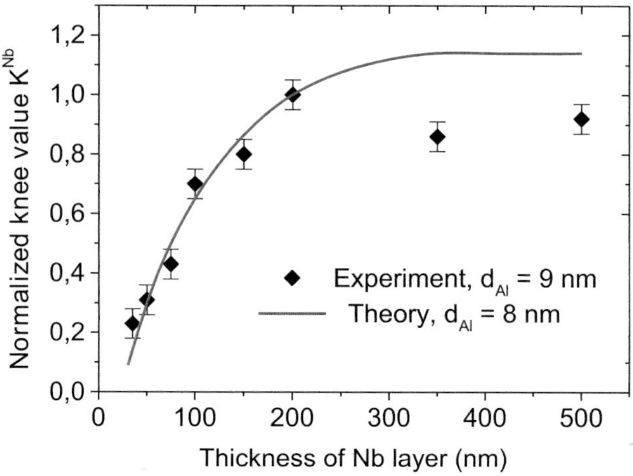

Figure 7.2 Calculated and measured values of the knee (normalized to the value at $d_{Nb} = 200$ nm) versus the thickness of the Nb base electrode. Reproduced with permission from Dmitriev et al. [31].

initial value in the calculations (see Fig. 7.2). The obtained data are summarized in Table 7.2.

Table 7.2 Parameters of Nb–AlO$_x$–Nb junctions ($A = 1700$ μ^2) for $d_{Nb} = 50$ nm [31]

d_{Al}, nm	I_k^*/I_g^* (theory)	I_k/I_g	V_g, mV
2	0	–	–
3	0.016	0	2.86
4	0.031	0.01	2.85
5	0.041	0.02	2.84
6	0.05	0.03	2.83
7	0.056	0.04	2.81
8	0.060	0.05	2.78
9	0.062	0.07	2.77
10	0.065	0.085	2.73
15	0.07	0.115	2.66

The SIS junctions with a thin Nb base electrode have almost ideal IVCs but are not suitable for high-frequency application since $d_{Nb} < \lambda_L^{Nb} = 90$ nm, which considerably increases the inductance of the microwave elements. To overcome this problem, an additional Al

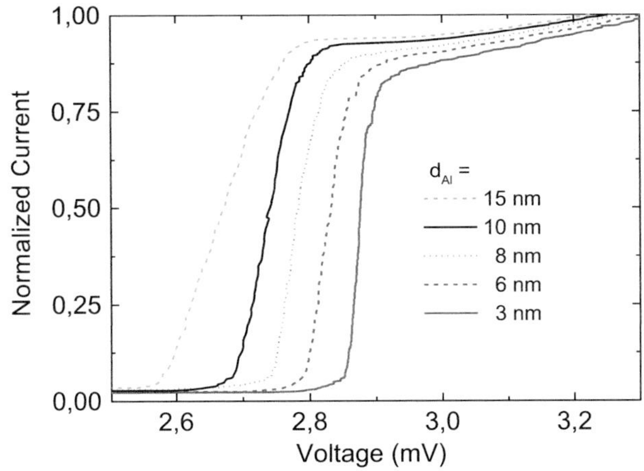

Figure 7.3 Experimentally measured IVCs at $d_{\mathrm{Nb}} = 50$ nm for five different thicknesses of the Al layer. Reproduced with permission from Dmitriev et al. [31].

layer was introduced in the Nb base electrode to realize a "knee-free" IVC for a reasonably thick base Nb [31]. The introduction of an additional Al interlayer into Nb/Al/AlO_x/Nb structures leads to a steeper IVC and disappearance of the knee structure. The reason is that with the introduction of such a layer, the order parameter in a thin Nb–Al bilayer near to the barrier becomes spatially homogeneous, and thus the density of states in this bilayer becomes BCS-like, with a smaller energy gap. The experimental IVC for Nb/$\mathrm{Al}^{\mathrm{add}}$/$\mathrm{Nb}^{\mathrm{add}}$–$\mathrm{Al}_2$/$\mathrm{AlO}_x$–Nb structure is shown in Fig. 7.4.

As a result of Nb–AlO_x–Nb technology optimization and extensive analysis of the SIS receiver designs, a number of low-noise receivers have been designed and tested [6, 15]. The results of a quasi-optical 500 GHz SIS receiver are presented below [6]. The receiver consists of a double dipole antenna SIS mixer, with integrated tuning elements. It was designed as a reference system to measure the ultimate performance of an integrated receiver [4, 5]. The tuning circuit consists of an end-loaded stripline connected to the double dipole antenna via a matching stripline transformer [6, 15]. The SIS junction is an Nb–AlO_x–Nb junction with an area of 1.5 $\mu\mathrm{m}^2$ and a normal resistance of $R_{\mathrm{n}} = 14\ \Omega$; $R_{\mathrm{j}}/R_{\mathrm{n}} = 30$. The Josephson effect is

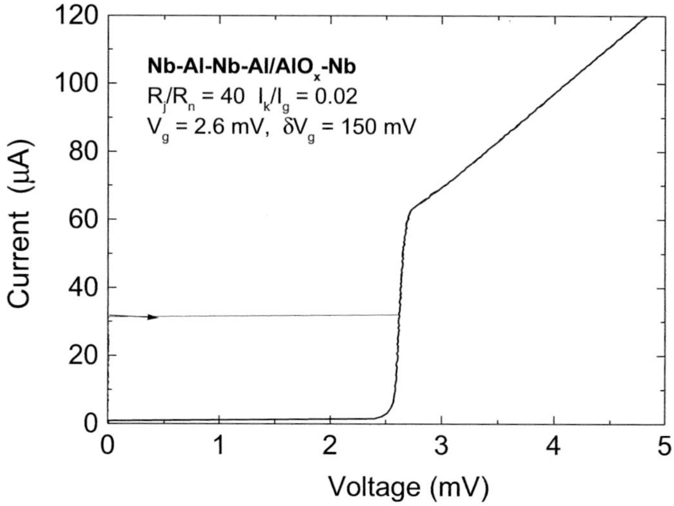

Figure 7.4 The IVC of the Nb/Al^a/Nb^a–Al/AlO$_x$–Nb junction at the thickness of the additional Al layer $d_{\text{Ala}} = 5$ nm and $d_{\text{Nba}} = 50$ nm. Reproduced with permission from Dmitriev et al. [31].

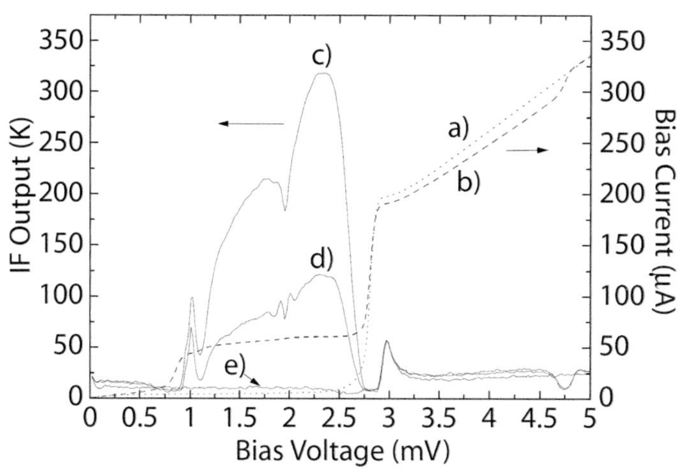

Figure 7.5 I–V curves and IF output of the receiver. (a) Unpumped I–V curve. (b) Pumped I–V curve, identical for hot and cold loads. (c) IF output for a 295 K load (at optimal LO power). (d) The same as in (c) for a 78 K load. (e) IF output for zero LO power [6].

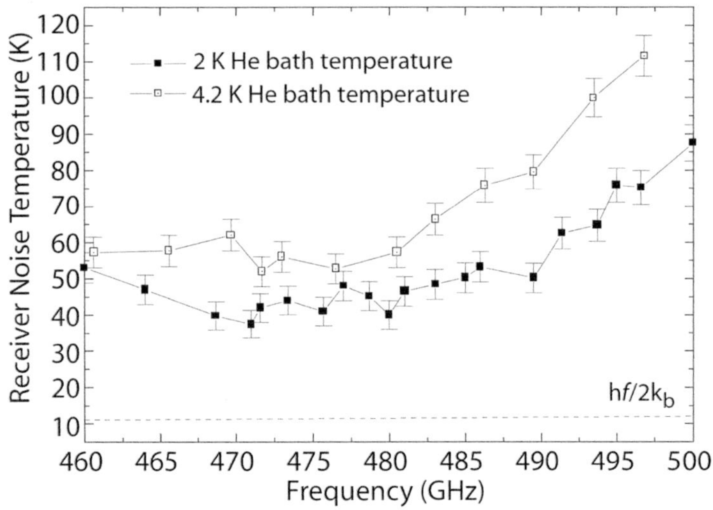

Figure 7.6 DSB receiver noise temperature for two different temperatures, uncorrected for the beam splitter. For comparison, the photon-noise floor of $hf/2k_B$ is included (represented by the dashed line) [6].

suppressed by an integrated magnetic field control line. The receiver chip ($4 \times 4 \times 0.5$ mm^3) is mounted on a silicon elliptical lens covered with an antireflection coating. A back reflector is mounted behind the antenna. The measured unpumped and pumped junction I–V curves, as well as the dependence of the IF output power on bias voltage, are shown in Fig. 7.5 [6]. The measured double-side-band (DSB) receiver noise temperature vs. the LO frequency is presented in Fig. 7.6 [6] and must be compared with the dashed line that indicates the photon noise limit $hf/2k_B$, with $hf/2$ being the photon energy. The best noise temperature, uncorrected for the beam splitter, of 40 ± 3 K was measured at 471 GHz. This is only approximately three times the zero-point fluctuation's noise level of 11.4 K.

7.1.2 Niobium-Based Tunnel Junctions with AlN Barrier

To realize the ultimate performance of the SIS mixers at sub-THz frequencies, tunnel junctions with very high tunnel barrier transparencies are required. Unfortunately, there is a limit for the increasing of the AlO_x barrier transparency (of approximately 10–15 kA/cm^2); at a higher current density, abrupt degradation of the junction's quality takes place. The idea of utilizing SIS tunnel junctions for heterodyne mixing at THz frequencies has received remarkable support due to developments of Nb–Al–AlN–Nb tunnel junctions with very high current densities of up to 100 kA/cm^2. That corresponds to low R_nS values down to 2 $\Omega \cdot \mu m^2$ (where R_n and S are junction the normal-state resistance and area, respectively) [34–38].

We produce Nb–Al–AlN–Nb tunnel junctions [39] in an oil-free UHV sputtering system with a base pressure of 10^{-6} Pa, which is provided by a combination of turbo-molecular and cryogenic pumps. This system is equipped with 5-inch DC and RF magnetron sources, an ion gun, and a grounded water-cooled substrate table. Wafers are fixed to the copper chucks using vacuum grease and are attached to the substrate table. The Nb–AlN–Nb junction fabrication procedure [39] follows the well-known recipe for conventional Nb–AlO_x–Nb junction production, and is described elsewhere [31]. The only difference in our case is the substitution of an oxidation step by a nitridation one. As in the case of the conventional Al oxide process, we deposited Nb and Al thin films using DC magnetron sputtering in an Ar atmosphere with a working gas pressure of 1 Pa. The dielectric layer for junction insulation consists of 250 nm SiO_2, defined in a self-aligned lift-off procedure. The wiring layer is defined by lift-off.

It is well known that by using a simple exposure of sputtered Al surface in an N_2 atmosphere, one cannot acquire a continuous AlN layer of sufficient thickness to be used as a tunnel barrier [35]. Several successful attempts of Al nitridation have been made using a glow discharge in a nitrogen atmosphere [35–38]. Following this idea, we grew an AlN tunnel barrier immediately after Al deposition by using RF magnetron discharge. Samples were attached to the grounded substrate table and maintained at 20°C. To obtain a density of nitrogen ions capable of producing an AlN tunnel barrier, the sample holder was positioned directly above a five-inch Al magnetron RF source with a holder-source distance of 14 cm. The electrical scheme of the nitridation process is presented in Fig. 7.7. We initiated a plasma discharge using a very small power density of 0.5–0.75 W/cm^2. The nitrogen pressure was kept constant at 4.5 Pa.

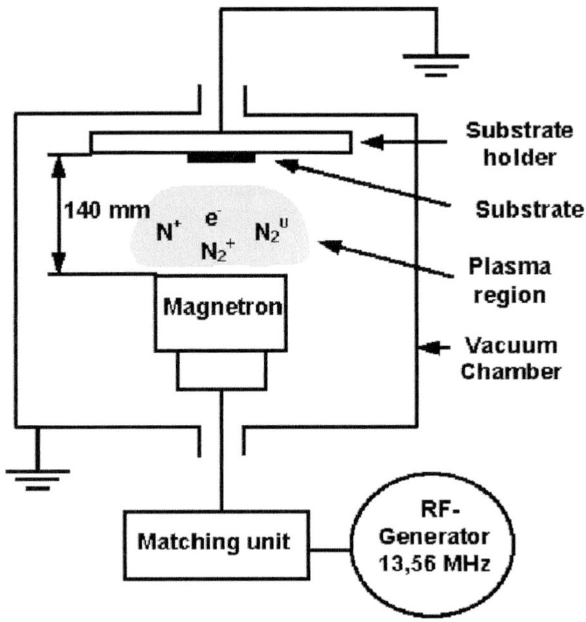

Figure 7.7 The electrical scheme of the nitridation process. Reproduced with permission from Dmitriev et al. [39].

The total duration of the nitridation process varied within the range of 100–300 seconds. The use of such conditions, the small power, large source-sample distance, and dense plasma, permitted us to avoid both exposure of the samples to an energetic flux of ions and significant sputtering of the Al target during AlN growth.

A set of Nb–AlN–Nb junction *IV* characteristics is presented in Fig. 7.8. The critical current is suppressed by a magnetic field. The R_nS value changes from 24 $\Omega \cdot \mu m^2$ for the curve (a) to 0.9 $\Omega \cdot \mu m^2$ for the curve (d). The increase of the subgap leakage follows the increase of the critical current density. Moreover, a self-heating in the junctions can be clearly seen in this figure. It causes both gap voltage reduction and back bending of the gap singularity. Figure 7.8b presents the *IV* characteristic of an Nb–AlN–Nb junction exposed to the nitrogen plasma for 300 seconds at 60 W of RF power. This junction with low R_nS of 10 $\Omega^*\mu m^2$ demonstrates excellent tunnel characteristics with $R_j/R_n = 16$. From other *IV*

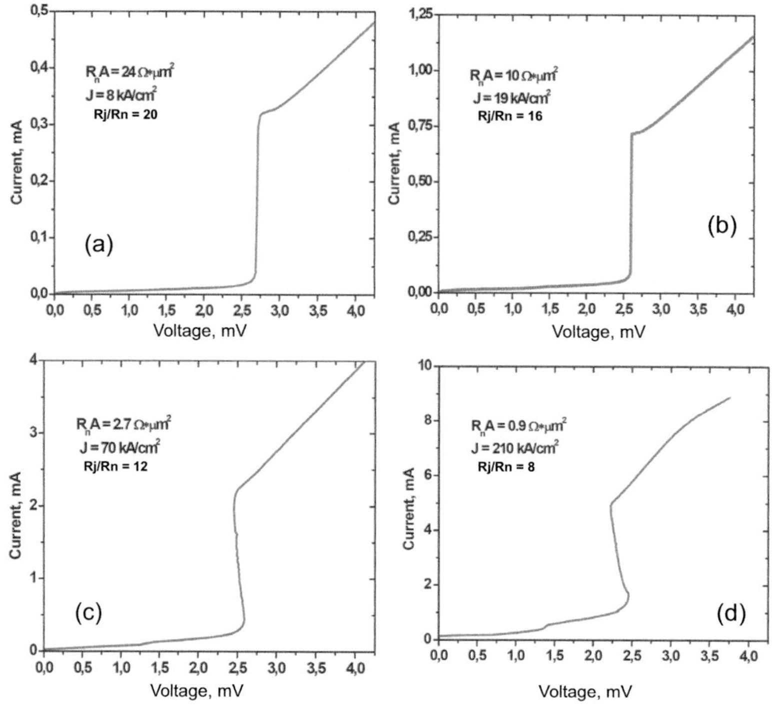

Figure 7.8 IVCs of Nb-AlN-Nb junctions with (a) 24 $\Omega \cdot \mu m^2$, (b) 10 $\Omega \cdot \mu m^2$, (c) 2.7 $\Omega \cdot \mu m^2$, (d) $R_n S = 0.9\ \Omega \cdot \mu m^2$; the critical current is suppressed by the magnetic field. Reproduced with permission from Dmitriev et al. [39].

curves presented in Fig. 7.8 it is clear that the $R_n S$ value can be easily lowered down to 5–7 without significant degradation. Dependences of the junction quality parameter R_j/R_n on tunnel current density for Nb/AlO$_x$/Nb junctions [39] are shown by asterisks in Fig. 7.9.

The implementation of an AlN tunnel barrier in combination with a NbN top superconducting electrode is expected to give a significant improvement in SIS THz mixer performance. To explore this idea we produced an Nb–AlN–NbN tunnel junction. NbN was deposited by DC reactive magnetron sputtering at ambient temperature with 1.8 W/cm^2 power density using Ar + 9% N_2 gas mixture. Otherwise the fabrication procedure was the same as described above for Nb–AlN–Nb junctions. The *IV* characteristic of this junction is presented in Fig. 7.10. $R_n S = 100\ \Omega \cdot \mu m^2$, $R_j/R_n = 32$, and the gap voltage

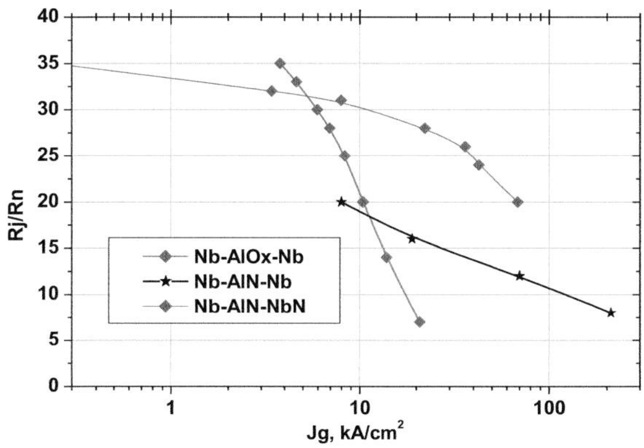

Figure 7.9 The dependencies of the R_j/R_n ratio vs. tunnel current density J_g for junctions fabricated at IREE using different techniques. Reproduced with permission from Dmitriev et al. [39].

is 3.7 mV. From this value and the voltage of the singularity corresponding to the difference of the superconducting gaps of the junction contacts $V_\delta = (\Delta_{NbN} - \Delta_{Nb})/e = 0.9$ mV we can estimate the gap voltage of our NbN film as $V_g^{NbN} = 2.3$ mV. The dependency of the ratio of subgap to normal state resistance (R_j/R_n) vs. tunnel

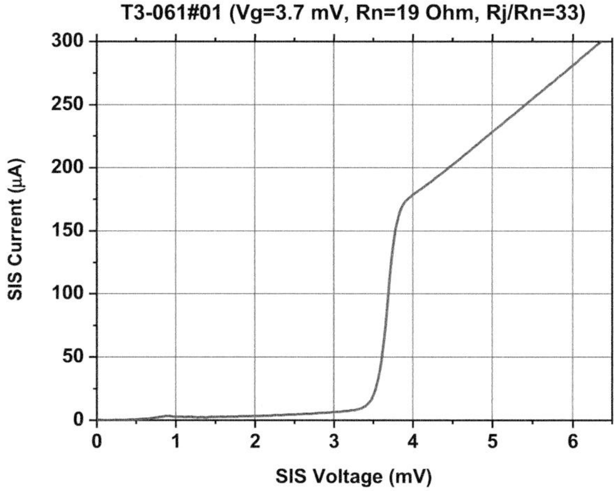

Figure 7.10 IV characteristic of the Nb–AlN–NbN junction ($S = 2\ \mu m^2$, $V_g = 3.7$ mV, $R_nS = 37\ \Omega \cdot \mu m^2$, $J_g = 6.5$ kA/cm^2); the Josephson supercurrent is suppressed by the magnetic field. Reproduced with permission from Torgashin et al. [40].

current density (J_g) for different types of the Nb-based junctions fabricated at IREE is presented in Fig. 7.9. One can see that the Nb–AlN–NbN junctions have very good quality at high current densities that is important for implementation in THz mixers.

High-quality Nb–AlN–NbN tunnel junctions were successfully used for development of the superconducting local oscillators and fully integrated superconducting receivers (see Sections 7.2 and 7.3). These junctions were also employed [41] to upgrade the 790–950 GHz CHAMP+ heterodyne array receiver [42] for the APEX (Atacama Pathfinder EXperiment) telescope [43]. The frequency of 950 GHz corresponds to a 3.9 mV photon step, which exceeds the V_g of "classical" SIS junctions with Nb electrodes ($V_g = 2.8$ mV, or even lower for extremely high current density junctions). Consequently, the voltage range available for the SIS mixer's operation is considerably wider for the Nb-AlN-NbN mixers as compared to the Nb ones. An additional interest is to gain experience with high-gap mixer technology for later use at frequencies above 1 THz, for example, in the Millimetron project [44] in the 950–1150 GHz range.

In order to make a wide band receiver, twin SIS junctions [45, 46] (each with an area of 0.5 μm^2) were used, which were coupled by a waveguide probe to the E field of a rectangular waveguide of 300×75 μm (Fig. 7.11). Due to the high current density of the produced AlN barrier, the lower R_n gives a higher $1/R_nC$ ratio for the

Figure 7.11 Photo of the SIS mixer (including the waveguide probe and filter structure) installed in a waveguide. The central part, with a SIS twin junction, is magnified. Reproduced with permission from Khudchenko et al. [41].

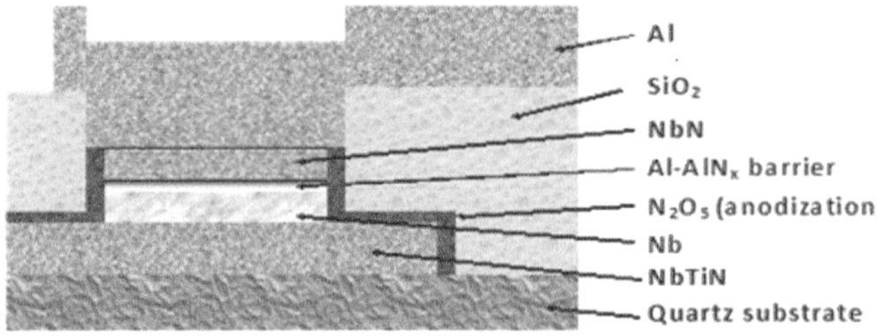

Figure 7.12 Cross section of the Nb/AlN/NbN tunnel junctions incorporated into a microstrip line consisting of a 300 nm thick bottom electrode (ground plane) made of NbTiN and a 500 nm thick top electrode made of Al.

junctions (C being the junction capacitance), thus providing a wider receiver band. The design of the tuning structure and waveguide mount is conceptually similar to the original design of the CHAMP+ high-band devices. The SIS junctions were embedded in a 4.5 μm wide microstrip line with a 6.5 μm interjunction distance, and coupled to the antenna by a 7 × 27 μm impedance transformer tuned for the high current density junctions.

The SIS mixer was based on high critical current density Nb/AlN/NbN tunnel junctions incorporated into a microstrip line consisting of a 300 nm thick bottom electrode (ground plane) made of NbTiN and a 500 nm thick top electrode made of Al [7]. The microstrip electrodes were separated by a 250 nm SiO_2 isolator (see Fig. 7.12). The Nb layer of the SIS junction was deposited onto the NbTiN film, while the NbN layer was in contact with the Al top electrode. First, an NbTiN film was deposited onto a fused quartz substrate at room temperature, using DC sputtering with an NbTi target in a nitrogen atmosphere. For the NbTiN film, the critical temperature T_c was measured to be 14.1 K, and the room temperature resistivity was estimated to be 85 $\mu\Omega \cdot$ cm. The geometry of the ground electrode was determined by a reactive ion etching (RIE) process. The tunnel junctions were fabricated from an Nb/AlN/NbN trilayer [40] with a normal state resistance–area product $R_n A = 7\ \Omega \cdot \mu m^2$, which corresponds to a current density of $J_c = 30\ kA/cm^2$; the Nb and NbN layers had thicknesses of 100 nm. An AlN tunnel barrier was grown immediately after deposition onto a 7 nm Al layer using an RF magnetron discharge. The samples were attached to the grounded substrate table maintained at 20°C, which was positioned directly above a 5-inch Al magnetron RF source with a holder-source distance of 14 cm. For fabrication of the high current density trilayer, we initiated a plasma discharge using a power density of 0.7 W/cm^2 at a nitrogen pressure of 4.5 Pa; the nitridation time was approximately 40 seconds. Afterwards, the NbN was deposited by DC reactive magnetron sputtering at an ambient temperature, with 1.8 W/cm^2 power density using an Ar + 9% N_2 gas mixture. Circular-shape junctions with an area of approximately 0.5 μm^2 were defined by deep ultraviolet photolithography. The SIS junctions were patterned from the Nb/AlN/NbN trilayer by successive RIE of the NbN layer using CF_4, by RF sputtering of AlN/Al film in Ar plasma and, finally, by RIE of the Nb layer using CF_4. The dielectric layer for junction insulation consists of 250 nm SiO_2, defined in a self-aligned lift-off procedure. In the final step, a 500 nm thick top microstrip electrode made of Al was deposited via DC magnetron sputtering. Afterwards, the thickness of the quartz substrate was reduced to 40 μm by mechanical polishing.

The Nb/AlN/NbN junctions fabricated onto an Si substrate with a 200 nm Nb layer and a 100 nm NbN film have a gap voltage $V_g = 3.7$ mV and a quality factor (the ratio of the subgap to the normal state resistance) $R_j/R_n > 30$ for R_nA approximately 30 $\Omega \cdot \mu m^2$, while showing $V_g = 3.4$ mV and $R_j/R_n = 23$ for the higher current density ($R_nA = 7\ \Omega \cdot \mu m^2$) and the submicrometer (0.5 μm^2) junction area. In contrast, the current SIS junctions with considerably thinner Nb electrodes (100 nm) and which were fabricated onto the NbTiN film have $V_g = 3.2$ mV for $R_nA = 7\ \Omega \cdot \mu m^2$ (see Fig. 7.13). We cannot completely explain yet the cause of such a decrease of the junction gap voltage.

7.1.3 NbN Tunnel Junctions with MgO Barrier

For many years, tunnel junctions based on niobium nitride (NbN) have been attracting interest as an alternative to Nb junctions for high-frequency applications because NbN has a large gap energy. There have been many reports on the development of NbN tunnel junctions using different tunnel barrier materials [47–50]. Initially, only NbN/MgO/NbN junctions exhibited reasonably good quality, because both NbN and MgO have the same crystal structure with

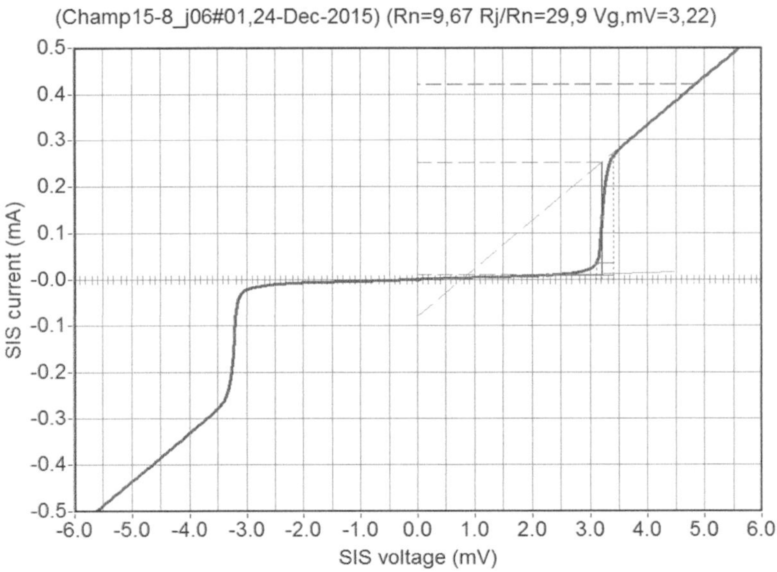

Figure 7.13 IVC of the twin Nb–AlN–NbN mixer element fabricated onto an SiO_2 substrate for APEX (area of each tunnel junction $S = 0.37\ \mu m^2$, $V_g = 3.22$ mV, $R_nS = 7\ \Omega \cdot \mu m^2$, J_g of approximately 34 kA/cm^2). The Josephson supercurrent is suppressed by the magnetic field.

a lattice mismatch of less than 5%. Recently, high-quality epitaxial NbN/AlN/NbN tunnel junctions with a wide range of current densities have been demonstrated [51–54]. Although previous works have proven the possibility of producing high-quality, all-NbN tunnel junctions, we have developed a new technique to fabricate NbN/MgO/NbN circuits. Our approach [17] resembles the "classical" technique proposed many years ago for the production of Nb/AlO_x/Nb junctions [23], which are the basic building blocks for most devices of modern superconducting electronics. According to our approach, a very thin Mg layer (approximately only 1.5 nm) is DC-sputtered on the NbN layer, and the Mg is then oxidized in the O_2 plasma (this is similar to the Al nitridization process used for the fabrication of Nb/AlN/NbN junctions [39, 55]).

The transmission electron microscopy (TEM) image (Fig. 7.14) shows the layered structure of the junction area of the sample. The MgO substrate, the bottom and top NbN electrodes, and the MgO

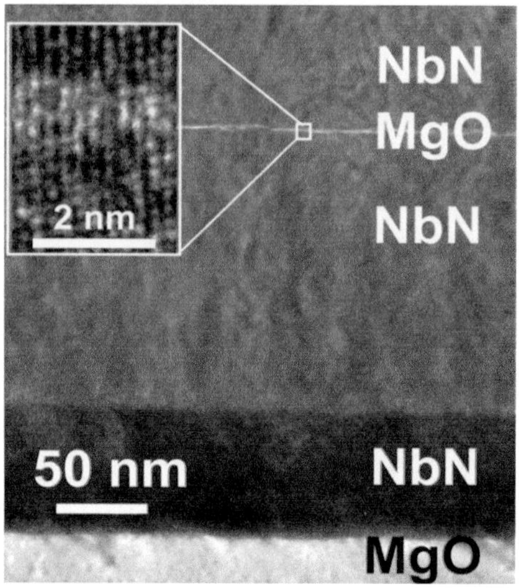

Figure 7.14 The transmission electron microscopy image of the layered NbN/MgO/NbN structure. Reproduced with permission from Koshelets et al. [17].

barrier are visible. The bottom electrode consists of the epitaxial 70 nm NbN monitor layer covered by a 150 nm thick NbN film, which is polycrystalline due to the lift-off structuring of this and all subsequent layers. The top NbN electrode is polycrystalline and has a thickness of 70 nm. A high-resolution TEM image of the 1.5 nm thick MgO barrier layer is shown in the insert of the picture. It was observed that the orientation of the crystal structure of NbN electrodes is maintained across the MgO barrier.

By using the developed technique, it is possible to fabricate high-quality junctions with a quasiparticle tunnel current density J_g in the range of 0.05–80 kA/cm^2. The IVC of the NbN/MgO/NbN junction ($J_g = 2$ kA/cm^2) is shown in Fig. 7.15; the dependencies of the gap voltage V_g and the quality factor R_j/R_n (the ratio of the leakage resistance R_j and the normal state resistance R_n) on the current density are presented in Fig. 7.16. To summarize, a new technique for fabrication of high-quality SIS tunnel junctions based on epitaxial NbN films with an MgO barrier has been developed; junctions with a gap voltage $V_g = 5.3$ mV and a quality barrier

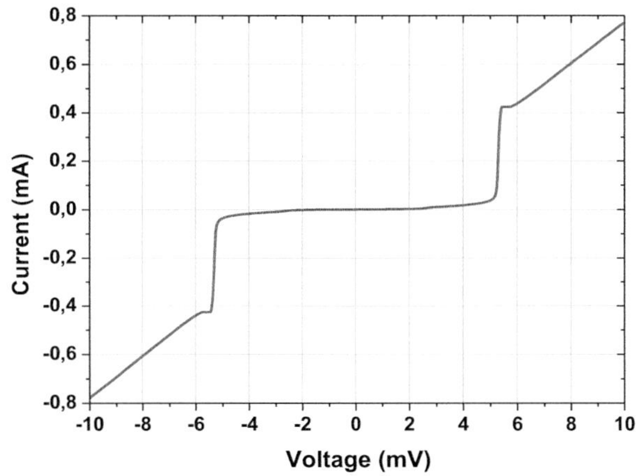

Figure 7.15 IVC of the NbN/MgO/NbN junction (I_c is suppressed by the magnetic field): $S = 18\ \mu\text{m}^2$, $J_g = 2\ \text{kA/cm}^2$, $V_g = 5.3$ mV, $R_n = 11.7\ \Omega$, $R_j(2\ \text{mV})/R_n = 80$, $R_j(4\ \text{mV})/R_n = 19.5$. Reproduced with permission from Koshelets et al. [17].

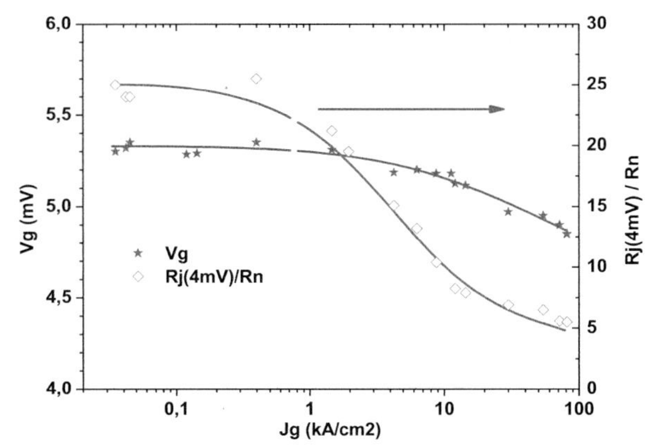

Figure 7.16 Dependencies of the NbN/MgO/NbN junction's parameters on the current density J_g. Experimental points are connected by lines as a guide for the eye. Reproduced with permission from Koshelets et al. [17].

parameter $R_j(4\ \text{mV})/R_n > 25$ have been fabricated. Such junction parameters are very promising for the development of an SIR for frequencies well above 1 THz.

7.2 Superconducting Terahertz Oscillators

Josephson junctions have been considered as natural terahertz oscillators for more than half a century, ever since Josephson discovered the effects named after him [56, 57]. Since that time, many quite different types of Josephson oscillators have been proposed and studied [58–68], but only a few of them were developed at level suitable for real applications. Let us consider one of the most attractive applications: the direct integration of a Josephson local oscillator (JLO) with the most sensitive heterodyne SIS mixer. There are a number of important requirements of the JLO's properties to make it suitable for application in the phase-locked superconducting integrated receiver (SIR). The continuous frequency tuning of the JLO over a wide frequency range (usually more than 100 GHz) and a possibility of the JLO's phase stabilization at any frequency in the operation range are required for most applications. The output power of the JLO should be sufficient to pump the matched SIS mixer within a wide frequency range and it can be electronically adjusted. Obviously, the JLO should emit enough power to pump an SIS mixer (of about 1 μW), taking into account a specially designed mismatch of about 5–7 dB between the JLO and the SIS mixer, which was introduced to avoid leakage of the input signal to the LO path. It is a challenge to realize the ultimate performance of the separate superconducting elements after their integration into a single-chip device. Another very important issue is the linewidth of the JLO. Even for wideband room-temperature PLL systems, the effective regulation bandwidth is limited by the length of the cables in the loop (about 10 MHz for a typical loop length of two meters). This means that the free-running JLO linewidth has to be well below 10 MHz to ensure stable JLO phase locking with a reasonably good spectral ratio (SR)—the ratio between the carrier and the total power emitted by the FFO [75].

7.2.1 Nb-Based Flux-Flow Oscillators

The Josephson flux flow oscillators (FFOs) [69–74] based on Nb–AlO_x–Nb and Nb–AlN–NbN junctions have proven [15, 16, 75] to be the most developed superconducting local oscillator for

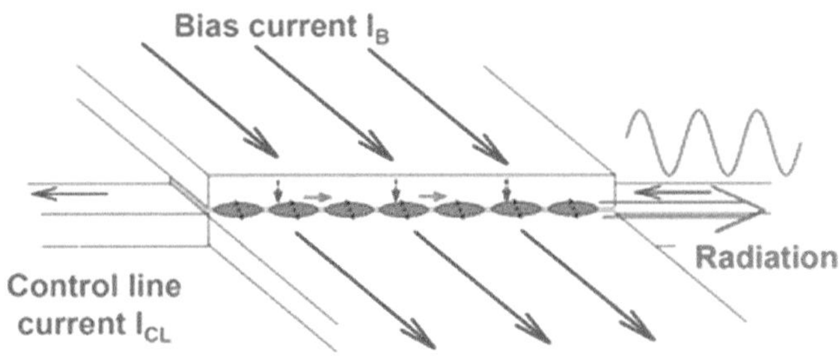

Figure 7.17 Schematic view of a flux-flow oscillator. Reproduced with permission from Koshelets et al. [16].

integration with an SIS mixer in a single-chip submicrometer-wave superconducting integrated receiver [14–21]. The FFO is a long Josephson tunnel junction of the overlapped geometry (see Fig. 7.17) in which an applied DC magnetic field and a DC bias current, I_B, drive a unidirectional flow of fluxons, each containing one magnetic flux quantum, $\Phi_0 = h/2e \approx 2{*}10^{-15}$ Wb. Symbol h represents Planck's constant and e is the elementary charge. An integrated control line with the current I_{CL} is used to generate the DC magnetic field that is applied to the FFO. According to the Josephson relation, the junction oscillates with a frequency $f = (1/\Phi_0)^* V$ (about 483.6 GHz/mV) if it is biased at voltage V. The fluxons repel each other and form a chain that moves along the junction. The velocity and density of the fluxon chain, and thus the power and frequency of the submicrometer-wave signal emitted from the exit end of the junction due to the collision with the boundary, may be adjusted independently by the appropriate settings of I_B and I_{CL}. The FFO differs from the other members of the Josephson oscillator family by the need for these two control currents, which in turn provides the possibility of an independent frequency and power tuning.

We experimentally investigated [16, 75] a large number of the FFO designs. The length, L, and the width, W, of the FFO used in our study were 300–400 μm and 4–48 μm, respectively. The value of the critical current density, J_C, was in the range of 4–8 kA/cm^2, giving a Josephson penetration depth of $\lambda_J \approx$ 6–4 μm. The corresponding value of the specific resistance was $R_n^* L^* W$ is ~50–25 $\Omega^*\mu m^2$. For the numerical calculations we used a typical value of the London penetration depth, $\lambda_L \approx 90$ nm, for all-Nb junctions, and a junction-specific capacitance $C_s \approx 0.08$ pF/μm^2. The active area of the FFO (i.e., the AlO_x or the AlN tunnel barrier) is usually formed as a long window in the relatively thick (200–250 nm) SiO_2 insulation layer, sandwiched between the two superconducting films (the base and wiring electrodes). The so-called idle region consists of the thick SiO_2 layer adjacent to the junction (on both sides of the tunnel region) between the overlapping electrodes. It forms a transmission line parallel to the FFO (not shown in Fig. 7.17). The width of the idle region (W_I = 2–14 μm) is comparable to the junction width. The idle region must be taken into account when designing an FFO with the desired properties. In our design, it is practical to use the flat-bottomed electrode of the FFO as a control line in which the current I_{CL} produces the magnetic field, which is mainly applied perpendicular to the long side of the junction.

Previously, the Nb–AlO_x–Nb or Nb–AlN–Nb trilayers were successfully used for the FFO's fabrication. Traditional all-Nb circuits are constantly being optimized but there seems to be a limit for linewidth optimizations at certain boundary frequencies due to the Josephson self-coupling (JSC) effect [76], as well as a high frequency limit, imposed by the Nb gap frequency ($\sim$700 GHz). This is the reason to develop novel types of junctions based on materials other than Nb. We reported on the development of the high-quality Nb–AlN–NbN junction-production technology [39]. The implementation of an AlN tunnel barrier in combination with an NbN top superconducting electrode provides a significant improvement in the quality of the SIS junction. The gap voltage of the junction $V_g = 3.7$ mV. From this value, and the gap voltage of the Nb film $\Delta_{Nb}/e = 1.4$ mV, we have estimated the gap voltage of our NbN film as $\Delta_{NbN}/e = 2.3$ mV [40]. The use of Nb for the top "wiring" layer is preferable due to smaller losses of Nb when compared to NbN below 720 GHz. Furthermore, the matching structures developed for the all-Nb SIRs can be used directly for the fabrication of receivers with Nb–AlN–NbN junctions. The general behavior of the new devices is similar to that of the all-Nb ones; even the control currents, necessary to provide magnetic bias for the FFO, were nearly the same for the FFOs of similar designs.

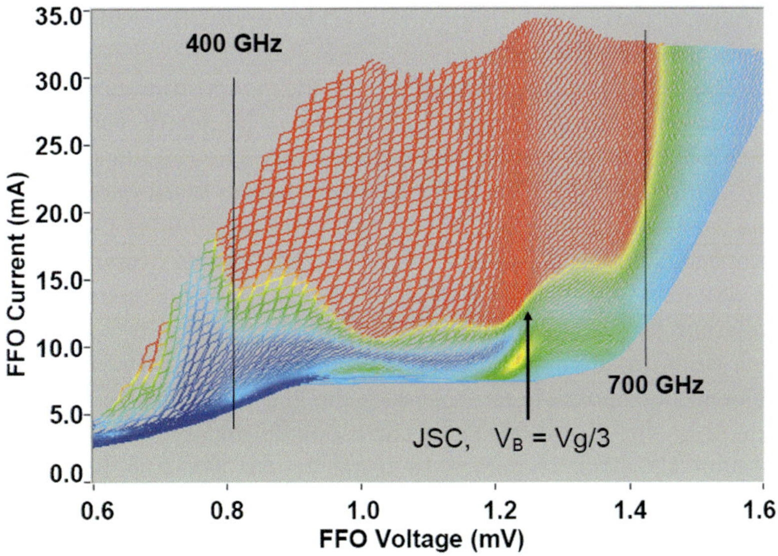

Figure 7.18 IVCs of the Nb–AlN–NbN FFO, measured at different magnetic fields produced by the integrated control line. The color scale shows the level of the DC current's rise at the HM induced by the FFO. The red area marks the region of the FFO's parameters where the HM current induced by the FFO exceeds 25% of the I_g. This level is well above the optimal value for an SIS-mixer operation. Reproduced with permission from Koshelets et al. [16].

A family of the Nb–AlN–NbN FFO IVCs, measured at different magnetic fields produced by the integrated control line, is presented in Fig. 7.18 ($L = 300$ μm, $W = 14$ μm, $W_I = 10$ μm). A single SIS junction with an inductive tuning circuit was employed as a harmonic mixer (HM) for the linewidth measurements. The tuning and matching circuits were designed to provide "uniform" coupling in the frequency range of 400–700 GHz. Measured values of the HM current induced by the FFO oscillations (HM pumping) are shown in Fig. 7.18 by the color scale. The HM pumping for each FFO bias point was measured at a constant HM bias voltage of 3 mV (pumping is normalized on the current jump at the gap voltage, $I_g = 140$ μA). From Fig. 7.18, one can see that an FFO can provide a large enough power over the wide frequency range, which is limited at higher frequencies only by the Nb superconducting gap in transmission line electrodes (base and wiring layers) and below 400 GHz by the design of the matching circuits.

The feature at approximately 600 GHz where the curves get denser is a JSC (Josephson self-coupling) boundary voltage. It was initially observed for all-Nb FFOs [76]. The JSC effect is the absorption of the FFO-emitted radiation by the quasi-particles in the cavity of the long junction. It considerably modifies the FFO's properties at the voltages $V \approx V_{\text{JSC}} = 1/3{*}V_{\text{g}}$ (V_{JSC} corresponds to 620 GHz for the Nb–AlN–NbN FFO). Just above this voltage, the differential resistance increases considerably; that results in an FFO-linewidth broadening just above this point. This, in turn, makes it difficult or impossible to phase-lock the FFO in that region. For an Nb–AlO_x–Nb FFO, the transition corresponding to $V_{\text{JSC}} = V_{\text{g}}/3$ occurs around 450 GHz. Therefore, by using the Nb–AlN–NbN FFOs we can cover the frequency gap from 450 to 550 GHz that is imposed by the gap value of all-Nb junctions. The feature in Fig. 4 around 1 mV is very likely due to a singularity in the difference between the superconducting gaps $\Delta_{\text{NbN}} - \Delta_{\text{Nb}}$.

Continuous frequency tuning at frequencies below 600 GHz for the Nb–AlN–NbN FFOs of moderate length is possible, although the damping is not sufficient to completely suppress the Fiske resonant structure at frequencies below $V_{\text{g}}/3$. For short junctions with a small α (wave attenuation factor), the distance between the steps in this resonant regime can be as large, so that it is only possible to tune the FFO within a certain set of frequencies. For a 300–400 μm long Nb–AlN–NbN junction, this is not the case—the quality factor of the resonator formed by a long Nb–AlN–NbN Josephson junction is not so high at frequencies > 350 GHz. Therefore, the resonance steps are slanting and the distance between them is not so large (see Fig. 7.18). This allows us to set any voltage (and any frequency) below V_{JSC}, but for each voltage, only a certain set of currents should be used. Therefore, in this case, we have the regions of forbidden bias-current values, which are specific for each voltage below V_{JSC}, instead of the forbidden voltage regions for the Fiske regime in Nb–AlO_x–Nb FFO [40]. Special algorithms have been developed for automatic working-point selection in flight.

The typical current–voltage characteristics (IVCs) of an Nb–AlN–NbN SIS junction of an area approximately 1 μm^2 is given in Fig. 7.19,

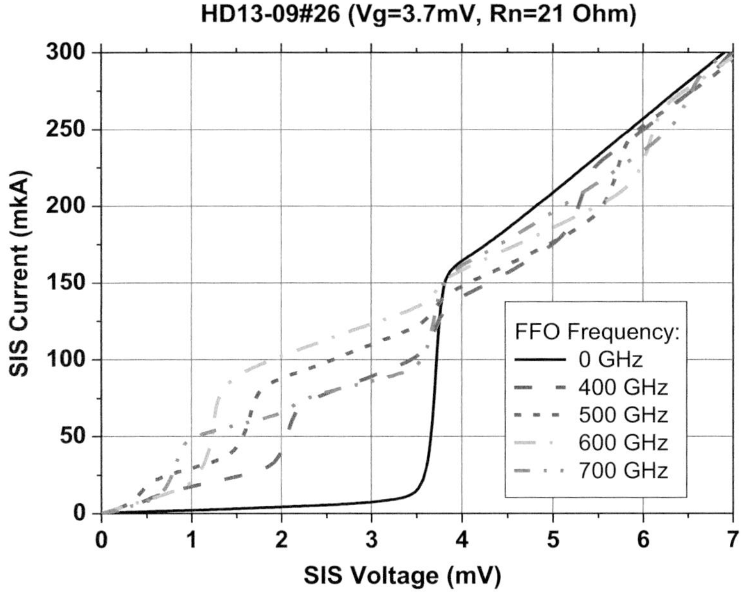

Figure 7.19 The IVCs of the SIS mixer: unpumped = solid curve; pumped at different frequencies = dashed and dotted lines (color online). Reproduced with permission from Koshelets et al. [16].

which represents both the unpumped IVC (the solid line) and the IVC when pumped by an Nb–AlN–NbN FFO at different frequencies (dotted lines). One can see that the FFO provides more than enough power for the mixer pumping. In this experiment, we used the test circuits with low-loss matching circuits tuned between 400 and 700 GHz. Even with the specially introduced 5 dB FFO/SIS mismatch (required for the SIR operation), the FFO delivered enough power for the SIS mixer's operation in the frequency range of 400–700 GHz. An important issue for the SIR's operation is a possibility to tune the FFO's power, while keeping the FFO frequency constant. This is demonstrated in Fig. 7.20, where the IVCs of an SIS mixer pumped at the FFO frequency of 500 GHz are shown, while they were being pumped at different FFO bias currents (different powers). Our measurements demonstrated [16, 75] that the FFO power can be adjusted in the range of 0–15 dB while keeping the same frequency, by proper adjustment of the FFO control line current.

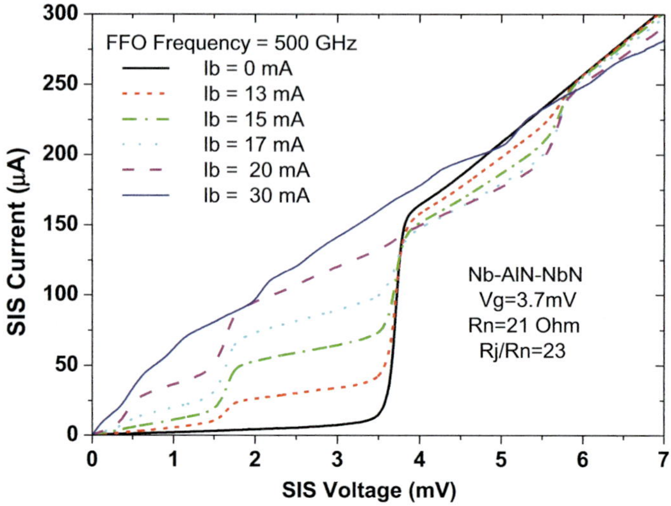

Figure 7.20 The IVCs of the SIS mixer: unpumped = black solid curve; pumped at different FFO bias currents (different powers) = lines with symbols; FFO frequency = 500 GHz (color online). Reproduced with permission from Koshelets et al. [16].

7.2.2 Linewidth of the Flux-Flow Oscillator and Its Phase-Locking

The FFO linewidth (LW) has been measured in a wide frequency range from 300 GHz up to 750 GHz by using a specially developed experimental technique [77–80]. A specially designed integrated circuit incorporates the FFO junction, the SIS harmonic mixer and the microwave matching circuits (see Fig. 7.21). Both junctions are fabricated from the same Nb/AlN/NbN or Nb/AlO$_x$/Nb trilayer. A block diagram of the set-up for the linewidth measurements is shown in Fig. 7.22 [78]. The FFO signal is fed to the harmonic mixer (a SIS mixer operated in Josephson or quasiparticle mode) together with a 17–20 GHz reference signal from a stable synthesizer. The required power level depends on the parameters of the HM; it is about of 1 μW for a typical junction area of 1 μm^2. The intermediate frequency (IF) mixer product ($f_{\mathrm{IF}} = \pm(f_{\mathrm{FFO}} - n \cdot f_{\mathrm{SYN}})$ at ~400 MHz is first boosted by a cooled HEMT amplifier ($T_{\mathrm{n}} \approx 5$ K, gain = 30 dB) and then by a high-gain room-temperature amplifier.

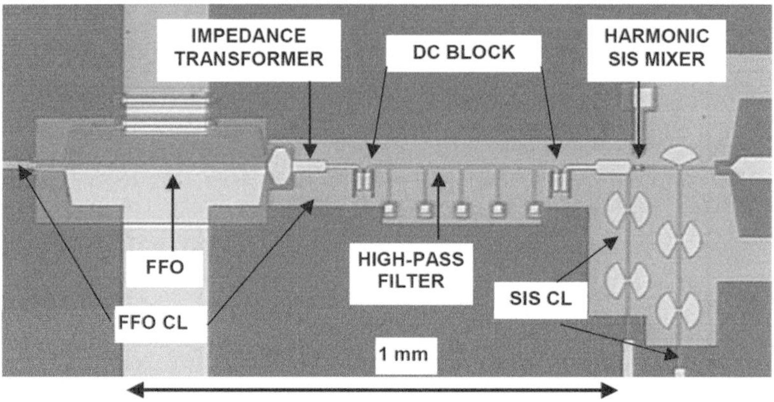

Figure 7.21 The central part of the microcircuits used for FFO linewidth measurements [79].

In order to accurately measure the FFO line shape, the IF signal must be time-averaged by the spectrum analyzer. To remove low-frequency drift and interference from the bias supplies, temperature drift, etc., we used a narrow band ($<$10 kHz) Frequency Discriminator (FD) system with a relatively low loop gain for the frequency locking of the FFO. With the FD narrow-band feedback system that stabilizes the mean frequency of the FFO (but which does not affect FFO's line shape), we can accurately measure the free-running FFO linewidth, which is determined by the much faster internal ("natural") fluctuations (see Fig. 7.23). The measured data are symmetrized relative to the center's frequency; these data are shown by diamonds. The profile of the FFO line recorded when biased at the steep Fiske step (FS), where the differential resistance is extremely small, can be different from the one measured on the smooth flux-flow step. Theoretically [58], the shape is Lorentzian for wideband fluctuations, while for narrow-band interference, at frequencies smaller than the autonomous FFO linewidth δf_{AUT}, the profile will be Gaussian; the theoretical curves are also shown in Fig. 7.23 for comparison. The theoretical lines providing the best fit near the peak are shown by the solid line and the dashed line for the Lorentzian and Gaussian profiles, respectively. The coincidence between the calculated curve and the symmetrized experimental

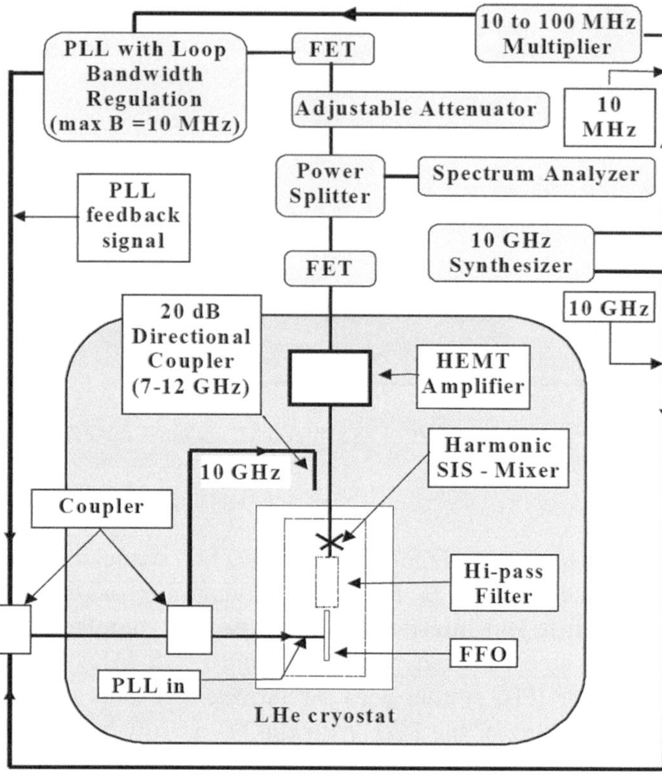

Figure 7.22 Block diagram of the PLL circuit and set-up for linewidth measurement. Reproduced with permission from Koshelets et al. [78].

data is excellent, and actually better than 5% in the emitted power, if a minor amplifier's nonlinearity of about 0.4 dB is taken into account.

The resulting IF signal is also supplied to the phase-locking loop (PLL) system. The phase-difference signal of the PLL is fed to the FFO control-line current [15, 16, 75, 78–81]. Wideband operation of the PLL (10–15 MHz full width) is obtained by minimizing the cable loop's length. A part of the IF signal is delivered to the spectrum analyzer via a power splitter (see Figs. 7.24 and 7.25). All instruments are synchronized to the harmonics of a common 10 MHz reference oscillator. Dependencies of the free-running FFO linewidth and the Spectral Ratio (SR) for the phase-locked FFO on

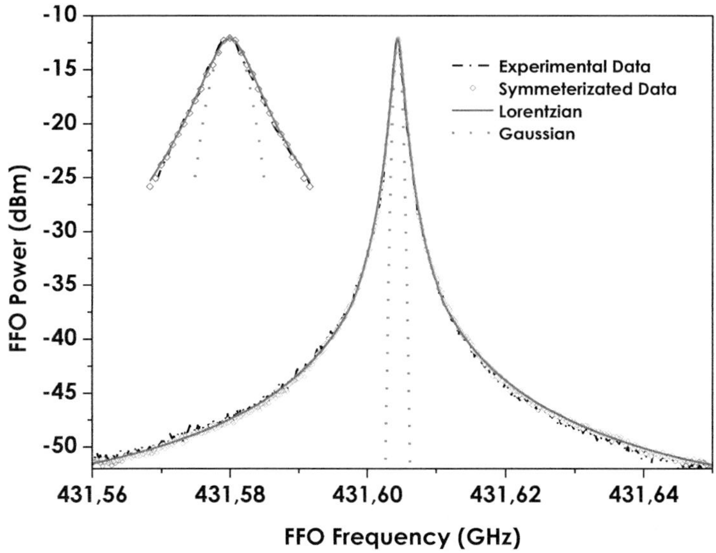

Figure 7.23 The FFO spectrum measured when biased on the Fiske step ($V_{FFO} = 893$ μV, $R_d = 0.0033\ \Omega$, $R_d^{CL} = 0.00422\ \Omega$, $\delta f_{AUT} = 1.2$ MHz) is represented by the dash-dotted line. The symmetrized experimental data are shown by diamonds. The fitted theoretical Lorentzian and Gaussian profiles are shown by solid and dotted lines, respectively. The inset shows a close-up view of the central peak with the frequency axis multiplied five times. Reproduced with permission from Koshelets et al. [80].

frequency for two different FFO technologies (Nb–O_x–Nb and Nb–AlN–NbN) are presented in Fig. 7.26. One can see that SR $>$ 70% can be realized for Nb–AlN–NbN FFO in the range of 250–750 GHz.

7.2.3 Sub-Terahertz Sound Excitation and Detection by Long Josephson Junctions

Interaction between phonons and electrons has attracted the attention of researchers over the years. Such interaction has been demonstrated by the use of an AC Josephson current for phonon generation in superconducting tunnel junctions [82, 83]; in these papers, a few possible mechanisms for phonon generation were considered. Direct generation of acoustic waves by the AC Josephson oscillations occurs when the tunnel barrier is piezoelectric. On the

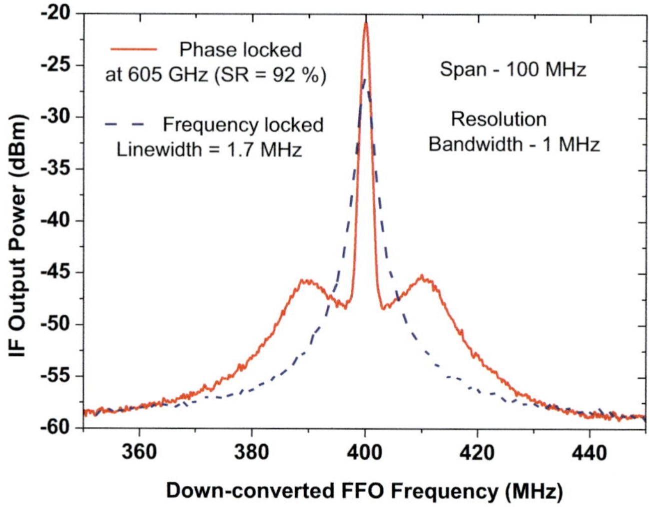

Figure 7.24 Spectra of the Nb–AlN–NbN FFO operating at 605 GHz (blue dashed line = frequency locked by FD; red solid line = phase-locked). Linewidth = 1.7 MHz; spectral ratio = 92% (color online).

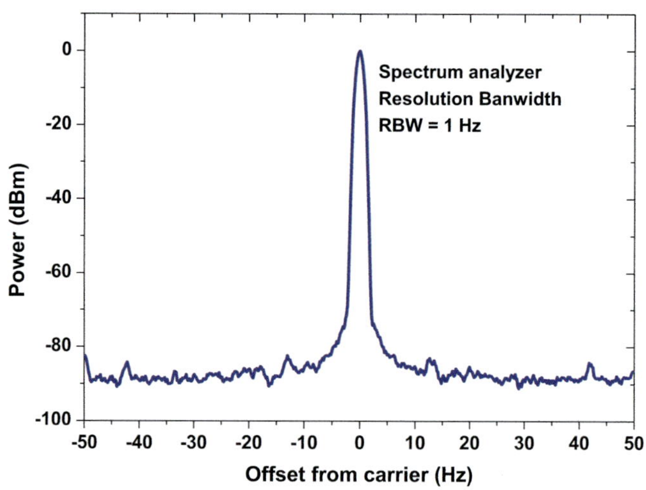

Figure 7.25 Spectra of the phase-locked Nb-AlN-NbN FFO operating at 605 GHz. Span = 100 Hz, RBW = 1 Hz, signal-to-noise ratio = 87 dB as measured in a bandwidth of 1 Hz.

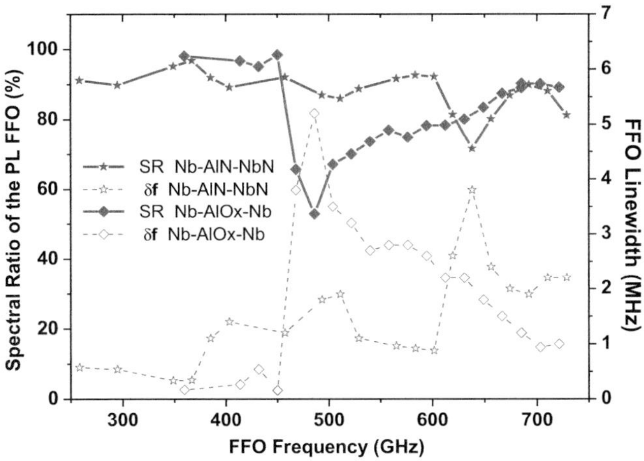

Figure 7.26 Dependencies of the free-running FFO linewidth and the spectral ratio for the phase-locked FFO on frequency. Data are presented for two different FFO technologies: Nb–O_x–Nb (represented by diamonds) and Nb–AlN–NbN (asterisks).

other hand, the need for an alternative explanation was suggested in reference [83] because the effect was observed in junctions with an amorphous barrier, and amorphous materials are typically not piezoelectric (although some materials may become piezoelectric in the amorphous state because amorphization can remove the inversion symmetry [84]). Therefore, it was suggested that, in the disordered material, the AC electrical field may instead act on uncompensated static charges with a finite dipole moment [83, 85, 86], resulting in the coherent generation of acoustic waves. The coupling strength of this process depends on the oxide properties; it can be comparable to (or even be well above the state of) Werthamer processes [87], which are basically the absorption of AC Josephson radiation energy by the quasiparticles and consequent photon-assisted tunneling—the so-called effect of Josephson self-coupling (JSC) [76, 87, 88].

The reverse effect—phonon-induced increase in the critical current of Josephson junctions [89] and appearance of the constant-voltage steps in the IVCs of the SNS junctions [90]—has been observed experimentally under low-frequency phonon excitation.

Interaction of the AC Josephson current and phonons was also found for intrinsic HT_c Josephson junctions [91–94]; it was observed as the occurrence of specific subgap structures in the form of current peaks (resonances) in the IVCs of these junctions. The obtained results have been explained [94–97] by the coupling between the intrinsic Josephson oscillations and phonons. Some time ago, a quite unusual superfine resonance structure (SFRS) was observed [98] in the IVCs of the Nb–AlO_x–Nb flux-flow oscillator (FFO); at that time, no reasonable explanation was proposed.

For comprehensive analysis of the superfine resonance structure [99], we have studied Nb–AlO_x–Nb and Nb–AlN–NbN FFOs. The length, L, and the width, W, of the tunnel junctions used in our study are 400 μm and 16 μm, respectively. The value of the critical current density, J_c, is in the range of 4–8 kA/cm^2—giving a Josephson penetration depth of $\lambda_J = 6$–4 μm. The active area of the FFO (i.e., the AlO_x or the AlN tunnel barrier) is formed as a long window in the relatively thick (200 nm) SiO_2 insulation layer that is sandwiched between two superconducting films (base and wiring electrodes). The FFOs were fabricated from a high-quality trilayer structure [39] on the monocrystalline silicon substrate of (001) orientation. We used commercially available double-side polished silicon wafers (room temperature resistivity > 10 kΩ*cm, thickness $d_S = 0.3 \pm 0.01$ mm).

For wideband measurements of the FFO spectra, a superconductor–insulator–superconductor (SIS) mixer has been integrated on the same chip with the FFO [77]—see the previous section; a simplified sketch of the device under the test is presented in the inset to Fig. 7.27a. The FFO and the SIS junction are connected by a specially designed microstrip circuit that provides RF coupling in the range of 300–800 GHz, while the break at DC gives us a possibility to bias and measure both devices independently; this circuit is presented schematically in the inset to Fig. 7.27a by a dashed line and the capacitor. Due to the strong nonlinearity of the SIS mixer, it was utilized as a high-number harmonic mixer (HM) [30], in which the FFO signal under investigation beats with the n-th harmonic of an applied reference signal (of about 20 GHz), which was fed to the SIS mixer via a coaxial cable from a synthesized signal generator. Signals at down-converted frequencies $f_{IF} = f_{FFO} - n \cdot f_{ref}$

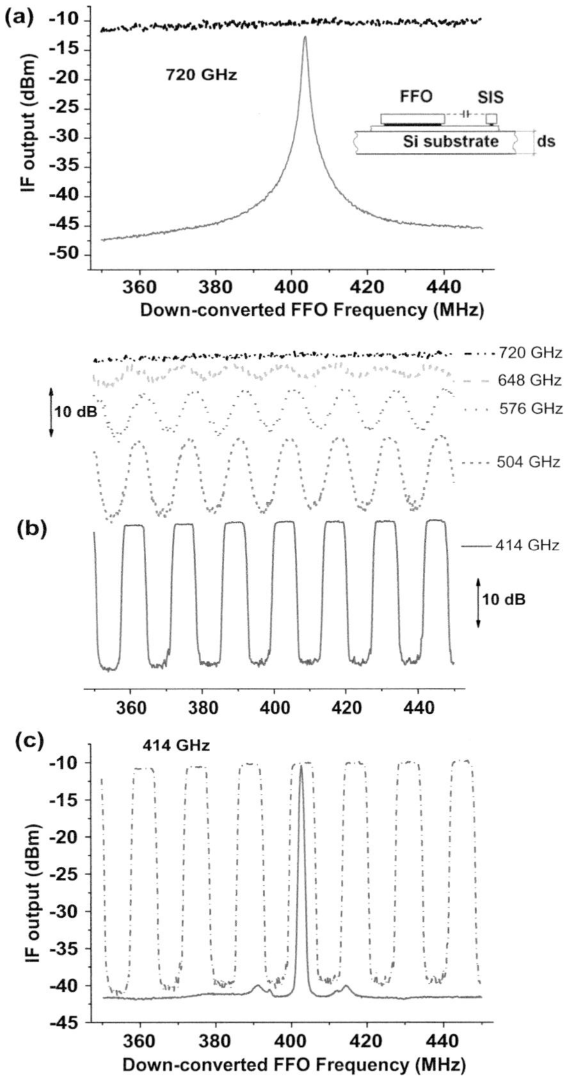

Figure 7.27 (Color online) [99]. Down-converted spectra of the Nb–AlO_x–Nb FFO measured at different FFO frequencies at $T = 4.2$ K by a spectrum analyzer in the regime "Max Hold" (see text) with resolution band width (RBW) = 1 MHz at fine tuning of the bias current; the frequency-locked FFO spectra measured at the FFO frequencies 720 and 414 GHz are presented by solid lines in graphs "a" and "c." A sketch of the experimental sample is shown as an inset to Fig. 7.27a (see text).

can be analyzed using a conventional spectrum analyzer. The down-converted signals measured at the FFO frequencies ranging from 414 to 720 GHz are presented in Fig. 7.27.

The spectrum of the frequency-locked FFO [77] operating at 720 GHz is shown in Fig. 7.1a by the solid line. The spectrum recorded at fine FFO frequency tuning in the range of 100 MHz is presented by the dash-dotted line. This spectrum was measured by using the so-called "Max Hold" regime, when the maximum value in each spectral channel of the analyzer (601 points per range) is memorized over a long enough period, providing that the FFO frequency is tuned by fine adjustment of the bias or CL current. The amplitude of the down-converted signal is almost constant for the FFO frequency 720 GHz (the dash-dotted line in Fig. 7.27a), while for the decreasing of the FFO frequency, a well-defined resonant structure appeared in the down-converted spectra (Fig. 7.1b).

At low FFO frequencies, the SFRS is very pronounced and the down-converted power in the dips is at least 1000 times (30 dB) smaller than in the case of the maximum level (see Fig. 7.1c); actually, the minimum level in this case is determined by the noise level of the HM. One can see that the FFO frequency can be continuously tuned only within a small range, while frequencies between these stable regions cannot be obtained. Even a small change of the bias current near the edge of the stable region results in a "jump" of the FFO voltage (frequency) to the next stable region. The distance between resonances is equal to 14.1 MHz; exactly the same resonance spacing was measured for all FFO frequencies and for all tested FFOs fabricated on the 0.3 mm-thick silicon substrates described above. Very similar behavior was also measured for the Nb–AlN–NbN junctions.

We attribute this superfine structure to the manifestation of resonant interaction of the acoustic waves with the Josephson oscillator. A few different mechanisms were proposed that may couple electron oscillations and phonons: (i) excitation of phonons in a tunnel barrier due to the electromagnetic interaction between the ionic charges and the charges of conduction electrons [82, 83, 85, 86, 93, 96, 97], or via the ac-Josephson effect in a tunnel barrier made from a piezo and ferroelectric materials; (ii) emission of phonons via non-equilibrium quasiparticle relaxation in the electrodes (not the barrier) caused by electron–phonon interaction [94, 100–102] and (iii) the dependence of the tunneling-matrix element on lattice displacements [31, 95]. The mechanism for the generation of phonons in Josephson junctions [103] is based on the excitation of the long wavelength acoustic resonance modes in the dielectric layer of the contact, which can influence the shape of the IVC of the junction in the same way as the excitation of electromagnetic cavity modes. This approach was extended [95] by including in consideration *all* optical phonons in superconductors (not only in the intermediate dielectric layer). To distinguish between these mechanisms, additional research is required, but that is outside the scope of this chapter.

According to our explanation, the experimentally measured IVCs are caused by excitation of the standing acoustic waves in the Si substrate. It is known that a considerable part of the power emitted by the FFO is reflected back; at low damping, these oscillations may reach the entry end of the FFO, raising the standing waves that manifest themselves as the Fiske steps. The standing electromagnetic waves of large amplitude (existing at least on the emitting end of the junction) excite acoustic waves. Note that even at higher voltages ($V > V_g/3, V_g$ is the gap voltage of the FFO), where the Fiske steps could be suppressed [76] due to higher dumping caused by the JSC effect, the standing electromagnetic waves still exist at the emitting end of the junction and can excite acoustic waves of considerable amplitude.

Since flatness (parallelism) of the Si substrates is fairly good (thickness variation is well below 5 μm over 100 mm wafer size), the acoustic wave is reflected back to the point of emission with an accuracy much better than 0.1 μm. At frequencies where the Si substrate's thickness is equal to the integer number of acoustic wavelengths, the reflected wave will arrive in phase with electromagnetic oscillations, resulting in an increase of the current amplitude, while in between these resonances, the oscillations will be suppressed.

At least two experimental facts lead to such a conclusion. Firstly, the frequency distance between two adjacent resonances coincides with the distance between sound resonances in the silicon substrate: $\Delta f = V_L/d_S^*2$, where V_L is the longitudinal speed of sound in Si along

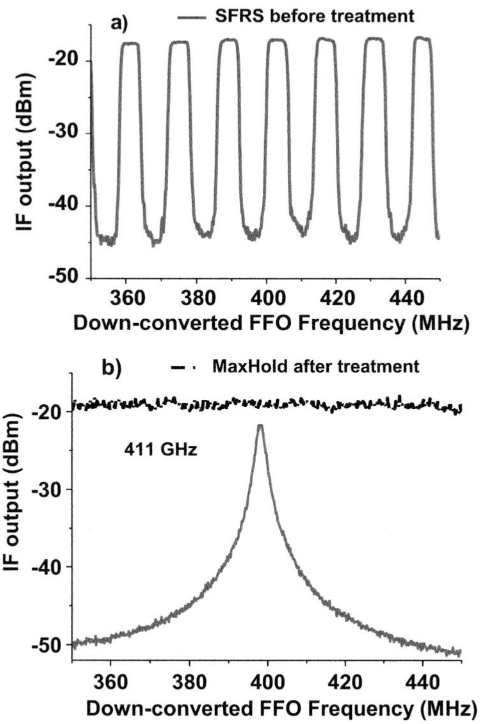

Figure 7.28 (Color online) [99]. Down-converted spectra of the Nb–AlO_x–Nb FFO fabricated on the 0.3 mm thick Si substrate before (a) and after abrasive treatment (b), measured at different FFO frequencies at T = 4.2 K using a spectrum analyzer in the "Max Hold" regime (see text). The frequency-locked FFO spectrum measured at the FFO frequency of 411 GHz is presented by solid lines in graph "b."

the [001] direction. For $V_L = 8480$ m/s [104] and $d_s = 0.3$ mm, the calculated resonance spacing is 14.1 MHz, which corresponds precisely to experimentally measured data. The frequency distance between two adjacent resonances is inversely proportional to the substrate thickness d_s; for the FFOs fabricated on a thicker Si substrate ($d_s = 0{,}525$ mm), the resonance spacing was about 8 MHz, which again corresponds well to the calculated value.

Secondly, after treatment of the opposite (bottom) substrate surface with an abrasive powder, the superfine resonant structure disappeared completely (see Fig. 7.28). We used a set of powders with particle sizes from 1 to 10 μm, resulting in a root mean square (RMS) surface roughness that was measured by an atomic force microscope (AFM) to be 30–280 nm, which is well above the wavelength of the acoustic wave at 500 GHz of about 13 nm (note that for a polished Si surface, the RMS is approximately 0.1 nm; see, e.g., [105]). It seems that the acoustic waves that are reflected from a rough Si surface arrive at the FFO plane in an arbitrary phase. That fact makes the establishment of standing acoustic waves impossible, giving us the possibility to phase lock the FFO [81] at *any* desirable frequency, which is vitally important for most practical applications. It was found that chemical etching of the bottom surface of the Si substrate (RMS roughness of approximately 250 nm) also completely eliminates the appearance of the SFRS. Such Si substrates with a chemically etched bottom surface are commercially available and were used for the fabrication of an integrated receiver with a phased-locked FFO [39, 40], which was successfully implemented for atmosphere monitoring from a high-altitude balloon [15, 16, 20]. On the other hand, the roughness of the etched Si substrate is negligibly small at sub-THz frequencies, and allows good RF coupling of the integrated receiver that is installed on the flat surface of the synthesized elliptical Si lens [15, 16, 20].

This explanation of the SFRS was confirmed by a preliminary theoretical consideration [106], in which coherent phonon radiation and detection due to the interaction of Josephson's electromagnetic oscillations with a mechanical displacement field were analyzed. The Josephson tunneling structure together with a silicon substrate constitute a high overtone composite resonator for bulk acoustic waves that propagate normally to the layers. Resonant generation of coherent acoustic waves revealed itself as a superfine structure in IVC, similar to the usual Fiske steps that are caused by reflection of electromagnetic waves in the junction resonance cavity.

An ability of the Josephson junction to generate and to detect the coherent acoustic waves has been demonstrated experimentally at frequencies up to 800 GHz. Frequency resolution well below 1 MHz can be realized for the frequency (or phase-locked) FFO, and that opens new possibilities for solid-state physics research and phonon spectroscopy. The superfine resonant structure in the FFO IVCs is attributed to acoustic wave generation by the FFO and the excitation of acoustic wave resonances in a thick Si substrate. The SFRS effect can be avoided by proper treatment of the bottom surface; on the other hand, this effect can be employed for high-resolution phonon spectroscopy at well-defined frequencies without additional FFO locking.

7.3 Superconducting Integrated Receivers

A superconducting integrated receiver (SIR) [14, 15] was proposed more than 10 years ago and has since then been developed up to the point of practical application [16–21]. Our approach [15, 16, 19] consists in developing a *single chip* heterodyne receiver, which is smaller and less complex than traditional devices. Typically, such a receiver consists of a number of main components (local oscillator, mixer, antenna structure, phase-lock circuit, etc.), which are usually built as separate units and are complex (and thus costly). According to our concept (see Fig. 7.29), we have integrated all these components into a single chip, reducing the overall system complexity in exchange for increased on-chip and lithographic fabrication complexity. An SIR comprises on a single chip all of the key elements needed for heterodyne detection: a low-noise SIS mixer with quasi-optical antenna, a flux-flow oscillator (FFO) [69–81] acting as a local oscillator (LO) and a second SIS harmonic mixer

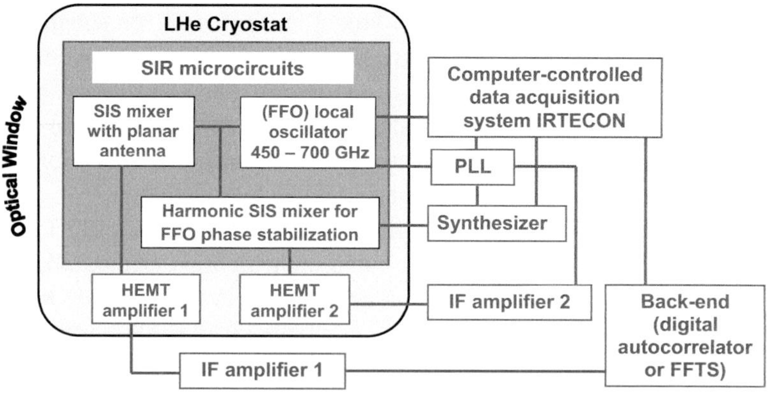

Figure 7.29 Block diagram of the superconducting integrated receiver.

(HM) for the FFO phase locking. The concept of the SIR looks very attractive for many practical applications due to its compactness and the wide tuning range of the FFO; a bandwidth up to 35% has been achieved with a twin-junction SIS mixer design. Recently, the frequency range of most practical heterodyne receivers was limited by the tunability of the local oscillator; nowadays, commercially available multipliers cover the band up to the 40% of the center's frequency [107], and the best SIS receivers (see, e.g., [108]) offer a bandwidth of 15–30%. Another potential advantage is the use of arrays of SIR channels within a single cryostat that could operate at the same or different LO frequencies.

One of the important practical applications of the SIR is TELIS (TErahertz and submillimeter LImb Sounder) [20, 109, 110]—a three-channel balloon-borne heterodyne spectrometer for atmospheric research that was developed in a collaboration between four institutes: Deutsches Zentrum für Luft- und Raumfahrt (DLR), Germany; Rutherford Appleton Laboratories (RAL), United Kingdom; and SRON—Netherlands Institute for Space Research, the Netherlands (in tight collaboration with Kotel'nikov Institute of Radio Engineering and Electronics, IREE, Moscow). All three receivers utilize state-of-the-art superconducting heterodyne technology and operate at 500 GHz (by RAL), at 480–650 GHz (by SRON + IREE) and at 1.8 THz (by DLR). TELIS is a compact, lightweight instrument capable of providing broad spectral coverage, high spectral resolution, and long flight duration. The TELIS instrument also serves as a test bed for many novel cryogenic technologies and as a pathfinder for satellite-based instrumentation.

TELIS is mounted on the same balloon platform (payload weight: 1200 kg) as the Fourier transform spectrometer MIPAS-B [111], developed by IMK (Institute of Meteorology and Climate Research of the University of Karlsruhe, Germany) and is operated in the mid-infrared range (680–2400 cm^{-1}). Both instruments observe the same air mass simultaneously, and together they yield an extensive set of stratospheric constituents that can be used for detailed analysis of atmospheric chemical models, such as ozone destruction cycles. In particular, the 480–650 GHz TELIS channel is able to measure vertical profiles of ClO, BrO, O_3, and its rare isotopologues, O_2, HCl, HOCl, H_2O, and its three rare isotopologues, HO_2, NO, N_2O,

NO_2, HNO_3, CH_3Cl, and HCN. In this paper, the design and technology of the 480–650 GHz channel as used in the flight configuration are presented in conjunction with the test results and the first preliminary scientific results.

7.3.1 The SIR Channel Design and Performance

A key element of the 480–650 GHz channel is the SIR [14–21] that comprises within a single chip (with a size of 4 mm*4 mm*0.5 mm, see Fig. 7.30) a low-noise SIS mixer with quasi-optical antenna, a superconducting FFO acting as an LO and a second SIS harmonic mixer (HM) for FFO phase-locking. To achieve the required instantaneous bandwidth of 480–650 GHz, a twin-SIS mixer with 0.8 μm^2 junctions and a new design of the FFO/SIS matching circuitry were implemented. A microscope photograph of the central part of the SIR chip with a double-dipole antenna is presented in Fig. 7.31. All components of the SIR microcircuits were fabricated from a high-quality Nb-based trilayer on an Si substrate [14, 15].

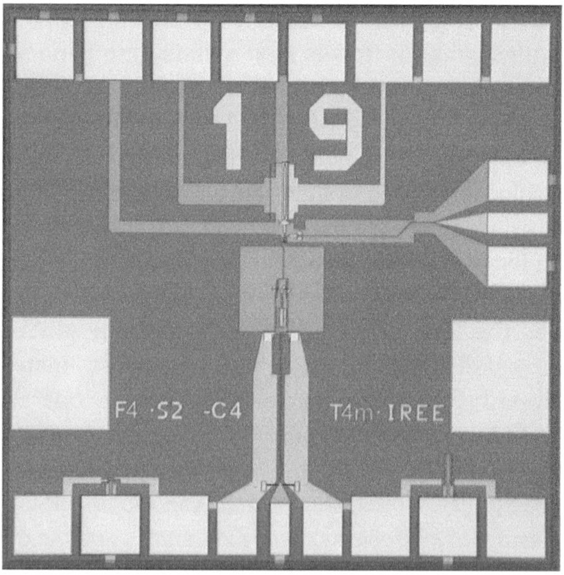

Figure 7.30 Photo of the SIR microcircuit with a double-slot antenna. Reproduced with permission from Koshelets et al. [16].

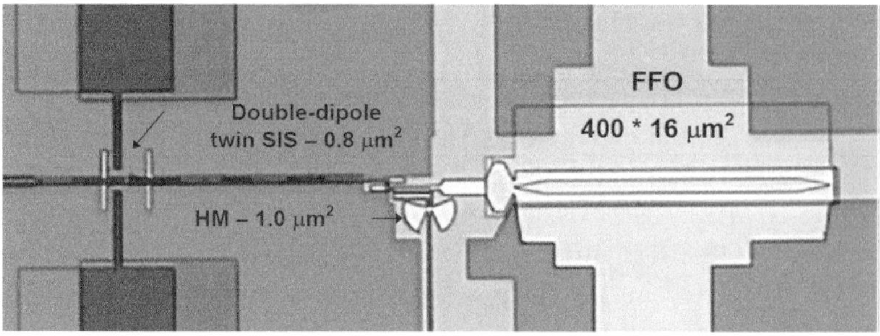

Figure 7.31 Central part of the SIR chip with a double-dipole antenna, a twin SIS-mixer, and a harmonic mixer for FFO phase locking. Reproduced with permission from Koshelets et al. [16].

The receiver chip was placed on the flat back surface of the elliptical silicon lens (forming an integrated lens-antenna) with accuracy of 10 μm, determined by the tolerance analysis of the optical system. As the FFO is very sensitive to external electromagnetic interferences, the SIR chip was shielded by two concentric cylinders: the outer cylinder was made from cryoperm and the inner one from copper with a 100 μm coating of superconducting lead. All SIR channel components (including input optical elements) were mounted onto a single plate inside a 240 mm × 180 mm × 80 mm box, which was cooled by the thermo-straps to a temperature of approximately 4.2 K.

The resolution of the TELIS back-end spectrometer is 2.160 MHz, sufficient to resolve the exact shape of atmospheric lines. The FFO line shape and spectral stability should ideally be much better than this. However, the free-running linewidth of the FFO can be up to 10 MHz; therefore, a phase-lock loop (PLL) has been developed to phase lock the FFO to an external reference oscillator [75–80]. For this, a small fraction of the FFO power was first directed to a so-called harmonic mixer (HM) placed on the SIR chip. The HM was pumped by an off-chip local oscillator source unit (LSU), which has a tunable reference frequency in the range of 19–21 GHz. The frequency of the LSU was chosen such that the difference frequency of the *n*th harmonic of the LSU, generated by the HM and the FFO, was about 4 GHz. This difference signal was then amplified by a cryogenic low-noise HEMT amplifier and down-converted to

400 MHz by using a second reference at 3.6 GHz. Finally, the frequency and phase of this 400 MHz signal was compared to yet another reference frequency of 400 MHz, and the resulting error signal was fed back to the FFO. The LSU and the reference signals at 3.6 GHz and at 400 MHz were all phase-locked to an internal, ultra-stable 10 MHz Master Oscillator.

The TELIS-SIR channel has been characterized in eight microwindows that have been selected for the flight in (Sweden). These microwindows have the following LO frequencies:

- 495.04 GHz for $H_2^{18}O$
- 496.88 GHz for HDO
- 505.60 GHz for BrO
- 507.27 GHz for ClO
- 515.25 GHz for O_2, pointing, and temperature
- 519.25 GHz for BrO and NO_2
- 607.70 GHz for ozone isotopes
- 619.10 GHz for HCl, ClO, and HOCl

Initial flight values for the parameters for the FFO, SIS, and HM mixers have been determined for each microwindow. Dedicated algorithms allowing for fast switching between LO frequencies and for in-flight optimization of the SIR have been developed. It takes approximately 1 minute of stabilization and optimization to switch between two LO settings. All experimental results presented here have been obtained using the SIR flight device. After optimization of the FFO design, the free-running linewidth between 7 and 0.5 MHz was measured in the frequency range of 350–750 GHz, which allows the phase-locking of 35–95% of the emitted FFO. Examples of the free-running (frequency-locked) and the phase-locked spectra of the FFO, measured for the flight SIR at the FFO frequency 605 GHz, are presented in Fig. 7.24.

The measured double sideband (DSB) receiver's noise temperature T_R, uncorrected for any loss, is presented in Fig. 7.32 as a function of LO frequency. As can be seen, the noise is well below 200 K at all frequencies of interest, with a minimum of 120 K at 500 and 600 GHz. The noise peak around 540–575 GHz is partially spurious, caused by the absorption of water vapor in the path between the calibration sources and the cryostat,

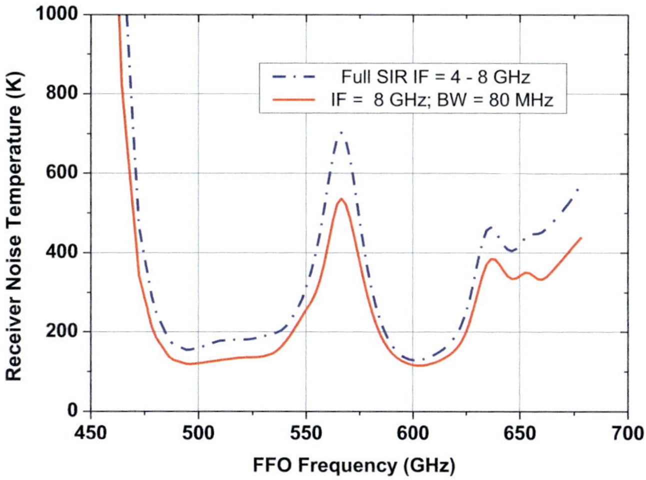

Figure 7.32 Measured DSB receiver noise temperature of the SIR device selected for flight at an 8 GHz IF frequency (red solid line) and integrated into the 4–8 GHz IF range (blue dashed line)—color online. Reproduced with permission from Koshelets et al. [16].

and partially real—due to the properties of the SIS-mixer tuning circuitry. The relatively high noise in this band is of no concern for scientific observations, since this part of the atmospheric spectrum is obscured by a highly saturated water-vapor line, rendering it virtually useless for atmospheric science. The noise as a function of IF is fairly flat in the frequency range of 4–8 GHz, as can be seen in Fig. 7.33, where the (DSB) receiver noise temperature is plotted as a function of IF frequency. The dependence of the receiver noise temperature on the SIS bias voltage is shown in Fig. 7.34 for the modified IF system; one can see that for Nb–AlN/NbN circuits there is very wide range of SIS bias voltages where T_R is almost constant.

The TELIS instrument had four successful flights: three in Kiruna, Sweden (2009, 2010, and 2011) and one in Timmins, Canada (2014); see Fig. 7.35. During all those flights, the shortest of which lasted 10 hours on float, thousands of limb spectra were measured. Examples of recently elaborated spectra measured by the SIR-TELIS channel at different LO frequencies are presented in Figs. 7.36 and 7.37 [17].

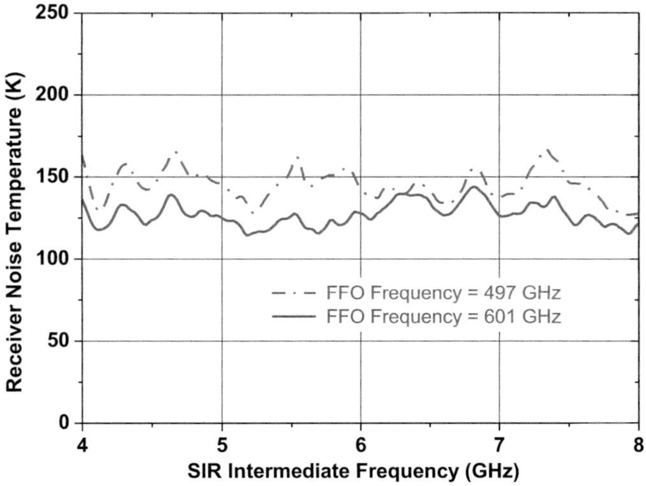

Figure 7.33 DSB receiver noise temperature as a function of the IF frequency, taken at two FFO frequencies: 497 and 601 GHz (color online). Reproduced with permission from Koshelets et al. [16].

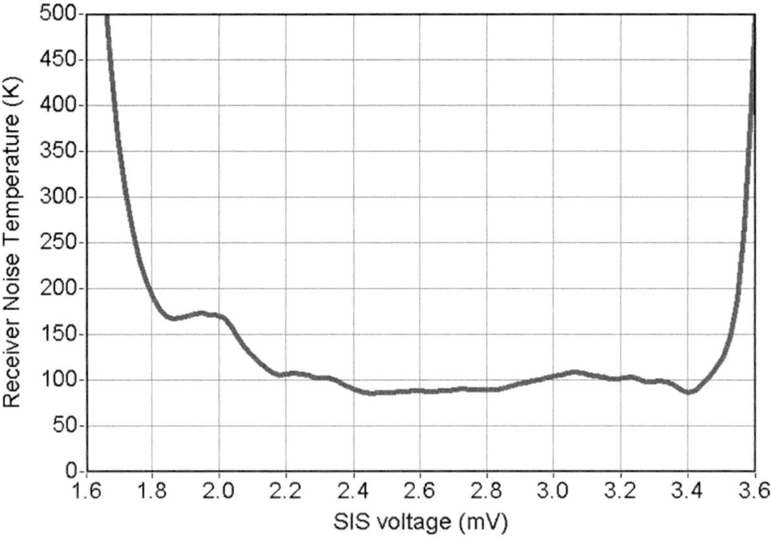

Figure 7.34 Receiver noise temperature as a function of the SIS bias voltage measured at the LO frequency 507 GHz in the full 4–8 GHz IF range. Reproduced with permission from Kiselev et al. [21].

Figure 7.35 TELIS-MIPAS launch at Esrange, Sweden; March 2009. Balloon size: 400,000 m^3.

The flights in northern Sweden focused on catalytic-ozone loss by halogens in the Arctic region, which is similar to processes causing the infamous ozone hole over the Antarctic. The spectra depicted in Figs. 7.36 and 7.37 [17] cover some core molecules to address these processes: namely, ozone itself and the Cl-bearing species HCl and ClO. ClO is the main form of active chlorine that is causing the catalytic destruction of ozone. HCl, on the other hand, is a so-called reservoir species as it is mostly inert in ozone chemistry. The ozone's destruction depends on the ClO concentration, which strongly depends on altitude. Limb sounding provides a tool to gain insight into the vertical distribution of these concentrations. A single recording of a spectrum mostly contains information on molecules at a tangent point, which is the lowest point in the atmosphere of the light path. By combining several spectra recorded at different tangent heights, a vertical profile can be constructed. The integration time of a single spectrum is 1.5 seconds and a typical limb scan typically contains 10–20 different recordings, covering tangent heights between 10 and 35 km, for a total measurement time, including calibration measurements, close to one minute. In the figures, only a few spectra from a given limb scan are shown to prevent cluttering.

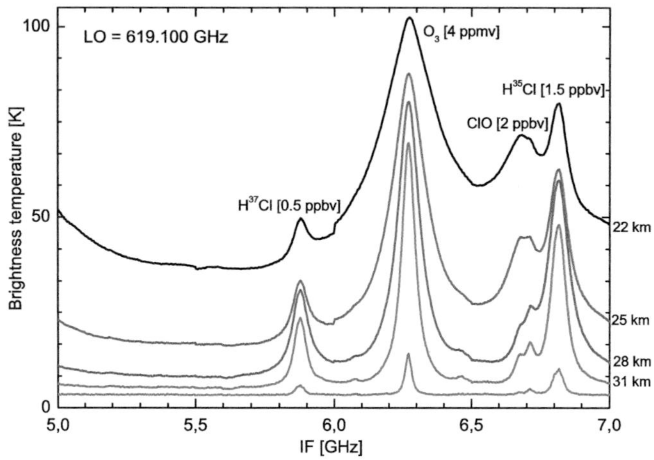

Figure 7.36 Spectra of two HCl isotopes, ozone and ClO. The LO frequency is 619.1 GHz. Spectra for tangent heights 22–31 km and upward looking 6° measured in Kiruna, 2010, are presented in the graph. Corresponding estimated concentrations of observed gases are shown. Reproduced with permission from Koshelets et al. [17].

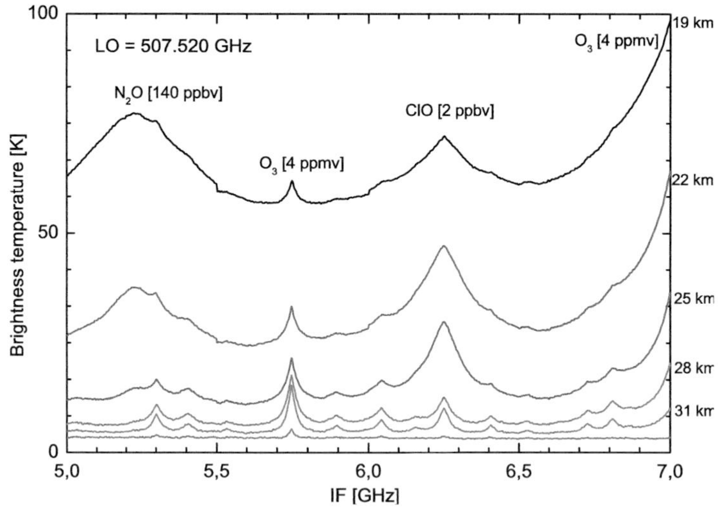

Figure 7.37 Spectra of ClO, ozone, and N_2O. The LO frequency is 507.52 GHz. Spectra for tangent heights 19–31 km and up-looking 6° measured in Kiruna, 2010, are presented in the graph. Reproduced with permission from Koshelets et al. [17].

For the results presented in Figs. 7.36 and 7.37 [17], limb scans' data were averaged 10 and 23 times, respectively. This calculation gives the total averaging time for one tangent height: about 16 seconds (in Fig. 7.36) and 35 seconds (in Fig. 7.37). To provide the same signal-to-noise ratio for the semiconductor-based receivers, which have noise temperatures not lower than 1800 K [112] in the same frequency range, the measurement time should be increased by two orders of magnitude, at least. The wideband coverage of the FFO gives the advantage that a huge number of molecules, whose absorption lines lie in the FFO-tuning range, can be measured during one campaign. The TELIS-SIR channel has been characterized in eight microwindows covering the FFO frequency range from 495.04 GHz (for $H_2^{18}O$) to 619.10 GHz (for HCl, ClO, and HOCl); in combination with a short integration time, this gives an opportunity to provide vertical profiles for many molecules for almost the same air mass.

The final product of those measurements is presented in several papers concerning atmospheric chemistry [113–116]. Analysis of all the flight data is an ongoing process, where post-flight characterization of the SIR (for example, precise laboratory measurement of the SIR sideband ratio) gave a new input to the processing of the flight data with higher accuracy. Already analyzed data proved the SIR to be one of the most sensitive sub-THz spectrometers, allowing us to measure concentrations of trace gases lower than 1 ppbv.

In summary, the capability of the Superconducting Integrated Receiver for high resolution atmospheric spectroscopy has been successfully proven during four scientific balloon flights. During these missions, the phase-locked SIR operation and frequency switching in the 480–650 GHz frequency range has been realized. An intrinsic spectral resolution of the SIR well below 1 MHz has been confirmed by CW signal measurements in the laboratory. An uncorrected double-sideband noise temperature below 120 K has been measured for the SIR when operated with a phase-locked FFO at an intermediate frequency bandwidth of 4–8 GHz. To ensure remote operation of the phase-locked SIR, several software procedures for automatic control have been developed and tested. The first tentative HCl profile has been presented, and its quality looks promising for future data reduction. Diurnal cycles of ClO and BrO have been observed at different viewing configurations (altitude), with a BrO line level of only approximately 0.5 K [16, 17, 19, 21].

The Superconducting Integrated Receiver can be considered as an operational device that is ready for many applications. The high sensitivity and spectral resolution of the integrated spectrometer enables the analysis of multi-component gas mixtures. Possibilities to use the SIR devices for analysis of exhaled air in medical surveys have been demonstrated [16, 19]. Recently the SIR was successfully implemented for the first spectral measurements of THz radiation emitted from intrinsic Josephson junction stacks (BSCCO mesa) at frequencies up to 750 GHz [17, 117]. Application of the SIR has allowed us to measure radiation emitted from intrinsic Josephson junction stacks in both regimes with a spectral resolution better than 1 MHz for the first time [117]. While at low bias, we found that linewidth is not smaller than 500 MHz, at high bias, the emission linewidth turned out to be in the range of 10–100 MHz. We attribute this to the hot spot acting as a synchronizing element; a linewidth as narrow as 7 MHz has been recorded at high bias. The phase-locked SIR has been used for the locking of the BSCCO oscillator during the test [17]. This first successful attempt to phase-lock a BSCCO oscillator to a stable microwave reference opens up prospects for its practical application.

7.4 Conclusions

The extremely high characteristic frequency, unique nonlinearity and very low intrinsic noise of niobium-based tunnel nanostructures, caused by their nature and cryogenic-operating temperatures, provide the possibility of developing unbeaten THz range detectors. The SIS mixers are certainly the best input devices in the sub-THz range; their noise temperatures are limited only by quantum effects. The technology of high-quality Nb-based tunnel junctions that were developed at Kotelnikov IREE primarily for the fabrication of low-noise SIS receivers, has been successfully used for a number of physical experiments in quite different areas. For example, the developed technology was implemented in the fabrication of superconducting metamaterials [118–120].

We proposed and developed the superconducting integrated receiver (SIR) that comprises in a single chip a planar antenna combined with a superconductor-insulator-superconductor (SIS) mixer, a superconducting flux-flow oscillator (FFO) acting as a local oscillator (LO), and a second SIS harmonic mixer (HM) for the FFO phase-locking. Improving on the fully Nb-based SIR, we developed and studied Nb–AlN–NbN circuits that exhibit extended operation frequency ranges. Continuous tuning of the phase-locked FFO was experimentally demonstrated at any frequency in the range of 250–750 GHz. The FFO free-running linewidth was measured as being between 1 and 5 MHz, which allows us to phase-lock up to 97% of the emitted FFO power. The output power of the FFO is sufficient to pump the matched SIS mixer. These achievements enabled the development of a 480–650 GHz integrated receiver for the atmospheric-research instrument TELIS (TErahertz and submillimeter LImb Sounder). We demonstrated for the first time the capabilities of the SIR technology for heterodyne spectroscopy in general, and atmospheric limb sounding in particular. At present, the SIR is probably the most functionally complex fully superconducting device that was already successfully implemented in practical applications. New techniques for the fabrication of high-quality SIS tunnel junctions with a gap voltage $V_g > 5$ mV, as well as an approach for phase-locking of the cryogenic oscillators with linewidths up to 50 MHz [17, 121], have been developed to extend the operation frequency of the SIR beyond 1 THz.

Acknowledgment

The authors thank the following colleagues for their support, help, and assistance in the development of the SIR technology: A Baryshev, A. B. Ermakov, M. I. Faley, H. Golstein, A. Golubov, T. de Graauw, L. de Jong, K. V. Kalashnikov, A. V. Khudchenko, N. V. Kinev, O. S. Kiselev, J. Kooi, M. Yu. Kupriyanov, A. de Lange, G. de Lange, I. L. Lapitskay, A. L. Pankratov, D. G. Paveliev O. M. Pylypenko, K. I. Rudakov, S. V. Shitov, M. Yu. Torgashin, A. V. Ustinov, V. L. Vaks, H. Wang, W. Wild, and P. Yagoubov.

References

1. Tucker, J. R. (1979). Quantum limited detection in tunnel junction mixers. *IEEE J. Quantum Electron.*, **15**, 1234–1258.
2. Tucker, J. R., and Feldman, M. J. (1985). Quantum detection at millimeter wavelengths. *Rev. Mod. Phys.*, **57**, 1055–1113.
3. Kooi, J. W., Chan, M., Phillips, T. G., Bumble, B., and Leduc, H. G. (1992). A low noise 230 GHz heterodyne receiver employing a 0.25 μm^2 Nb/AlOx/Nb tunnel junction. *IEEE Trans. Microwave Theory Tech.*, **40**, 812–815.
4. Karpov, A., Blondell, J., Voss, M., and Gundlach, K. H. (1995). Four photons sensitivity heterodyne detection of submillimeter radiation with superconducting tunnel junctions. *IEEE Trans. Appl. Supercond.*, **5**, 3304–3307.
5. Gaidis, M. C., Leduc, H. G., Bin, M., Miller, D., Stern, J. A., and Zmuidzinas, J. (1996). Characterization of low noise quasi-optical SIS mixers for the submillimeter band. *IEEE Trans. Microwave Theory Tech.*, **44**, 1130–1139.
6. Baryshev, A. M. (2005). Superconductor-insulator-superconductor THz mixer integrated with a superconducting flux-flow oscillator. PhD thesis, Delft University of Technology, ISBN 90-9019220-4.
7. Jackson, B. D., de Lange, G., Zijlstra, T., Kroug, M., Klapwijk, T. M., and Stern, J. A. (2005). Niobium titanium nitride-based superconductor-insulator-superconductor mixers for low-noise terahertz receivers. *J. Appl. Phys.*, **97**, 113904.
8. de Graauw, Th., Whyborn, N., Caux, E., Phillips, T. G., Stutzki, J., Tielens, X., Güsten, R., Helmich, F. P., Luinge, W., Pearson, J., Roelfsema, P., Schieder, R., Wildeman, K., and Wavelbakker, K. (2006). The Herschel-heterodyne instrument for the far-infrared (HIFI). Available at: herschel.esac.esa.int/Publ/2006/SPIE2006 HIFI paper.pdf
9. de Lange, G., Jackson, B. D., Jochemsen, M., Laauwen, W. M., De Jong, L., Kroug, M., Zijlstra, T., and Klapwijk, T. M. (2006). Performance of the flight model HIFI band 3 and 4 mixer units. *Proc. SPIE*, **6275**, 627517.
10. Honigh, C. E., Haas, S., Hottgenroth, D., Jacobs, K., and Stutzki, J. (1997). Low-noise broadband fixed tuned wave-guidemixers at 660 and 800 GHz. *IEEE Appl. Phys.*, **7**(2), 2582–2586.
11. Kooi, J. W., Kovács, A., Sumner, M. C., Chattopadhyay, G., Ceria, R., Miller, D., Bumble, B., Le Duc, R., Stern, J. A., and Phillips, T. G. (2007). A 275–425 GHz tunerless waveguide receiver based on AlN SIS technology. *IEEE Trans. Microwave Theory Tech.*, **55**(10), 2086–2096.

12. Kooi, J. W. (2008). Advanced receivers for submillimeter and far infrared astronomy. PhD thesis, Groningen State University, ISBN 978-90-367-3653-4.

13. Uzawa, Y., Fujii, Y., Gonzalez, A., Kaneko, K., Kroug, M., Kojima, T., Miyachi, A., Makise, K., Saito, S., Terai, H., and Wang, Z. (2015). Tuning circuit material for mass-produced terahertz SIS receivers. *IEEE Trans. Appl. Supercond.*, **25**(3), 2401005.

14. Koshelets, V. P., Shitov, S. V., Filippenko, L. V., Baryshev, A. M., Golstein, H., de Graauw, T., Luinge, W., Schaeffer, H., and van de Stadt, H. (1996). First implementation of a superconducting integrated receiver at 450 GHz. *Appl. Phys. Lett.*, **68**, 1273.

15. Koshelets, V. P., and Shitov, S. V. (2000). Integrated superconducting receivers. *Supercond. Sci. Technol.*, **13**, R53–R69.

16. Koshelets, V. P., Birk, M., Boersma, D., Dercksen, J., Dmitriev, P. N., Ermakov, A. B., Filippenko, L. V., Golstein, H., Hoogeveen, R., de Jong, L., Khudchenko, A. V., Kinev, N. V., Kiselev, O. S., Kudryashov, P. V., van Kuik, B., de Lange, A., de Lange, G., Lapitsky, I. L., Pripolzin, S. I., van Rantwijk, J., Selig, A., Sobolev, A. S., Torgashin, M. Yu., Vaks, V. L., de Vries, E., Wagner, G., and Yagoubov, P. A. (2011). Integrated submm wave receiver: development and applications, in *Nanoscience Frontiers - Fundamentals of Superconducting Electronics*, ed. Sidorenko, A. Springer Series: NanoScience and Technology (Springer Berlin Heidelberg), pp. 263–296.

17. Koshelets, V. P., Dmitriev, P. N., Faley, M. I., Filippenko, L. V., Kalashnikov, K. V., Kinev, N. V., Kiselev, O. S., Artanov, A. A., Rudakov, K. I., de Lange, A., de Lange, G., Vaks, V. L., Li, M. Y., and Wang, H. B. (2015). Superconducting integrated terahertz spectrometers. *IEEE Trans. Terahertz Sci. Technol.*, **5**, 687– 694.

18. Koshelets, V. P., Ermakov, A. B., Filippenko, L. V., Khudchenko, A. V., Kiselev, O. S., Sobolev, A. S. Torgashin, M. Yu., Yagoubov, P. A., Hoogeveen, R. W. M., and Wild, W. (2007). Integrated submillimeter receiver for TELIS. *IEEE Trans. Appl. Supercond.*, **17**, 336–342.

19. Koshelets, V. P., Ermakov, A. B., Filippenko, L. V., Kinev, N. V., Kiselev, O. S., Torgashin, M. Yu., de Lange, A., de Lange, G., Pripolzin, S. I., and Vaks, V. L. (2010). *Proc. SPIE*, **7854**, 78540J.

20. de Lange, G., Boersma, D., Dercksen, J., Dmitriev, P. N., Ermakov, A. B., Filippenko, L. V., Golstein, H., Hoogeveen, R., de Jong, L., Khudchenko, A. V., Kinev, N. V., Kiselev, O. S., van Kuik, B., de Lange, A., van Rantwijk, J., Sobolev, A. S., Torgashin, M. Yu., de Vries, E., Yagoubov, P. A., and Koshelets, V. P. (2010). Development and characterization of the

superconducting integrated receiver channel of the TELIS atmospheric sounder. *Supercond. Sci. Technol.*, **23**(4), 045016 (8pp).

21. Kiselev, O., Birk, M., Ermakov, A., Filippenko, L., Golstein, H., Hoogeveen, R., Kinev, N., van Kuik, B., de Lange, A., de Lange, G., Yagoubov, P., and Koshelets, V. (2011). Balloon-borne superconducting integrated receiver for atmospheric research. *IEEE Trans. Appl. Supercond.*, **21**, 612–615.
22. Rowell, J. M., Gurvitch, M., and Geerk, J. (1981). Modification of tunneling barriers on Nb by a few monolayers of Al. *Phys. Rev. B*, **24**, 2278–2281.
23. Gurvitch, M., Washington, W. A., and Huggins, H. A. (1983). High quality refractory Josephson tunnel junctions utilizing thin aluminum layers. *Appl. Phys. Lett.*, **42**, 472–474.
24. Huggins, H. A., and Gurwitch, M. (1985). Preparation and characteristics of Nb-AlOxide-Nb tunnel junctions. *J. Appl. Phys.*, **57**, 2103–2109.
25. Kroger, H., Smith, L. N., and Jillie, D. W. (1981). Selective niobium anodisation process for fabrication Josephson tunnel junctions. *Appl. Phys. Lett.*, **39**, 280–282.
26. Morohashi, S., and Hasuo, S. (1987). Experimental investigations and analysis for high-quality Nb/Al-AlO_x/Nb Josephson junctions. *J. Appl. Phys.*, **61**, 4835–4849.
27. Morohashi, S., Shinoki, F., Shoji, A., et al. (1985). High quality Nb/Al-AlOx/Nb Josephson junction. *Appl. Phys. Lett.*, **46**, 2103–2109.
28. Imamura, T., Shiota, T., and Hasuo, S. (1992). Fabrication of high quality Nb/AlOx-Al/Nb Josephson junctions: I - Sputtered Nb films for junction electrode. *IEEE Trans. Appl. Supercond.*, **2**, 1.
29. Imamura, T., and Hasuo, S. (1992). Fabrication of high quality Nb/AlOx-Al/Nb Josephson junctions: II - Deposition of thin Al layers on Nb layers. *IEEE Trans. Appl. Supercond.*, **2**, 84.
30. Imamura, T., and Hasuo, S. (1992). Fabrication of high quality Nb/AlOx-Al/Nb Josephson junctions: III – Annealing stability of AlOx tunneling barriers. *IEEE Trans. Appl. Supercond.*, **2**, 222.
31. Dmitriev, P. N., Ermakov, A. B., Kovalenko, A. G., Koshelets, V. P., Iosad, N. N., Golubov, A. A., and Kupriyanov, M. Yu. (1999). Niobium tunnel junctions with multi-layered electrodes. *IEEE Trans. Appl. Supercond.*, **9**, 3970–3973.
32. Koshelets, V. P., Kovtonyuk, S. A., Serpuchenko, I. L., Filippenko, L. V., and Shchukin, A. V. (1991). High quality Nb-AlOx-Nb tunnel junctions for microwave and SFQ logic devices. *IEEE Trans. Magn.*, **27**, 3141–3144.

33. Golubov, A. A., Houwman, E. P., Gijsbertsen, J. G., Krasnov, V. M., Lokstra, J. F., Rogalla, H. (1995). Proximity effect in superconductor-insulator-superconductor Josephson tunnel junctions: theory and experiment. *Phys. Rev. B*, **51**, 1073–1089.

34. Shiota, T., Imamura, T., and Hasuo, S. (1992). Nb Josephson junctions with an AlNx barrier made by plasma nitridation. *Appl. Phys. Lett.*, **61**, 1228–1230.

35. Kleinsasser, A. W., Mallison, W. H., and Miller, R. E. (1995). Nb/AlN/Nb Josephson-junctions with high critical current density. *IEEE Trans. Appl. Supercond.*, **5**, 2318–2321.

36. Kawamura, J., Miller, D., Chen, J., Zmuidzinas, J., Bumble, B., Le Duc, H. G., and Stern, J. A. (2000). Very high-current-density Nb/AlN/Nb tunnel junctions for low-noise submillimeter mixers. *Appl. Phys. Lett.*, **76**, 2119.

37. Bumble, B., Le Duc, H. G., Stern, J. A., and Megerian, K. G. (2001). Fabrication of Nb/AlNx/NbTiN junctions for SIS mixer applications. *IEEE Trans. Appl. Supercond.*, **11**, 76–79.

38. Iosad, N. N., Ermakov, A. B., Meijer, F. E., Jackson, B. D., and Klapwijk, T. M. (2002). Characterization of the fabrication process Nb/Al-AlN_x/Nb tunnel junctions with low R_nA values up to 1 $\Omega\mu m^2$. *Supercond. Sci. Technol*, **15**, 945–951.

39. Dmitriev, P. N., Lapitskaya, I. L., Filippenko, L. V., Ermakov, A. B., Shitov, S. V., Prokopenko, G. V., Kovtonyuk, S. A., and Koshelets, V. P. (2003). High quality Nb-based tunnel junctions for high frequency and digital applications. *IEEE Trans. Appl. Supercond.*, **13**(2), 107–110.

40. Torgashin, M. Yu., Koshelets, V. P., Dmitriev, P. N., Ermakov, A. B., Filippenko, L. V., and Yagoubov, P. A. (2007). Superconducting integrated receivers based on Nb-AlN-NbN circuits. *IEEE Trans. Appl. Supercond.*, **17**, 379–382.

41. Khudchenko, A., Baryshev, A. M., Rudakov, K., Koshelets, V., Dmitriev, P., Hesper, R., and de Jong, L. (2016). High gap Nb-AlN-NbN SIS junctions for frequency band 790–950 GHz. *IEEE Trans. Terahertz Sci. Technol.*, **6**(1), 127–132.

42. Kasemann, C., Gu'sten, R., Heyminck, S., Klein, B., Klein, T., Philipp, S. D., Korn, A., Schneider, G., Henseler, A., Baryshev, A., and Klapwijk, T. M. (2006). CHAMP+: a powerful array receiver for APEX. *Proc. SPIE*, **6275**, 62750N.

43. Gu'sten, R., Booth, R., Cesarsky, C., and Menten, K., et al., (2006). Ground-based and airborne telescopes. *Proc. SPIE*, **6267**, 626714-1.

44. Smirnov, A. V., Baryshev, A. M., de Bernardis, P., Vdovin, V. F., Goltsman, G. N., Kardashev, N. S., Kuzmin, L. S., Koshelets, V. P., Vystavkin, A. N., Lobanov, Yu. V., Ryabchun, S. A., Finkel, S. A., and Khokhlov, D. R. (2012). The current stage of development of the receiving complex of the Millimetron space obseratory. Radiophys. Quantum Electron., **54**(8–9), 557–568.

45. Zmuidzinas, J., LeDuc, H. G., Stern, J. A., and Cypher, S. R. (1994). Two-junction tuning circuits for submillimeter SIS mixers. *IEEE Trans. Microwave Theory Tech.*, **42**(4), 698–706.

46. Belitsky, V. Yu., Jacobsson, S. W., Filippenko, L. V., and Kollberg, E. L. (1995). Broadband twin-junction tuning circuit for submillimeter SIS mixers. *Microwave. Opt. Technol. Lett.*, **10**, 74–78.

47. Shoji, A., Kosaka, S., Shinoki, F., Aoyagi, M., and Hayakawa, H. (1983). All refractory Josephson tunnel junctions fabricated by reactive ion etching. *IEEE Trans. Magn.*, **19**, 827–830.

48. Shoji, A., Aoyagi, M., Kosaka, S., Shinoki, F., and Hayakawa, H. (1985). Niobium nitride Josephson tunnel junctions with magnesium oxide barriers. *Appl. Phys. Lett.*, **46**, 1098–1100.

49. Blaugher, R. D., Przybysz, J. X., Talvacchio, J., and Battyan, J. (1987). Processing of all-NbN tunnel junction series arrays. *IEEE Trans. Magn.*, **23**(2), 673–675.

50. Talvacchio, J., and Braginski, A. I. (1987). Tunnel junctions fabricated from coherent NbN/MgO/NbN and NbN/Al_2O_3/NbN structures. *IEEE Trans. Magn.*, **23**, 859–862.

51. Wang, Z., Uzawa, Y., and Kawakami, A. (1997). High current density NbN/AlN/NbN tunnel junctions for submillimeter wave SIS mixers. *IEEE Trans. Appl. Supercond.*, **7**(2), 2797–2800.

52. Uzawa, Y., Wang, Z., and Kawakami, A. (1998). Terahertz NbN/AlN/NbN mixers with Al/SiO/NbN microstrip tuning circuits. *Appl. Phys. Lett.*, **73**(5), 680–682.

53. Wang, Z., Terai, H., Kawakami, A., and Uzawa, Y. (1999). Interface and tunneling barrier heights of NbN/AlN/NbN tunnel junctions. *Appl. Phys. Lett.*, **75**, 701.

54. Wang, Z., Terai, H., Qiu, W., Makise, K., Uzawa, Y., Kimoto, K., and Nakamura, Y. (2013). High-quality epitaxial NbN/AlN/NbN tunnel junctions with a wide range of current density. *Appl. Phys. Lett.*, **102**(14), 142604-1–142604-5.

55. Zijlstra, T., Lodewijk, C. F. J., Vercruyssen, N., Tichelaar, F. D., Loudkov, D. N., and Klapwijk, T. M. (2007). Epitaxial aluminum nitride tunnel

barriers grown by nitridation with a plasma source. *Appl. Phys. Lett.*, **91**, 233102–233105.

56. Josephson, B. D. (1962). Possible new effects in superconductive tunneling. *Phys. Lett.*, **1**, 251–253.
57. Josephson, B. D. (1964). Coupled superconductors. *Rev. Mod. Phys.*, **36**, 216–220.
58. Likharev, K. K. (1986). *Dynamics of Josephson Junctions and Circuits*, Gordon and Breach: London.
59. Varmazis, C., Sandell, R. D., Jain, A. K., and Lukens, J. E. (1978). Generation of coherent tunable Josephson radiation at microwave frequencies with narrowed linewidth. *Appl. Phys. Lett.*, **33**, 357.
60. Jain, A. K., Likharev, K. K., Lukens, J. E., and Sauvageau, J. E. (1984). Mutual phase-locking in Josephson junction arrays. *Phys. Rep.*, **109**, 309–426.
61. Dueholm, B., Levring, O. A., Mygind, J., Pedersen, N. F., Soerensen, O. H., and Cirillo, M. (1981). Multisoliton excitations in long Josephson junctions. *Phys. Rev. Lett.*, **46**, 1299.
62. Joergensen, E., Koshelets, V. P., Monaco, R., Mygind, J., Samuelsen, M. R., and Salerno, M., (1982). Thermal fluctuations in resonant motion of fluxon on a Josephson transmission line: theory and experiment. *Phys. Rev. Lett.*, **49**, 1093–1096.
63. Cirillo, M., and Lloyd, F. (1987). Phase lock of a long Josephson junction to an external microwave source. *J. Appl. Phys.*, **61**, 2581.
64. Booi, P. A. A., and Benz, S. P. (1994). Emission linewidth measurements of two-dimensional array Josephson oscillators. *Appl. Phys. Lett.*, **64**, 2163.
65. Barbara, P., Cawthorne, A. B., Shitov, S. V., and Lobb, C. J. (1999). Stimulated emission and amplification in Josephson junction arrays. *Phys. Rev. Lett.*, **82**, 1963–1965.
66. Salez, M., and Boussaha, F. (2010). Fluxon modes and phase-locking at 600 GHz in superconducting tunnel junction nonuniform arrays. *J. Appl. Phys.*, **107**, 013908.
67. Song, F., Müller, F., Scheller, T., Semenov, A. D., He, M., Fang, L., Hübers, H.-W., and Klushin, A. M. (2011). Compact tunable subterahertz oscillators based on Josephson junctions. *Appl. Phys. Lett.*, **98**, 142506.
68. Galin, M. A., Klushin, A. M., Kurin, V. V., Seliverstov, S. V., Finkel, M. I., Goltsman, G. N., Müller, F., Scheller, T., and Semenov, A. D. (2015). Towards local oscillators based on arrays of niobium Josephson junctions. *Supercond. Sci. Technol.*, **28**, 055002 (7pp).

69. Nagatsuma, T., Enpuku, K., Irie, F., and Yoshida, K. (1983). Flux-flow type Josephson oscillator for millimeter and submillimeter wave region. *J. Appl. Phys.*, **54**, 3302.

70. Nagatsuma, T., Enpuku K., Sueoka, K., Yoshida, K., and Irie, F. (1984). Flux-flow type Josephson oscillator for millimeter and submillimeter wave region. II. Modeling. *J. Appl. Phys.*, **56**, 3284.

71. Nagatsuma, T., Enpuku, K., Yoshida, K., and Irie, F. (1985). Flux-flow type Josephson oscillator for millimeter and submillimeter wave region. III. Oscillation stability. *J. Appl. Phys.*, **58**, 441.

72. Qin, J., Enpuku, K., and Yoshida, K. (1988). Flux-flow type Josephson oscillator for millimeter and submillimeter wave region. IV. Thin-film coupling. *J. Appl. Phys.*, **63**, 1130.

73. Zhang, Y. M., and Wu, P. H. (1990). Numerical calculation of the height of velocity-matching step of flux-flow type Josephson oscillator. *J. Appl. Phys.*, **68**, 4703–4709.

74. Zhang, Y. M., Winkler, D., and Claeson, T. (1993). Detection of mm and submm wave radiation from soliton and flux-flow modes in a long Josephson junction. *IEEE Trans. Appl. Supercond.*, **3**, 2520–2523.

75. Koshelets, V. P., Dmitriev, P. N., Ermakov, A. B., Sobolev, A. S., Torgashin, M. Yu., Kurin, V. V., Pankratov, A. L, Mygind, J. (2005). Optimization of the phase-locked flux-flow oscillator for the submm integrated receiver. *IEEE Trans. Appl. Supercond.*, **15**, 964–967.

76. Koshelets, V. P., Shitov, S. V., Shchukin, A. V., Filippenko, L. V., Mygind, J., and Ustinov, A. V. (1997). Self-pumping effects and radiation linewidth of Josephson flux flow oscillators. *Phys. Rev. B*, **56**, 5572–5577.

77. Koshelets, V. P., Shitov, S. V., Shchukin, A. V., Filippenko, L. V., and Mygind, J., (1996). Linewidth of submillimeter wave flux-flow oscillators. *Appl. Phys. Lett.*, **69**, 699–701.

78. Koshelets, V. P., Shitov, S. V., Shchukin, A. V., Filippenko, L. V., Dmitriev, P. N., Vaks, V. L., Mygind, J., Baryshev, A. M., Luinge, W., and Golstein, H. (1999). Flux flow oscillators for sub-mm wave integrated receivers. *IEEE Trans. Appl. Supercond.*, **9**, 4133–4136.

79. Koshelets, V. P., and Mygind, J., (2001). Flux flow oscillators for superconducting integrated submm wave receivers, in *Studies of High Temperature Superconductors*, ed. Narlikar, A. V., NOVA Science Publishers: New York, **39**, pp. 213–244.

80. Koshelets, V. P., Ermakov, A. B., Dmitriev, P. N., Sobolev, A. S., Baryshev, A. M., Wesselius, P. R., and Mygind, J. (2001). Radiation linewidth of flux flow oscillators. *Supercond. Sci. Technol.*, **14**, 1040–1043.

81. Koshelets, V. P., Shitov, S. V., Filippenko, L. V., Vaks, V. L., Mygind, J., Baryshev, A. B., Luinge, W., and Whyborn, N. (2000). Phase locking of 270-440 GHz Josephson flux flow oscillator. *Rev. Sci. Instrum.*, **71**(1), 289–293.

82. Berberich, H., Buemann, R., and Kinder, H. (1982). Monochromatic phonon generation by the Josephson effect. *Phys. Rev. Lett.*, **49**, 1500.

83. Kinder, H., Berbctich, P., and Schick, A. (1984). High resolution phonon spectroscopy using the Josephson effect. *Physica B+C*, **127**, 210–213.

84. Ehre, D., Lyahovitskaya, V., Tagantsev, A., and Lubomirsky, I. (2007). Amorphous piezo- and pyroelectric phases of $BaZrO_3$ and $SrTiO_3$. *Adv. Mater.*, **19**, 1515–1517.

85. Schlomann, E. (1964). Dielectric losses in ionic crystals with disordered charge distributions. *Phys. Rev.*, **135**, A413–A419.

86. Kozub, V. I. (1983). Possible explanation of monochromatic phonon emission by the Josephson junction. *Phys. Rev. Lett.*, **51**, 1498.

87. Werthamer, N. R. (1966). Nonlinear self-coupling of Josephson radiation in superconducting tunnel junctions. *Phys. Rev.*, **147**, 255–263.

88. Hasselberg, L.-E., Levinsen, M. T., and Samuelsen, M. R. (1974). Theories of subharmonic gap structures in superconducting junctions. *Phys. Rev. B*, **9**, 3757–3765.

89. Tredwell, T. J., and Jacobsen, E. H. (1975). Phonon-induced enhancement of the superconducting energy gap. *Phys. Rev. Lett.*, **35**, 244–246; Tredwell, T. J., and Jacobsen, E. H. (1976). Phonon-induced increase in the energy gap of superconducting films. *Phys. Rev. B*, **13**, 2931–2942.

90. Volodin, A. P., Logvenov, G. Yu., Ryazanov, V. V., and Fal'kovskii, I. V. (1987). Acoustoelectric effect at a Josephson SNS junction. *Sov. Phys. JETP Lett.*, **46**, 48–51.

91. Schlenga, K., Hechtfischer, G., Kleiner, R., Walkenhorst, W., and Muller, P. (1996). Subgap structures in intrinsic Josephson junctions of $Tl_2Ba_2Ca_2Cu_3O_{10+\delta}$ and $Bi_2Sr_2CaCu_2O_{8+\delta}$. *Phys. Rev. Lett.*, **76**, 4943–4946.

92. Yurgens, A., Winkler, D., Zavaritsky, N., and Claeson, T. (1996). Gap and sub-gap structures of intrinsic Josephson tunnel junctions in $Bi_2Sr_2CaCu_2O_x$ single crystals. *Proc. SPIE*, **2697**, 433–451.

93. Helm, Ch., Preis, Ch., Forsthofer, F., Keller, J., Schlenga, K., Kleiner, R., and Müller, P. (1997). Coupling between phonons and intrinsic Josephson oscillations in cuprate superconductors. *Phys. Rev. Lett.*, **79**, 737–740.

94. Krasnov, V. M. (2006). Quantum cascade phenomenon in $Bi_2Sr_2CaCu_2O_{8+\delta}$ single crystals. *Phys. Rev. Lett.*, **97**, 257003(1–4).
95. Maksimov, E. G., Arseev, P. I., and Maslova, N. S. (1999). Phonon assisted tunneling in Josephson junctions. *Solid State Commun.*, **111**, 391–395.
96. Helm, Ch., Preis, Ch., Walter, Ch., and Keller, J. (2000). Theory for the coupling between longitudinal phonons and intrinsic Josephson oscillations in layered superconductors. *Phys. Rev. B*, **62**, 6002–6014.
97. Katterwe, S.-O., Motzkau, H., Rydh, A., and Krasnov, V. M. (2011). Coherent generation of phonon-polaritons in $Bi_2Sr_2CaCu_2O_{8+x}$ intrinsic Josephson junctions. *Phys. Rev. B*, **83**, 100510R (1–4).
98. Koshelets, V. P., Ermakov, A. B., Shitov, S. V., Dmitriev, P. N., Filippenko, L. V., Baryshev, A. M., Luinge, W., Mygind, J., Vaks, V. L., and Pavel'ev, D. G. (2001). Superfine resonant structure on IVC of long Josephson junctions and its influence on flux flow oscillator linewidth. *IEEE Trans. Appl. Supercond.*, **11**, 1211–1214.
99. Koshelets, V. P. (2014). Subterahertz sound excitation and detection by long Josephson junction. *Supercond. Sci. Technol.*, **27**, 065010 (7pp).
100. Eisenmenger, W., and Dayem, A. H. (1967). Quantum generation and detection of incoherent phonons in superconductors. *Phys. Rev. Lett.*, **18**, 125–127.
101. Eisenmenger, W. (1976). Superconducting tunnel junctions as phonon generators and detectors, in *Physical Acoustics: Principles and Methods*, Academic Press: New York, pp. 79–153.
102. Bron, W. E. (1980). Spectroscopy of high-frequency phonons. *Rep. Prog. Phys.*, **43**, 301–352.
103. Ivanchenko, Yu. M., and Medvedev, Yu. V. (1971). Exitation of sound in a Josephson junction. *Sov. Phys. JETP*, **33**, 1223–1228.
104. McSkimin, H. J., and Andreatch, P. (1964). Elastic moduli of silicon vs hydrostatic pressure at 25.0°C and −195.8°C. *J. Appl. Phys.*, **35**, 2161–2165.
105. Teichert, C., MacKay, J. F., Savage, D. E., Lagally, M. G., Brohl, M., and Wagner, P. (1995). Comparison of surface roughness of polished silicon wafers measured by light scattering topography, soft x-ray scattering, and atomic force microscopy. *Appl. Phys. Lett.*, **66**, 2346–2348.
106. Polzikova, N. I., Mansfeld, G. D., Tokpanov, Y. S., and Koshelets, V. P. (2011). Resonant subterahertz coherent acoustic waves excitation by Josephson junction. *Proc. 2011 Joint Conference of the IEEE International Frequency Control Symposium & European Frequency and Time Forum*, ISSN: 1075-6787, 483–485.

107. http://vadiodes.com/index.php/en/products/full-band-multipliers-wr-series

108. http://www.almaobservatory.org/en/about-alma/how-does-alma-work/technology/front-end

109. Hoogeveen, R. W. M., Yagoubov, P. A., de Lange, A., Selig, A. M., Koshelets, V. P., Ellison, B. N., and Birk, M. (2005). *Proc. SPIE*, **5978**, 440.

110. Hoogeveen, R. W. M., Yagoubov, P. A., de Lange, G., de Lange, A., Koshelets, V., Birk, M., and Ellison, B. (2007). *Proc. SPIE*, **6744**, 67441U-1.

111. Friedl-Vallon, F., Maucher, G., Seefeldner, M., Trieschmann, O., Kleinert, A., Lengel, A., Keim, C., Oelhaf, H., and Fischer, H. (2004). *Appl. Opt.*, **43**, 3335.

112. Sobis, P., Drakinskiy, V., Wadefalk, N., Karandikhar, Y., Hammar, A., Emrich, A., Zhao, H., Bryllert, T., Tang, A.-Y., Nilsson, P.-Å., Schleeh, J., Kim, H., Jacob, K., Murk, A., Grahn, J., and Stake, J. (2014). Low noise GaAs Schottky TMIC and InP Hemt MMIC based receivers for the ISMAR and SWI instruments. Micro-and Millimetre Wave Technology and Techniques Workshop 2014, ESA-ESTEC, Noordwijk, The Netherlands.

113. de Lange, A., Birk, M., de Lange, G., Friedl-Vallon, F., Kiselev, O., Koshelets, V., Maucher, G., Oelhaf, H., Selig, A., Vogt, P., Wagner, G., and Landgraf, J. (2012). HCl and ClO in activated Arctic air; first retrieved vertical profiles from TELIS submillimetre limb spectra. *Atmos. Meas. Tech.*, **5**, 487–500.

114. Xu, J., Schreier, F., Vogt, P., Doicu, A., and Trautmann, T. (2013). A sensitivity study for far infrared balloon-borne limb emission sounding of stratospheric trace gases. *Geosci. Instrum. Methods Data Syst. Discuss.*, **3**, 251–303.

115. Kasai, Y., Sagawa, H., Kreyling, D., Dupuy, E., Baron, P., Mendrok, J., Suzuki, K., Sato, T. O., Nishibori, T., Mizobuchi, S., Kikuchi, K., Manabe, T., Ozeki, H., Sugita, T., Fujiwara, M., Irimajiri, Y., Walker, K. A., Bernath, P. F., Boone, C., Stiller, G., von Clarmann, T., Orphal, J., Urban, J., Murtagh, D., Llewellyn, E. J., Degenstein, D., Bourassa, A. E., Lloyd, N. D., Froidevaux, L., Birk, M., Wagner, G., Schreier, F., Xu, J., Vogt, P., Trautmann, T., and Yasui, M. (2013). Validation of stratospheric and mesospheric ozone observed by SMILES from International Space Station. *Atmos. Meas. Tech.*, **6**, 2311–2338.

116. Sagawa, H., Sato, T., Baron, P., Dupuy, E., Livesey, N., Urban, J., von Clarmann, T., de Lange, A., Wetzel, G., Kagawa, A., Murtagh, D., and

Kasai, Y. (2013). Comparison of SMILES ClO profiles with other satellite and balloon-based measurements. *Atmos. Meas. Tech. Discuss.*, **6**(1), 613–663.

117. Li, M., Yuan, J., Kinev, N., Li, J., Gross, B., Gu'enon, S., Ishii, A., Hirata, K., Hatano, T., Koelle, D., Kleiner, R., Koshelets, V. P., Wang, H., and Wu, P. (2012). Linewidth dependence of coherent terahertz emission from $Bi_2Sr_2CaCu_2O_8$ intrinsic Josephson junction stacks in the hot-spot regime. *Phys. Rev. B*, **86**, 060505(R).
118. Butz, S., Jung, P., Filippenko, L. V., Koshelets, V. P., and Ustinov, A. V. (2013). A one-dimensional tunable magnetic metamaterial. *Opt. Express*, **21**(19), 22540–22548.
119. Vidiborskiy, A., Koshelets, V. P., Filippenko, L. V., Shitov, S. V., and Ustinov, A. V. (2013). Ultra-compact tunable split-ring resonators. *Appl. Phys. Lett.*, **103**, 162602.
120. Jung, P., Butz, J., Marthaler, M., Fistul, M. V., Leppäkangas, J., Koshelets, V. P., and Ustinov, A. V. (2014). A multi-stable switchable metamaterial. *Nat. Commun.*, **5**, 3730.
121. Kalashnikov, K. V., Khudchenko, A. V., and Koshelets, V. P. (2013). Harmonic phase detector for phase locking of cryogenic terahertz oscillator. *Appl. Phys. Lett.*, **103**, 102601-1–102601-4.

Chapter 8

Application in Superconducting Quantum Interference Devices SQUIDs

D. Drung and J. Beyer
Physikalisch-Technische Bundesanstalt (PTB), Abbestr. 2–12, 10587 Berlin, Germany
dietmar.drung@ptb.de, joern.beyer@ptb.de

The superconducting quantum interference device (SQUID) is a highly sensitive detector for magnetic flux or any quantity that can be efficiently converted into flux. Comprehensive overviews of the fundamentals, technology, and applications of SQUIDs and SQUID systems are found in the literature (for example [1–6]). In this chapter, a short introduction into the basic function of a SQUID, its operation, and its design for magnetic field and current sensing is given. It is shown by several examples that the device is very versatile and important for a variety of applications such as biomagnetism, astronomy, low-temperature physics, and metrology. Due to the focus of the book on Nb-based devices, the large field of devices with high critical temperature (high-T_c SQUIDs) will not be considered here.

Josephson Junctions: History, Devices, and Applications
Edited by Edward Wolf, Gerald Arnold, Michael Gurvitch, and John Zasadzinski

ISBN 978-981-4745-47-5 (Hardcover), 978-1-315-36452-0 (eBook)
www.panstanford.com

8.1 SQUID Fundamentals

A SQUID basically consists of a superconducting loop interrupted by one or more Josephson junctions. Depending on the number of junctions, it is named rf SQUID (one junction) or dc SQUID (two junctions), respectively. This classification results from the basic mode of operation: the rf SQUID is driven by a high-frequency signal (tens or hundreds of megahertz typically) applied to a tank circuit magnetically coupled to the SQUID, whereas the dc SQUID is biased with a direct current. The dc SQUID was discovered by Jaklevic et al. [7], only a few years after Josephson's predictions of superconducting tunneling. The rf SQUID was introduced shortly afterwards [8]. In those days, it was extremely difficult to fabricate reliable Josephson junctions with thin-film techniques. As a result, early SQUIDs were equipped with "point contacts" consisting of a Nb screw pressed against a Nb body, thereby forming an adjustable Josephson junction.

It is obviously much more demanding to obtain two point contacts with similar parameters than to adjust only one. Therefore, although the dc SQUID generally allows a lower overall noise, the rf SQUID became the standard device in the following years. The situation changed after reliable thin-film processes were developed in the 1970s and 1980s, and the dc SQUID began to replace the rf SQUID. Nowadays, the dc SQUID is dominant in the field of Nb-based sensors, and the rf SQUID is practically obsolete. Therefore, this chapter deals with the dc SQUID only; the rf SQUID is extensively discussed in literature (see for example [4]).

8.1.1 Basic SQUID Function

The dc SQUID is based on two effects: flux quantization and superconducting tunneling. Its function is schematically illustrated in Fig. 8.1. The simplest circuit is assumed: a superconducting loop interrupted by two Josephson junctions. The junctions have parallel-connected resistors R to eliminate hysteresis in their current–voltage characteristics [9, 10]. We first consider the case in Fig. 8.1a, where a quasi-static current I is passed through the SQUID but no magnetic field is applied. Due to symmetry,

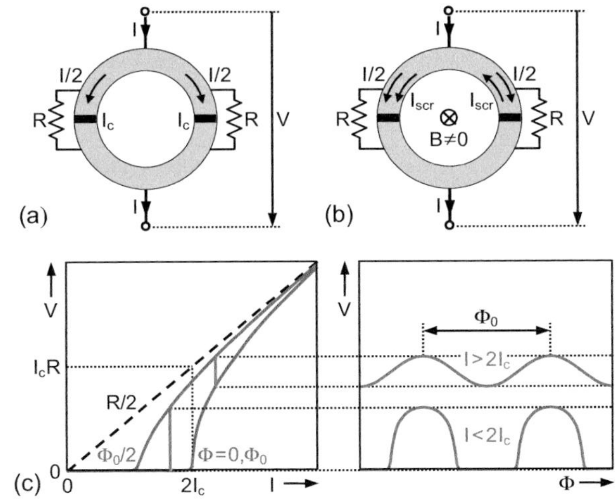

Figure 8.1 Basic SQUID circuit with (a) zero and (b) nonzero applied magnetic flux density B. The resulting voltage–current and voltage–flux characteristics are schematically shown in (c). The dashed line shows the resistance that would be obtained without tunneling. To remove hysteresis in their characteristics, the two Josephson junctions (black regions incorporated into the superconducting loop) are shunted by resistors R connected in parallel.

the current splits into two equal halves $I/2$ that flow through the two Josephson junctions. The corresponding current–voltage characteristic is shown in Fig. 8.1c. Neglecting noise rounding, the total critical current of the SQUID $2I_c$ is the sum of the two junction critical currents I_c.

If a magnetic field is applied perpendicular to the SQUID loop, the resulting flux in the loop $\Phi = BA$ (assuming that the flux density B is constant over the area A of the loop) will cause a screening current I_{scr} to circulate. As we will see later, Josephson junctions are commonly made as small as possible, so that the effect of the applied field on the junction critical currents can be neglected. For example, in Fig. 8.1b, the screening current is added to $I/2$ in the left junction, but subtracted from $I/2$ in the right one. Therefore, the critical current of the left junction is already reached at $I/2 < I_c$ and hence the critical current of the SQUID is reduced with applied flux. If the magnetic flux caused by the screening current exceeds $\pm\Phi_0/2$, the flux state of the SQUID changes by one flux quantum $\Phi_0 \approx 2.068 \times 10^{-15}$ V·s and the screening current changes its direction because this is energetically more favorable than a further increase in the screening current. This way, the total flux in the loop is always kept equal to an integer number of flux quanta (flux quantization) and the critical current of the SQUID changes periodically with the applied flux.

As illustrated in Fig. 8.1c, the flux dependence of the critical current leads to a periodic voltage–flux characteristic if the SQUID is biased at constant current I. The period is exactly equal to one flux quantum Φ_0. For $I \leq 2I_c$, the zero-voltage state is included in the voltage–flux characteristic, whereas for $I > 2I_c$ the SQUID is always in the voltage state with a finite dc voltage across the device. Note that in the voltage state, due to the ac Josephson effect, a high-frequency ac voltage is always superimposed with a fundamental of 483.6 MHz per microvolt of dc voltage. In Fig. 8.1 it was assumed that this ac Josephson voltage is outside the measurement bandwidth and filtered out by the measurement setup (which is always the case in practice). Finally, in the above discussion we have neglected thermal noise. In practice, the characteristics in Fig. 8.1 are rounded near zero voltage due to thermal noise.

The SQUID function is determined by four basic quantities: the inductance of the SQUID loop L, the junction critical current I_c, the shunt resistance per junction R, and the parasitic junction capacitance C. The tunnel junctions may be intrinsically shunted or equipped with parallel-connected resistors. Strictly speaking, in the latter case the quasiparticle tunneling would slightly contribute to the effective shunt resistance R; however, this effect is commonly negligible. The specific junction capacitance is relatively high because of the plate-capacitor-like structure with the very thin (a few nanometers typically) tunnel barrier between the electrodes. For window-type junctions, C includes the effect of the overlap area around the tunnel barrier. For cross-type junctions, the overlap contribution is practically eliminated [11]. Extremely small junctions are used in nanometer-sized SQUIDs, the so-called nanoSQUIDs [12]. Here, the parasitic capacitance from the vicinity of the junctions can dominate the total capacitance.

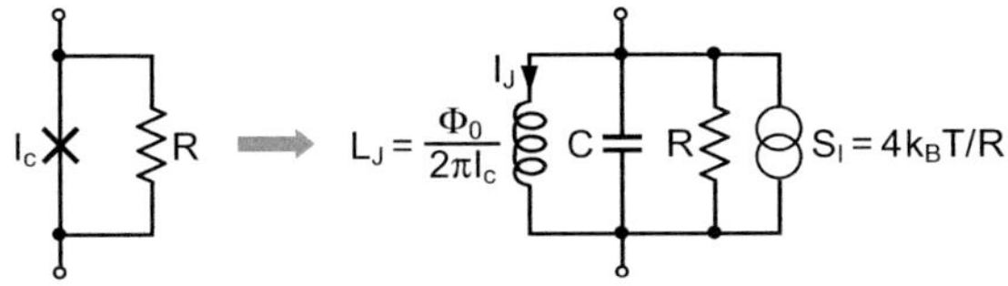

Figure 8.2 Simplified equivalent circuit of a resistively shunted junction with small Josephson tunnel current $I_J \ll I_c$. The effect of superconducting tunneling was approximated by an inductance $L_J = \Phi_0/(2\pi I_c)$.

The SQUID is a strongly nonlinear device. Analytical solutions for the corresponding mathematical equations are available in a few special cases only, but unfortunately for parameters less suitable for practical devices. Intensive computer simulations were performed beginning in the mid-1970s to understand the device and to find design rules for optimum SQUID function [1, 13–16]. These simulations yielded conditions for the three major SQUID parameters that will be discussed and made plausible here on the basis of the simplified junction circuit depicted in Fig. 8.2. For small supercurrents $I_J \ll I_c$ through the tunnel junction, the sine function in the dc Josephson effect can be approximated by the linear term of its Taylor series. As a result, in this case the effect of superconducting tunneling can be taken into account by an equivalent inductance $L_J = \Phi_0/(2\pi I_c)$. For larger tunnel currents I_J approaching $\pm I_c$, the higher-order terms in the sinusoidal current–phase relation become dominant. Therefore, the effective inductance increases nonlinearly with I_J, becomes infinite at $I_J = \pm I_c$ and even negative if I_J exceeds $\pm I_c$. This clearly shows that the device dynamics are very complicated. However, the simplified circuit in Fig. 8.2 is helpful for understanding some basic relationships.

Thermal noise in the shunt resistance can be taken into account by a current source in parallel to the junction having a power spectral density $S_I = 4k_B T/R$, where $k_B \approx 1.38 \times 10^{-23}$ J/K is the Boltzmann constant and T is the absolute temperature. Integrating over frequency, the total rms noise current flowing through L_J is found to be $I_{J,rms} = (k_B T/L_J)^{1/2}$. Obviously, to avoid thermal noise "wiping out" the superconducting tunnel effect, it is required that $I_{J,rms} \ll I_c$. Thus we obtain a condition for the noise parameter

$$\Gamma = (I_{J,rms}/I_c)^2 = 2\pi k_B T/(\Phi_0 I_c) \ll 1. \tag{8.1}$$

The parameter Γ describes the rounding of the junction characteristics due to thermal noise and is a measure for the apparent reduction of the critical current in the presence of thermal noise [17]. For a typical critical current $I_c = 10\ \mu A$, one obtains $L_J = 33$ pH and an rms noise current $I_{J,rms} = 1.33\ \mu A$ at liquid helium temperature ($T = 4.2$ K), resulting in $\Gamma \approx 0.018$. Equation 8.1 is fulfilled at 4.2 K for critical currents above about 1 μA.

Due to the parasitic capacitance C, the Josephson junction forms a parallel resonant circuit. For the simplified circuit in Fig. 8.2 we find a quality factor $Q_J = R(C/L_J)^{1/2}$. Generally, it is advisable to keep the quality factor of resonant circuits in nonlinear systems below about unity to minimize excess noise from down-mixing effects. In the case of a Josephson junction, hysteresis occurs in the current–voltage characteristic for Q_J above about unity (without noise the hysteresis limit is $Q_J \approx 0.84$ [1]). Q_J can be set for given I_c and C by selecting R appropriately. However, R should not be chosen too low because this would result in a small peak-peak output voltage V_{pp} of the SQUID (typically $V_{pp} \approx 0.4\ I_c R$). Therefore, in practice a good compromise is

$$\beta_C = Q_J^2 = 2\pi I_c R^2 C/\Phi_0 \approx 1. \tag{8.2}$$

At 4.2 K and relatively low critical currents $I_c \approx 3.5\ \mu A$ (corresponding to $\Gamma \approx 0.05$) minimum SQUID noise is obtained for β_C between 1 and 2 [1]. In contrast, for very low $\Gamma < 0.01$ typically obtained at millikelvin temperatures, it is advisable to keep $\beta_C < 0.5$ to improve damping. This is particularly important if the effective junction capacitance is higher than expected due to parasitic capacitance in the SQUID layout [16]. Although a slight hysteresis in the SQUID characteristics might not be visible due to thermal noise, it can degrade the noise performance substantially at low values of Γ.

As discussed above, the flux dependence of the critical current of the SQUID results from the screening current I_{scr} interacting with the Josephson junctions. Obviously, I_{scr} decreases with increasing SQUID inductance L. Thus, the modulation depth of the critical current and the resulting output voltage modulation ΔV are maximized for small SQUID inductance $L \ll L_J$. In the limit $L/L_J \to 0$, the critical current of the SQUID as a function of applied flux becomes $2I_c|\cos(\pi\Phi/\Phi_0)|$, i.e., it is completely suppressed at

$\Phi = (n + 1/2)\Phi_0$ (n is an integer). However, simulations show that for too low L/L_J the SQUID noise increases, leading to the design rule [13]

$$\beta_L = L/(\pi L_J) = 2LI_c/\Phi_0 \approx 1. \qquad (8.3)$$

Combining Eqs. 8.3 with Eq. 8.1, one obtains a practical limit for the SQUID inductance $L_{max} \approx 1$ nH at 4.2 K.

The design rules Eqs. 8.2 and 8.3 are helpful guides in practice. One first selects the SQUID inductance according to the intended application (typically $L \approx 100$ pH) and assumes the smallest junction size to minimize capacitance (typically $C \approx 0.4$ pF). Next, Eq. 8.3 is used to determine I_c from the given L. With the help of Eq. 8.2 one obtains the required R and the resulting number of squares in the shunt resistor layout for the nominal sheet resistance of the shunt material (e.g., 4 Ω per square for 70 nm thick AuPd). The final optimization is preferably done during routine fabrication by fine-tuning the critical current density of the Josephson junctions and the film thickness of the shunt resistors.

8.1.2 SQUID Noise

Achieving a low noise level is the most important issue in the field of SQUID sensors. As discussed above, the period in the voltage–flux characteristic is exactly equal to the flux quantum Φ_0. Therefore, the flux sensitivity of the SQUID is automatically "calibrated" and the measured output noise can easily be converted into $\Phi_0/\sqrt{\text{Hz}}$. However, in most cases the signal applied to the SQUID is not magnetic flux directly, but rather magnetic field or current, the latter being passed through a coil inductively coupled to the SQUID loop. In any case, it is necessary to specify the noise figure of the sensor in units of the quantity to be measured, for example the flux density in the pickup coil B or the input current I_i. The corresponding power spectral densities are related to the flux noise density S_Φ by

$$S_B = S_\Phi/A_{eff}^2 \text{ or } S_I = S_\Phi/M_i^2. \qquad (8.4)$$

Here, A_{eff} is the effective field-sensitive area of the SQUID magnetometer; often, the inverse of A_{eff} in units of T/Φ_0 is referred to as field sensitivity. In the case of current sensing applications, M_i is the mutual inductance between the input coil and the SQUID loop.

The flux noise in the SQUID is caused by thermal noise in the shunt resistors, that is accounted for in Fig. 8.2 by the current source $S_I = 4k_BT/R$. As a result, the dc voltage V across the SQUID shows fluctuations V_N that are interpreted as fluctuations in flux $\Phi_N = V_N/V_\Phi$ ($V_\Phi = \partial V/\partial\Phi$ is the transfer coefficient at the chosen working point in the voltage–flux characteristic). It is important to note that the flux noise $\sqrt{S_\Phi}$ is a superposition of a "true" flux noise component (i.e., fluctuations in the screening current $I_{N,scr}$) and a voltage noise component that represents "apparent" flux noise only. When operating the SQUID as a high-frequency amplifier by coupling an input coil inductively to the SQUID loop, the noise in the screening current induces a noise voltage in the input coil [18, 19]. This causes backaction of the SQUID on the signal source, makes the noise analysis more complicated, and can degrade the signal-to-noise ratio. Fortunately, in most SQUID applications this effect is small or even completely negligible, so that it is commonly sufficient to know the total flux noise density S_Φ only.

A useful figure of merit for characterizing SQUIDs is the noise energy per bandwidth ε referred to the SQUID inductance L. To derive it, we substitute the fluctuations in flux Φ_N by equivalent current fluctuations $I_N = \Phi_N/L$ and calculate the noise energy $LI_N^2/2 = \Phi_N^2/2L$. We then replace Φ_N^2 by the spectral density S_Φ and obtain the noise energy per bandwidth $\varepsilon = S_\Phi/2L$ that is often quoted in units of Planck's constant $h \approx 6.63 \times 10^{-34}$ J/Hz. For the near-optimum case $\beta_L \approx 1$ and $\beta_C \approx 1$ numerical simulations [13, 14] yield the white noise level

$$\varepsilon_w \approx 9k_BTL/R \text{ or } \varepsilon_w \approx 16k_BT(LC)^{1/2}. \tag{8.5}$$

Due to the strongly nonlinear SQUID characteristics and the "inherent local oscillator" (the Josephson ac voltage at typically 5 GHz with a rich spectrum of harmonics), thermal noise in the shunt resistors is mixed down from the microwave regime into the signal frequency range. A small-signal analysis without nonlinear effects yields $\varepsilon_w \approx 2k_BTL/R$ [20], i.e., more than three-fourths of the noise energy in Eq. 8.5 is caused by down-mixing. Equation 8.5 was obtained for a "bare" SQUID without including parasitic capacitance in the SQUID layout—for example, due to a multiturn input coil coupled to the SQUID or the transmission line connecting the SQUID with the bonding pads. If parasitic high-Q resonant circuits in the SQUID design are driven by thermal noise, excess noise from down-mixing can become a severe problem and Eq. 8.5 can substantially underestimate the noise energy. Fortunately, in practice this excess noise can be strongly reduced by proper resonance damping, although the final noise energy will be always higher than without parasitic capacitance due to thermal noise in the damping resistors.

The general design rule for minimum noise energy is that parasitic capacitance should be kept as small as possible [16] and that, if parasitic resonant circuits cannot be avoided, appropriate damping by extra resistors or resistor–capacitor series shunts should be implemented [21–23]. This is particularly important when operating a SQUID at millikelvin temperatures because of the reduced noise level. At higher temperatures, hysteresis effects caused by parasitic capacitance are "wiped out" to a certain extend by thermal noise, which can somewhat relax the issue of resonance damping.

The white noise energy ε_W has a minimum when β_C is varied via the shunt resistance R for fixed SQUID inductance L and junction capacitance C. Below the optimum value of R, the noise energy degrades due to the increased thermal current noise $4k_B T/R$, while above optimum the down-mixing noise rises due to insufficient damping. The second expression in Eq. 8.5 shows a practical limit if L and C are fixed by SQUID layout and fabrication process. It assumes that R is optimally selected, typically near $\beta_C \approx 1$. The first expression in Eq. 8.5 can be used if R and L are given. It implies that the β_L is optimally chosen via the critical current I_c and that the junctions are sufficiently well damped. It is also applicable for strongly overdamped junctions with $\beta_C \ll 1$.

The white noise of well-designed Nb-based SQUIDs is typically in fair agreement with theory. Figure 8.3 shows an example of a SQUID with strongly overdamped Josephson junctions ($\beta_C \approx 0.2$ and $\beta_L \approx 0.5$). The measured white noise energy of 28 h at 4.2 K is a factor of 1.75 above the value of 16 h calculated with Eq. 8.5 for $L = 110$ pH and $R = 5.4\ \Omega$, corresponding to a deviation of 32% in rms flux noise $\sqrt{S_\Phi}$. Cooling the SQUID to 310 mK improves the white noise energy by a factor of 12 to 2.3 h. Note that for

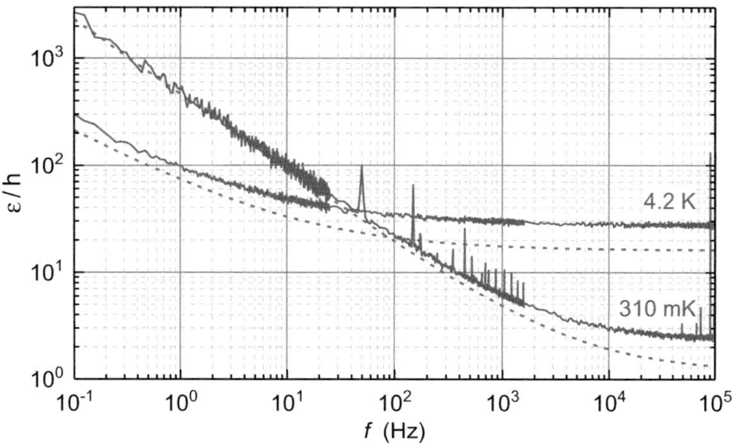

Figure 8.3 Example of noise spectra for a SQUID with strongly overdamped Josephson junctions (device C214G05 in [24], reproduced with permission). Solid lines show the measured noise at 4.2 K and 310 mK, and dashed lines are calculated from Eqs. 8.5 and 8.6 for $L = 110$ pH and $R = 5.4\ \Omega$. The exponent α was determined from the measured noise spectra: $\alpha = 0.53$ at 4.2 K and $\alpha = 0.7$ at 310 mK, respectively. The good agreement between measured and calculated low-frequency data at 310 mK is coincidental.

operation at very low temperatures, the noise temperature of the shunt resistors (and correspondingly the SQUID noise) saturates at typically 300 mK due to self-heating (hot-electron effect [25]). Large cooling fins attached to the resistors improve the situation, but it is commonly difficult to reduce the effective resistor noise temperature below about 100 mK.

As observed in virtually all electronic devices (semiconducting as well as superconducting), also the SQUID noise rises at low frequencies. This additional low-frequency noise is called flicker noise or $1/f$ noise, the latter name resulting from the typical scaling of the power spectral density $S_\Phi \propto 1/f$. In contrast to the white noise, the low-frequency noise is less well understood and can generally not be predicted from the SQUID design. There are noise contributions from low-frequency critical current fluctuations that can be suppressed by special modulation schemes involving bias reversal (for a review see Chapter 4 in [4]). Fortunately, in contrast to high-T_c junctions, modern Nb–AlO_x–Nb trilayer-based junctions show very small levels of critical current fluctuations and bias reversal schemes are commonly not required. This simplifies the readout electronics, which is helpful in particular for multichannel systems. However, there is another low-frequency noise component which reveals as a "true" flux noise and cannot be eliminated by bias reversal. Unfortunately, this excess low-frequency flux noise increases when the operation temperature is reduced below about 2 K, in contrast to the effect of critical current fluctuations that decreases with temperature [26].

A special peculiarity of the excess flux noise is a weaker scaling with frequency, $S_\Phi \propto 1/f^a$ with α typically around 0.6 for low-noise devices [24, 26, 27]. The increase in the noise at low temperatures is accompanied by a rise in the exponent α. Recently, the noise energy of a large variety of SQUIDs at 4.2 K and <320 mK was reported, and an approximate equation for the noise energy including low-frequency excess noise was empirically found for α ranging between about 0.5 and 0.9 [24]

$$\varepsilon \approx \varepsilon_\mathrm{w} + 0.09\,h \times (f/200\ \mathrm{kHz})^{-\alpha}. \tag{8.6}$$

Equation 8.6 gives an estimate of the excess flux noise between about 1 Hz and 100 kHz. It is applicable to SQUIDs when other sources of low-frequency noise (e.g., due to critical current fluctuations or picked-up environmental noise) do not noticeably contribute. At very low frequencies $\lesssim$0.1 Hz, the common $1/f$ scaling ($\alpha \approx 1$) is typically observed. The dashed lines in Fig. 8.3 are calculated with Eq. 8.6 for the experimentally observed α values of 0.53 at 4.2 K and 0.7 at 310 mK, respectively. The agreement between the calculated and measured noise spectra is adequate considering that Eqs. 8.5 and 8.6 are approximate formulas only. To conclude, the white noise decreases with temperature as expected. Cooling the SQUID to $\lesssim$300 mK helps to improve the noise energy by typically about one order of magnitude compared to operation at 4.2 K. However, it is generally observed that the low-frequency noise degrades when lowering the temperature below about 2 K. For the particular device in Fig. 8.3, the noise at 310 mK exceeds that at 4.2 K for frequencies below about 40 Hz. Therefore, operation at millikelvin temperatures is usually not favorable for low-frequency applications.

8.1.3 Inductance and Effective Area

The basic equations for designing a SQUID sensor are summarized in Chapter 5 of [4]. In this section, we present equations for calculating the inductance and the effective field-sensitive area of two representative superconducting structures. The polygonal structure in Fig. 8.4a is commonly used for thin-film pickup coils or for the SQUID loop [28]. A narrow slit, indicated in Fig. 8.4a by a vertical solid line, interrupts the loop to enable connection with other elements, for example the two Josephson junctions. For simplicity, we first neglect the contribution of the slit to the total inductance and effective area of the polygonal loop. In practice, the slit's contribution can be made small by covering it with a superconducting plate, however, at the expense of additional stray capacitance [29]. The coplanar line in Fig. 8.4b typically serves as an interconnect line, for example between pickup coils in planar thin-film gradiometers [30], or as a "spoke" in multiloop magnetometers [20]. It has a larger parasitic inductance and effective area than the microstrip geometry (where one strip is placed on top of the other), but a substantially reduced capacitance. The latter is beneficial because stray capacitance should always be minimized in SQUID design.

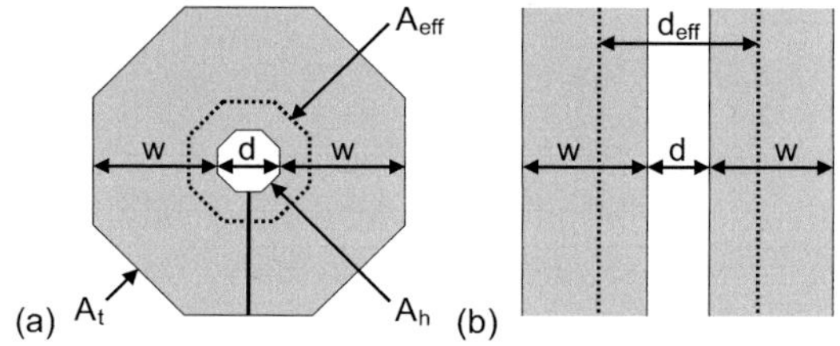

Figure 8.4 Top view of two basic superconducting structures: (a) polygonal loop, (b) coplanar line. Superconducting films are marked in gray. In (a), the hole area A_h is defined by the inner perimeter of the polygonal loop and A_t is the total area (gray region plus A_h). The effective area A_{eff} and effective slit width d_{eff} are indicated by dotted lines for the depicted case $w = 2d$.

When applying a homogeneous flux density B to the superconducting loop in Fig. 8.4a, magnetic flux is focused into the hole, in particular if the hole diameter d is much smaller than the linewidth w [31]. The effective area A_{eff} is defined by $A_{eff} = \Phi_h/B$ where Φ_h is the total flux focused into the hole area A_h. In other words, A_{eff} is the area that a superconducting loop with $w \to 0$ should have to collect the same total flux for given B. Similarly, flux is focused into the slit of the coplanar line in Fig. 8.4b. The effective slit width d_{eff} is equal to the slit width d of an equivalent coplanar line with $w \to 0$.

Unfortunately, analytical equations for L_h and A_{eff} of the polygonal loop in Fig. 8.4 exist in special cases only [31]. At PTB, we are using the following approximate equations that were deduced from numerical calculations under the assumption of idealized superconducting structures (i.e., film thickness t and London penetration depth λ_L are much smaller than the linewidth w and spacing d) [32]. The equations were derived for a regular polygon with N corners, but are useful for other shapes as well (e.g., the optimized multiloop magnetometer in [33]). For the inductance of the polygonal loop we find

$$L_h = \gamma_L \mu_0 c/\pi \text{ with } \gamma_L = [\ln(d/w+2.8)+0.2+2.7d/c]/2.07, \quad (8.7)$$

where $\mu_0 = 4\pi \times 10^{-7}$ H/m is the vacuum permeability and c is the perimeter of the hole. The inductance is proportional to the hole perimeter c multiplied by a geometry factor γ_L. This geometry factor depends on the ratio d/w, but also on the shape of the loop due to the term $2.7d/c$ in Eq. 8.7. For a regular polygon one obtains the perimeter

$$c = N \tan(\pi/N)d. \quad (8.8)$$

For a square loop ($N = 4$) in the limit $d/w \to 0$, the hole inductance $L_h = 1.25\ \mu_0 d$ numerically calculated by Jaycox and Ketchen [29] is about 7% larger than the result $L_h = 1.17\ \mu_0 d$ obtained from Eq. 8.7. However, the numerical result $L_h = 1.19\ \mu_0 d$ reported in [34] is in good agreement with Eq. 8.7. For an octagonal loop in the limit $d/w \to 0$, a hole inductance $L_h \approx 1.05\ \mu_0 d$ was quoted in [28], which agrees well with $L_h = 1.04\ \mu_0 d$ resulting from Eqs. 8.7 and 8.8.

The effective area of the polygonal loop is given by

$$A_{eff} = \gamma_A (A_h A_t)^{1/2} \text{ with } \gamma_A = 1 - 0.68/(d/w + 2.07)^{1.75}. \quad (8.9)$$

The geometry factor γ_A depends on the ratio d/w, but is independent of the shape of the loop. A_h and A_t are the area of the hole and the total area of the loop, respectively. For a regular polygon we obtain

$$(A_h A_t)^{1/2} = (N/4) \tan(\pi/N)d(d + 2w). \quad (8.10)$$

In the limiting cases of square ($N = 4$) or circular ($N \to \infty$) shapes, Eq. 8.10 simplifies to $(A_h A_t)^{1/2} = d(d+2w)$ or $d(d+2w)\pi/4$, respectively. For a circular loop, the effective area was analytically calculated in the limit $d/w \to 0$ [31]. Equation 8.9 differs from the analytical result $\gamma_A = 8/\pi^2$ by less than 0.12%. For a square loop, the numerical result $\gamma_A = 0.81$ in [34] is in excellent agreement with Eq. 8.9. However, the experimental value $\gamma_A \approx 1.1$ reported by Ketchen et al. [31] is 36% higher than Eq. 8.9 predicts. In Ketchen's experiments, the slit had been covered by a superconducting plate. The increase in effective area was probably caused by the residual stray flux coupled into the loop via the slit.

The inductance per length L' and effective slit width (effective area per length) d_{eff} of an infinitely long coplanar line according to Fig. 8.4b can be analytically calculated, yielding $L' = \mu_0 K(k)/K(k')$ [35] and $d_{\text{eff}} = 0.5\pi(d+2w)/K(k')$. Here, $K(k)$ is the complete elliptic integral of the first kind with modulus $k = d/(d+2w)$ and $k' = (1 - k^2)^{1/2}$. Simplified but still accurate approximate formulas without elliptic integral [32] are given here:

$$L' = \gamma_{L'}\mu_0 \text{ with}$$

$$\gamma_{L'} = [\ln(4d/w + 22)/\ln(8w/d + 4.9)]/1.98, \quad (8.11)$$

$$d_{\text{eff}} = \gamma_d(d + 2w) \text{ with } \gamma_d = 0.5\pi/\ln(8w/d + e^{\pi/2}). \quad (8.12)$$

The term $e^{\pi/2}$ ensures that Eq. 8.12 yields the correct result $d_{\text{eff}} \to d$ in the limit $w/d \to 0$. Neglecting the influence of the line ends, the total inductance and effective area of a coplanar line of finite length l are given by $L = L'l$ and $A_{\text{eff}} = d_{\text{eff}}l$, respectively.

The geometry factors according to Eqs. 8.7 to 8.12 are plotted in Fig. 8.5 versus the aspect ratio w/d in a wide range, 10^{-3} to 10^2. Due to the logarithmic terms in the equations, the dependence on w/d is relatively weak. For large aspect ratios $w/d \gg 1$, the geometry factors of the polygonal loop become independent of w/d as reported in literature [29, 31]. For comparison, the geometry factor $\gamma_w = 0.5\ln(8d/w) - 1$ for the inductance $L_w = \gamma_w \mu_0 d$ of a circular wire-wound coil in the limit $w/d \ll 1$ is shown in Fig. 8.5 as green line [36]. In this case, w denotes the wire diameter and d the coil diameter.

The polygonal loop is commonly contacted at the outer edge, for example to the Josephson junctions or to an interconnect line. So far

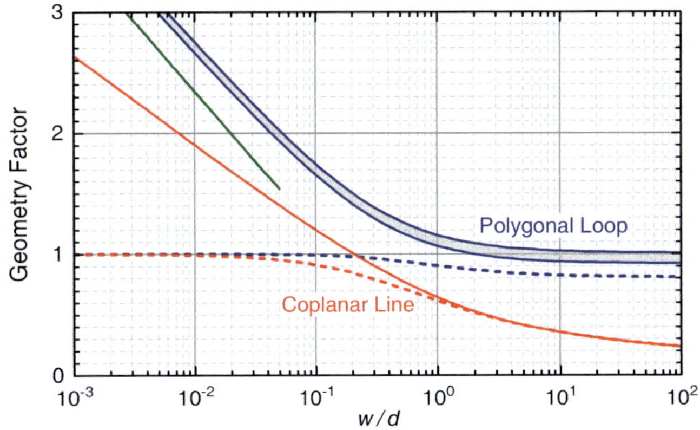

Figure 8.5 Calculated geometry factors of the structures in Fig. 8.4 plotted versus the aspect ratio w/d. Solid lines show the inductance (γ_L and $\gamma_{L'}$), dashed lines the effective area and slit width (γ_A and γ_d). The inductance of the polygonal loop also depends on the shape due to the term $2.7d/c$ in Eq. 8.7; here, the two limiting cases of square ($N = 4$, lower trace) and circular ($N \to \infty$, upper trace) shape are shown, respectively. For comparison, the geometry factor of a circular wire-wound coil is indicated as green line [36].

we have neglected the effect of the slit; that is, we have assumed an infinitely narrow slit. A finite slit width will add parasitic inductance L_{sl} and increase the effective area by $A_{sl,eff}$. This can be estimated by approximating the slit by a piece of coplanar line and using Eqs. 8.11 and 8.12 to calculate L_{sl} and $A_{sl,eff}$. As L' and d_{eff} of the coplanar line depend only weakly on the aspect ratio w/d, the actual choice of w/d is not crucial. For example, a typical value for the inductance of a slit $L' = 300$ nH/m was quoted in [28]. This value is obtained from Eq. 8.11 for $w/d = 86$. Decreasing or increasing w/d by a factor of 2 changes L' by only +12% or –10%, respectively.

Finally, the presented equations can also be applied to calculate an elongated loop with $d_{long} > d_{short}$. For this, the loop is approximated by a combination of a coplanar line of length d_{long}–d_{short}, "capped" on both ends by half a polygonal loop with inner dimension d_{short}. The total inductance and effective area are estimated from Eqs. 8.7–8.12 as the sum of the contributions from coplanar line and polygonal loop. This approach is used at PTB to estimate the inductance and effective area of the SQUIDs shown as examples in the following sections.

8.2 Making the SQUID a Practical Device

8.2.1 The Bare SQUID

So far we have discussed a "bare" SQUID, i.e., a device without coupling structures that just consists of a superconducting loop with two resistively-shunted Josephson junctions as schematically shown in Fig. 8.1a. Figure 8.6a shows an implementation of such a device that was intended as a miniature magnetometer for analyzing the residual magnetic field in cryogenic setups. The SQUID loop was designed to achieve a field sensitivity of 2 $\mu T/\Phi_0$ corresponding to an effective area of 1034 μm^2. A single-turn coil on top of the loop allows one to apply a feedback current I_F to the device (this is required for operation; see Section 8.3). The mutual inductance between this coil and the SQUID loop is $M_F = 47.5$ pH, corresponding to a current sensitivity $1/M_F = 43.5\ \mu A/\Phi_0$.

The lower trace in Fig. 8.6b shows the flux noise measured with a superconducting shield enclosing the device. The white noise

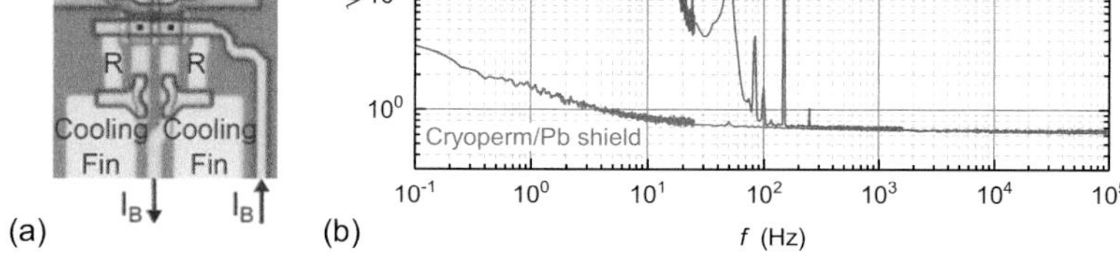

Figure 8.6 (a) Micrograph and (b) flux noise spectrum of a basic SQUID with a nominal field sensitivity of 2 $\mu T/\Phi_0$ (corresponding to an effective area of 1034 μm^2). The two Josephson junctions are indicated by black squares. The noise was measured at $T = 4.2$ K with and without a superconducting shield enclosing the device. The increased noise level without shield is due to pickup of environmental magnetic noise. The roll-off above about 10 Hz results from the screening effect of the metal liquid-helium transport dewar used for the measurement.

level of 0.65 $\mu\Phi_0/\sqrt{\text{Hz}}$ corresponds to a flux density noise level of 1.3 pT/$\sqrt{\text{Hz}}$, which is just comparable to a low-noise flux-gate magnetometer. Although this noise level is achieved with a very small device (which can be advantageous in some applications) one sees that the bare SQUID is not really a sensitive magnetometer. In Section 8.2.4 we will discuss how the magnetic field sensitivity of the SQUID can be improved by three to four orders of magnitude to achieve noise levels down to below 1 fT/$\sqrt{\text{Hz}}$.

Besides magnetometry, the other main application of SQUIDs is current sensing. For this, the signal current to be measured is passed through the single-turn feedback coil (now used as an input coil) and the resulting flux change in the SQUID loop is sensed via the SQUID voltage. The figure of merit for a current sensor is the current noise referred to the input coil. With the measured current sensitivity one obtains a current noise level of 28.3 pA/$\sqrt{\text{Hz}}$. This is too high for most applications. Furthermore, as shown by the upper trace in Fig. 8.6b, the low-frequency noise rises by orders of magnitude when operating the SQUID without magnetic shield. This results from the SQUID's sensitivity to magnetic fields. Although being too small for magnetic field sensing applications, it is by far too high for unshielded operation in current sensing applications. In the following section we will show, how the current noise can be improved while making the device less sensitive to environmental magnetic noise. Note that in Fig. 8.6b the flux noise without shield of about 300 $\mu\Phi_0/\sqrt{\text{Hz}}$ at 1 Hz corresponds to a flux density noise of 0.6 nT/$\sqrt{\text{Hz}}$. Values around 1 nT/$\sqrt{\text{Hz}}$ at 1 Hz are quite common at the PTB site located in an urban area (Berlin). This is about six orders of magnitude larger than the intrinsic noise level of an optimized SQUID magnetometer. The peak-to-peak power-line interference typically lies in the range of 100 nT to 1 μT.

8.2.2 Low-Inductance Current Sensors

The noise of a SQUID-based current sensor can be reduced by connecting a large number N_S of SQUIDs in series [37]. Provided that all devices in a SQUID series array (SSA) are identical and that the signal current is equally well coupled to all of them, the SSA behaves like a single SQUID with increased output voltage. As the noise

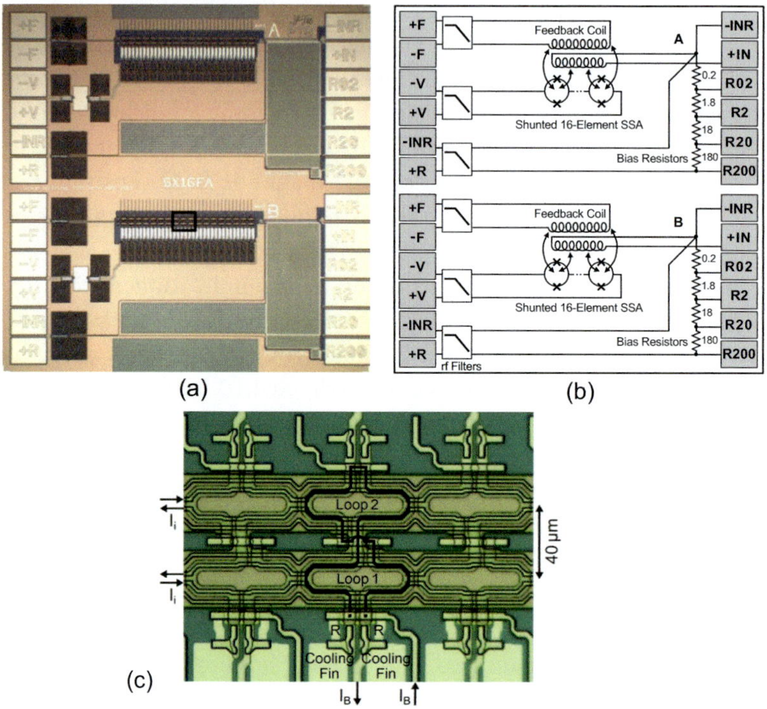

Figure 8.7 (a) Micrograph and (b) simplified equivalent circuit of a sensor chip with two separate 16-element SSAs and integrated bias resistors R_b between nominally 0.2 mΩ and 200 mΩ (PTB type X16FA). The displayed area is 3 mm × 3 mm. In (b) the SQUIDs are drawn as circles with two crosses indicating the Josephson junctions, and nominal resistance values are quoted in mΩ. In (c), a magnification is shown with three individual SQUID cells. The displayed region is marked in (a) by a black frame. The SQUIDs are configured as first-order series gradiometers as indicated by a bold black line. The different colors in (a) and (c) result from the microscopes used.

voltages across the individual SQUIDs of the array are uncorrelated, the total rms noise voltage scales with $\sqrt{N_S}$. In contrast, the voltage changes caused by the input signal add coherently and, hence, the total output voltage of the SSA increases linearly with N_S. Therefore, the signal-to-noise ratio of the SSA improves with $\sqrt{N_S}$, i.e., the effective rms flux and current noise levels scale with $1/\sqrt{N_S}$.

To make the SSA insensitive to magnetic fields, the individual SQUID loops may be configured as so-called first-order gradiometers, i.e., two equally large loops with different orientation are connected in series to get zero net flux if a homogeneous magnetic field is applied. An example of a chip containing two independent arrays of 16 SQUIDs each is depicted in Fig. 8.7 along with a simplified circuit diagram. The chip is an improved variant of the initial version described in [38]. First-order gradiometers are implemented as indicated by a bold black line in the magnification Fig. 8.7c. Elongated SQUID loops are used to obtain a good magnetic coupling with the single-turn input and feedback coils. A total SQUID inductance $L \approx 145$ pH is estimated using the analysis described in Section 8.1.3. The input inductance of the array is <3 nH. Inductor-resistor filters between the individual SQUID cells and shunt resistors across the input coils were implemented to obtain smooth and well-behaved array characteristics, in particular at millikelvin temperatures. Integrated bias resistors R_b between nominally 0.2 mΩ and 200 mΩ are intended for the readout of superconducting detectors (Section 8.4.4). The required resistor can be selected by wire-bonding to the corresponding pad. All lines to the room temperature electronics (left side of the chip) are passed through on-chip rf filters.

A critical issue for SSAs is flux trapping during cool-down. If the background flux in the individual SQUIDs differs due to the stray field of vortices trapped in the films, the voltage–flux characteristics do no longer add coherently and the overall characteristic can be severely distorted. Therefore, the linewidth in the critical parts of the array should be chosen such that vortices cannot enter the film during cool-down, and closed superconducting loops should be avoided which can trap flux due to screening currents or noise. The latter issue implies that arrays of parallel gradiometer SQUIDs [39] have strongly reduced cooling fields compared to series gradiometers and are therefore not recommended. It was shown [40] that complete vortex expulsion from narrow superconducting strips of width w occurs if the cooling field B_{cool} is kept in the range

$$B_{\mathrm{cool}} \lesssim \Phi_0/w^2. \tag{8.13}$$

Thus, to reliably cool down a SQUID array in the Earth's magnetic field ($\sim$50 μT) a maximum linewidth of about 5 μm should be used. For a 2.5 μm technology this means that only a single-turn input coil can be realized. Therefore, the input coils of the devices in Fig. 8.7 are just 2.5 μm wide lines on top of the narrow SQUID loops. The feedback lines (where the magnetic coupling needs not to be maximized) are located on both sides of SQUID the loops in a coplanar structure.

In the above considerations, we have assumed that for optimum performance all SQUIDs in the array should be identical. However, varying the loop size or input coil mutual inductance intentionally, one can obtain a nonperiodic transfer characteristic with a unique peak at zero flux [41–43]. These irregular arrays can be used to measure the absolute magnetic field in contrast to single SQUIDs or regular arrays that detect field changes only due to the Φ_0 periodicity in their characteristics. However, both regular and irregular SQUID arrays suffer from flux trapping which is an important issue for absolute field sensors. In this chapter, only regular arrays are discussed because they are much more common than irregular ones.

Figure 8.8 shows the flux noise obtained at 4.2 K with a 32-element SSA. This device is similar to the one in Fig. 8.7, but involves twice the number of SQUIDs and an extra circuit to reduce the current noise contribution of the room temperature preamplifier. A low white noise level of 0.17 $\mu\Phi_0/\sqrt{\text{Hz}}$ is found, corresponding to about 1 $\mu\Phi_0/\sqrt{\text{Hz}}$ in the individual SQUIDs. This is a typical value for well-designed SQUIDs with $L \approx 145$ pH and implies a noise energy $\varepsilon \approx 22\,h$. Due to the low flux noise, the input-referred current noise of about 4.5 pA/$\sqrt{\text{Hz}}$ is a factor of 6 lower than that of the bare SQUID in Fig. 8.6. Furthermore, the effective area is only about 1 μm^2 in all three spatial directions (in-plane and perpendicular to the chip), corresponding to a field sensitivity of about 2 mT/Φ_0. This is a factor of 1000 improvement over the bare SQUID in Fig. 8.6, and leads to a substantially smaller noise degradation when operating the SSA unshielded (upper trace in Fig. 8.8). Note that SSA current sensors are generally more susceptible to environmental noise pickup than single SQUIDs. This results from the fact that environmental noise (when homogeneous over the area of the SSA)

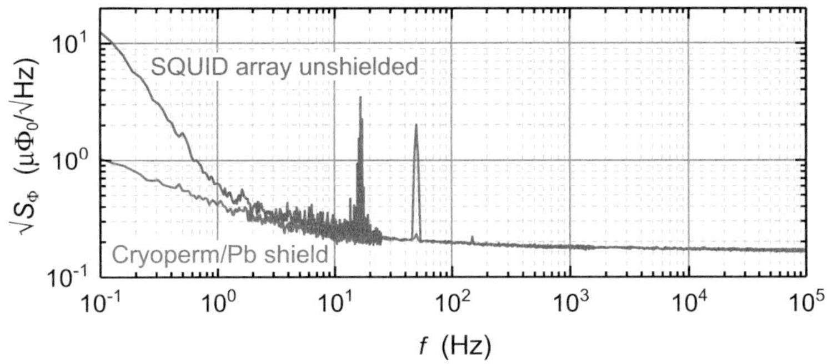

Figure 8.8 Flux noise spectrum of a 32-element SSA (PTB type X216FB) measured at $T = 4.2$ K with and without a superconducting shield enclosing the device. The SSA was optimized for current sensing applications and involves on-chip current feedback to reduce the effect of preamplifier current noise (see Section 8.3.3). The individual SQUIDs are of the same design as those in Fig. 8.7.

is added coherently, so that the resulting total excess flux noise is identical to that of a single SQUID. In contrast, the intrinsic flux noise of the SSA improves with increasing N_S compared to a single SQUID and, hence, the signal-to-noise ratio of the SSA degrades more strongly by noise pickup.

So far we have implied the common way of SQUID operation called current bias, where a constant current is passed through the device and the voltage across it is measured. Alternatively, the SQUID can be operated with voltage bias by connecting a voltage source in parallel to the device and sensing the current change through the SQUID caused by the applied magnetic flux. For sensing this output current, the SSAs described above are well suited because they have an adequate noise level, a low magnetic field sensitivity, and zero dc input impedance due to the superconducting input coil. The voltage source can be implemented by a low-value resistor through which a current is passed. Integrating all components on a single chip, a compact two-stage sensor can be realized that has overall characteristics like a single SQUID [38]. An example of such a sensor will be shown in the following Section 8.2.3. The SQUID bias modes and the dimensioning of a two-stage setup are discussed in detail in literature (for example in Chapter 2 of [3]).

8.2.3 High-Inductance Current Sensors

The SSAs described in the previous section are well suited for the readout of superconducting detectors (see Section 8.4.4) or as low-noise preamplifiers in two-stage setups. However, the current noise is in the pA/$\sqrt{\text{Hz}}$ range, and the input inductance is far below 1 μH. A large input inductance and a considerably increased current sensitivity can be achieved by using a multiturn input coil. Early SQUID devices in the 1960s and 1970s were equipped with wire-wound Nb coils. At around 1980, thin-film coupling schemes were introduced [29, 44–46]. The most widely used scheme is depicted in Fig. 8.9. A SQUID loop with a large linewidth w serves a so-called "washer" onto which a spiral multiturn coil is placed. As shown in Fig. 8.5, the inductance of a polygonal loop becomes independent of the aspect ratio w/d if the linewidth w is made much larger than the hole dimension d. Thus the design of the structure in Fig. 8.9 is relatively straight-forward: One selects the hole dimension to obtain the desired hole inductance L_h, and expands the outer dimension of the washer to accommodate the required number of turns N_i for the intended input coil inductance L_i.

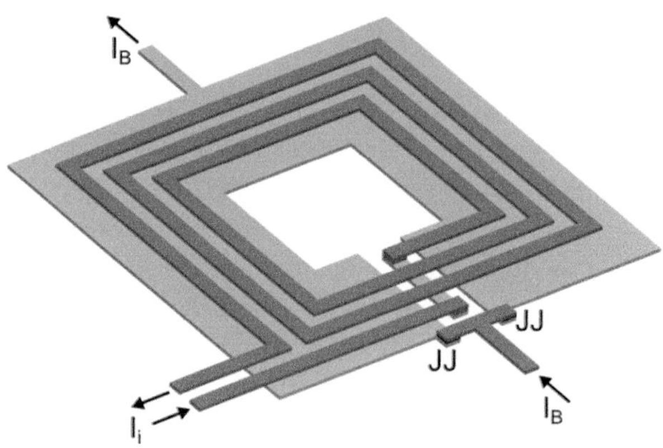

Figure 8.9 Thin-film coupling scheme developed by Ketchen and Jaycox [29]. In the original design the slit was covered by a superconducting plate, requiring three superconducting layers. Here, a simplified variant with two superconducting layers is shown. An input coil with three turns is drawn; in practice, up to about 100 turns are common. The shunt resistors in parallel to the Josephson junctions (JJs) are omitted for clarity.

For typical parameters, a high coupling constant $k = M_\mathrm{i}/(L_\mathrm{i}L)^{1/2} \approx 0.9$ is easily achieved. Neglecting coupling losses ($k \approx 1$), the mutual inductance between input coil and SQUID M_i and the input coil inductance L_i are given by

$$M_\mathrm{i} \approx N_\mathrm{i}L,\ L_\mathrm{i} \approx N_\mathrm{i}^2 L. \tag{8.14}$$

More detailed and accurate equations are given in [29]; however, Eq. 8.14 is often sufficient for dimensioning the SQUID. In practice, there is not much degree of freedom in the design. Once the hole size is fixed, one can basically only adjust the number of turns for obtaining the desired input inductance.

In Fig. 8.9 the tunnel junctions are located at the outer edge of the washer where the magnetic fields are low. The layout involves two superconducting layers only, which is achieved by wiring the return line of the input coil through the slit. This increases the inductance contribution of the slit, but the effect on the overall performance is modest as the input coil partially couples flux into the slit, thereby increasing the mutual inductance M_i together with the total SQUID inductance L. The slit can be covered by a superconducting plate to minimize its inductance contribution, but this introduces significant parasitic capacitance and requires a third superconducting layer [29]. Generally, microwave resonances in the structure are a severe problem. They can strongly distort the SQUID characteristics and increase the noise level substantially by down-mixing. As mentioned in Section 8.1.2, proper resonance damping is crucial to achieve a low noise level in practice [21–23]. Although low-noise operation was reported even without resonance damping (e.g., in [46]), resonance damping is strongly recommended because it makes the device more robust and tolerant against parameter variations.

To give an example, 71 turns are required according to Eq. (8.14) for coupling a 1 μH input inductance to a 200 pH SQUID loop. For a conventional fabrication process with 2.5 μm minimum linewidth and spacing, a 5 μm pitch in the spiral input coil and a total width $w \approx 355$ μm are obtained. Assuming $L' \approx 0.3$ pH/μm for the slit [28], the total contribution to the SQUID inductance $L_\mathrm{sl} \approx 107$ pH is about half the intended SQUID inductance. This example shows that with a 2.5 μm fabrication technology, the SQUID inductance cannot be lowered below about 100 pH when using the design in Fig. 8.9. As mentioned above, the slit could be covered, but this increases the parasitic capacitance and requires an extra superconducting layer. The increase in capacitance can easily undermine the benefit of a reduced SQUID inductance. When connecting several washers in parallel (typically two or four) and their input coils in series, the total SQUID inductance is reduced and the total input inductance increases [28, 47]. This relaxes the problem of large inductance ratios L_i/L. Gradiometric configurations with multiple washers are beneficial to reduce the sensitivity to external magnetic fields.

An elegant way to couple a large input inductance to a small SQUID inductance is to use an extra input transformer [23, 38, 48]. In the double-transformer scheme, the spiral coil on top of the SQUID washer is connected to the secondary (low-inductance) side of the input transformer, whose primary (high-inductance) side is used for the input coil. Typically, the input transformer is designed in the same way as the SQUID transformer, i.e., its low-inductance side acts as a "washer" for the multiturn input coil. Without coupling losses in the two individual transformers, the overall coupling is also ideal, $k = 1$. However, for finite coupling losses in the transformers, the overall coupling degrades relatively strongly. Unfortunately, the equations for the overall coupling constant are somewhat cumbersome. Assuming for simplicity that both transformers have the same coupling constant k_0 and that the inductance of the spiral coil on the SQUID washer is equal to the low-inductance side of the input transformer, the overall coupling constant is given by $k = k_0^2/(2 - k_0^2)$ [38]. Thus, for relatively high individual coupling constants k_0 of 90% or 80% the overall coupling constant k degrades to 68% or 47%, respectively. On the other hand, in the double transformer scheme the transformers require fewer turns, leading to reduced parasitic capacitance. The degradation in coupling is (partially) compensated by smooth, well-behaved SQUID characteristics and a low noise level.

An example of a practical SQUID with double-transformer coupling is shown in Fig. 8.10. It is an improved version of the device described in [38], and was fabricated using a Nb–AlO_x–Nb

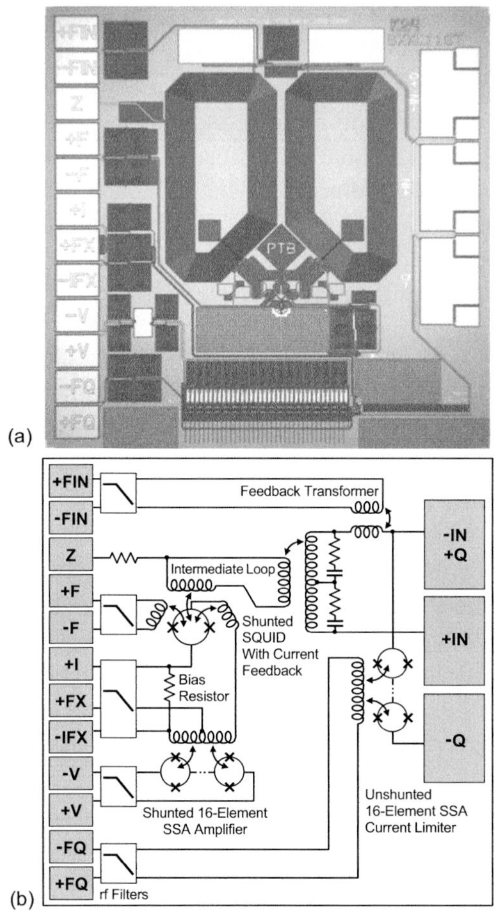

Figure 8.10 (a) Micrograph and (b) simplified equivalent circuit of a current sensor with an input inductance of about 1.8 μH (PTB type XXL116T). The integrated two-stage sensor involves double-transformer coupling and is equipped with an optional input current limiter (reproduced with permission from Drung et al. [38]). The displayed area is 3 mm × 3 mm.

trilayer process with two superconducting layers and a minimum lithographic feature size of 2.5 μm. It has a high input inductance $L_i \approx 1.8$ μH which is coupled to the SQUID inductance $L \approx 80$ pH with an overall coupling constant $k \approx 0.72$. Two large input transformers (each with a 40-turn input coil shunted by a resistor–capacitor series circuit) are wired as a first-order series gradiometer in order to reduce the sensitivity to homogeneous magnetic fields. Variants with fewer turns (down to 4 on each input transformer) were also implemented to cover the input inductance range down to about 24 nH. An additional feedback transformer in series to the input coil is intended for applications where the feedback current is applied to the input circuit (nulling the input current minimizes crosstalk in multichannel magnetometer systems [49]).

To protect the input against large currents in magnetic resonance experiments, an optional on-chip current limiter (Q spoiler [50]) may be used. It is realized in Fig. 8.10 by a series array of 16 unshunted 20 pH SQUIDs connected in series to the input coil. All lines connecting the sensor chip with the room temperature readout electronics are passed through on-chip rf filters. The SQUID is designed with four parallel loops, arranged as a second-order gradiometer. A 16-element SSA is integrated on the chip to act as a low-noise preamplifier. Current feedback between the SQUID sensor and the SSA is utilized to increase the overall gain (see Section 8.3.3). A typical white flux noise of 0.8 $\mu\Phi_0/\sqrt{\text{Hz}}$ is achieved at 4.2 K, resulting in a current noise level (referred to the input coil) of $\sim$0.2 $\text{pA}/\sqrt{\text{Hz}}$ for the nominal current sensitivity $1/M_\text{i} = 0.24\ \mu\text{A}/\Phi_0$. At $\lesssim$300 mK, the white noise typically drops to 0.25 $\mu\Phi_0/\sqrt{\text{Hz}}$.

An important figure of merit for superconducting current sensors is the coupled noise energy ε_c referred to the input inductance L_i rather than to the SQUID inductance L. Using the current noise density referred to the input coil $S_\text{I} = S_\Phi/M_\text{i}^2$ one obtains

$$\varepsilon_\text{c} = S_\text{I} L_\text{i}/2 = S_\Phi/(2k^2 L) = \varepsilon/k^2. \qquad (8.15)$$

The input-referred noise energy ε_c of the device in Fig. 8.10 is typically 50 h at 4.2 K and 5 h at $\lesssim$300 mK, respectively. Note that at 4.2 K single-stage sensors (devices without the SSA preamplifier) achieve nearly the same noise level; however, when operating a current sensor at millikelvin temperatures, a two-stage readout is required to avoid the noise being dominated by the room-temperature readout electronics (see Section 8.3).

8.2.4 Magnetic Field Sensors

The "traditional" way of realizing a sensitive magnetic field sensor is to connect a superconducting wire-wound pickup coil to the input coil of a SQUID current sensor. This scheme was introduced soon after the invention of the SQUID and is still widely used. Figure 8.11 shows two examples, a magnetometer and a first-order gradiometer. In the case of a magnetometer, the magnetic flux in the pickup coil, $\Phi_P = BA_P$, causes a screening current $\Phi_P/(L_P + L_i)$ which flows through the input coil, thereby generating a change in the magnetic flux Φ in the SQUID. The flux transfer coefficient is

$$\Phi/\Phi_P = M_i/(L_P + L_i) \approx 0.5k(L/L_P)^{1/2}. \qquad (8.16)$$

For the optimization we assume that for given SQUID inductance L the input coil inductance L_i is varied by the number of input coil turns N_i while the coupling constant $k = M_i/(L_i L)^{1/2}$ remains constant. Under this condition, the flux transfer Φ/Φ_P is maximized for matched inductances $L_i = L_P$. The approximation on the right side of Eq. (8.16) is obtained for the optimum case $L_i \approx L_P$. Note that in practice it is not very crucial to exactly keep the matching condition because the optimum is quite wide. Being a factor of two away from optimum ($L_i/L_P = 0.5$ or 2), reduces the flux transfer by 5.7% only. Furthermore, when considering the resulting field noise, one has to include the fact that the effective SQUID inductance is reduced in the presence of the pickup coil by screening effects. The detailed optimization is rather complicated [23]. Fortunately, in practice the straight-forward condition $L_i \approx L_P$ is generally sufficient for dimensioning a magnetometer.

Knowing the flux transfer coefficient, the effective area of the magnetometer $A_{eff} = A_P \times \Phi/\Phi_P$ can be calculated from the pickup coil area A_p. The resulting noise is given by

$$S_B = S_\Phi/A_{eff}^2 \approx \varepsilon_c 8 L_P/A_P^2. \qquad (8.17)$$

The approximation on the right side is valid for the matched case $L_i \approx L_P$. For fixed pickup coil parameters L_P and A_P, the noise scales with the coupled noise energy ε_c. The self-inductance of the wire-wound pickup coil can be calculated with the approximation $L_P \approx \mu_0 d_P[0.5\ln(8d_P/w_P) - 1]$ given in [36] (d_P and w_P are the diameters of the coil and the wire, respectively). The inductance

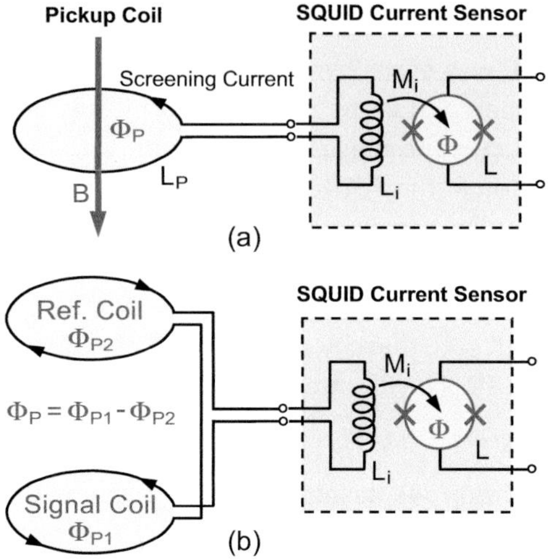

Figure 8.11 Coupling a wire-wound pickup coil to a SQUID current sensor: (a) Magnetometer and (b) axial first-order gradiometer.

contribution of the interconnect lines between pickup coil and SQUID can be included in the total pickup coil inductance L_P. Assuming for example a circular single-turn pickup coil with 20 mm coil diameter, 0.1 mm wire diameter, and a 16 cm twisted pair with $L' = 4$ nH/cm between pickup coil and SQUID, the total pickup coil inductance amounts to 132 nH. Coupling this to a SQUID with $\varepsilon_c \approx 50\ h$ will result in a noise level $\sqrt{S_B} \approx 0.6$ fT/$\sqrt{\text{Hz}}$ according to Eq. 8.17. In this example, the dimensioning was suboptimal because the contribution of the twisted pair was relatively high (about a quarter of the total inductance $L_P + L_i$). Using a multiturn pickup coil and a correspondingly increased input coil inductance would reduce the noise even further. Thus we conclude that with modern dc SQUIDs it is relatively straight-forward to achieve noise levels well below 1 fT/$\sqrt{\text{Hz}}$. In practice, the system noise level is usually limited by thermal noise currents in the superinsulation of the dewar containing the magnetometer (see Section 8.4.2).

Magnetometers are useful in extremely well shielded environments only. With moderate shielding, the effect of environmental interference can be reduced by wiring two identical pickup coils to a first-order gradiometer as shown in Fig. 8.11b. The source to be measured (e.g., the human brain or heart) is located as close as possible to one of the coils, the signal coil. Due to the strong decrease of the source's magnetic field with distance, the other coil (the reference coil) will "see" only a weak signal, i.e., the net flux will be only slightly reduced compared to a magnetometer. In contrast, the effect of a homogeneous magnetic field is suppressed because the flux contributions Φ_{P1} and Φ_{P2} in the two pickup coils cancel each other out. Thus a gradiometer strongly reduces the effect of remote noise sources that have small spatial derivatives compared to those of the local signal source. The SQUID itself is commonly housed in a well shielded package, sufficiently far away from the pickup coils to avoid distortion of the magnetic fields. An example of a practical realization is given in Fig. 8.12. The remote location of the SQUID package is beneficial in applications where the object under investigation is exposed to large magnetic fields that would distort the SQUID function (for example in magnetic resonance experiments).

First-order gradiometers are often adequate in moderate magnetic shielding (for example a chamber with two layers of high-permeability material plus one eddy-current screen of Al). Higher order gradiometers (second or third) are required for magnetically unshielded measurements. Generally, an "ideal" gradiometer of n-th order is sensitive to the n-th and higher spatial derivatives of the applied field, but suppresses all lower spatial derivatives including the homogeneous field component. In practice, imbalance occurs due to slightly different coil areas or tilt angles, so that a "real" gradiometer is also sensitive to the homogeneous field component in all three spatial directions. In multichannel systems, the imbalance can be compensated by adding a set of reference channels (magnetometers and lower-order gradiometers).

Gradiometers cannot only be realized "in hardware" as depicted in Fig. 8.11, but also electronically by combining the analog outputs of different channels [51] or "in software" after digitizing the output signals of the SQUID readout electronics. The latter approach is very powerful and efficient, and typically used in large biomagnetic multichannel systems. Considering the dynamic range of the associated electronics, a suitable combination of passive and active methods (i.e., shielded room and gradiometric configurations)

Figure 8.12 Example of a commercial SQUID package. The superconducting connection to the input coil is realized via screw contacts. For low-noise operation, the SQUID carrier is enclosed by a Nb shield. On the left side, the socket for the wiring to the room temperature electronics is visible. Picture courtesy of Magnicon GmbH, Hamburg.

yields the best overall performance and can suppress environmental interference by about five orders of magnitude. A comprehensive review of shielding and noise cancellation issues with the focus on biomagnetic applications is found in Chapter 3 of [3].

The wire-wound pickup coil can be substituted by a thin-film coil in order to get a compact single-chip magnetometer or gradiometer. For integrated devices, the sensitivity can be maximized by applying the multiloop concept rather than the transformer coupling of Fig. 8.9. The basic idea is quite simple: the SQUID loop is directly used for collecting flux, but a large number of loops is connected in parallel to reduce the effective SQUID inductance to an acceptable level. The effective area of the complete device is given by the area of the individual loops. A theoretical description and several device examples are given in [20]. The approach of fractional-turn loops was first implemented in bulk Nb SQUIDs in the early 1970s [52]. It was later attempted as an alternative to the washer structure [45], but has not become established for SQUID current sensors. In the early 1990s, the concept was successfully applied to thin-film magnetometers [53]. A few years later, record noise levels of 1.13 $\mathrm{fT}/\sqrt{\mathrm{Hz}}$ were obtained with a device implemented on a 7.2 mm $\times$ 7.2 mm chip [33]. The lowest noise reported so far is 0.33 $\mathrm{fT}/\sqrt{\mathrm{Hz}}$ for a 12.5 mm $\times$ 12.5 mm device involving sub-micrometer cross-type tunnel junctions [54].

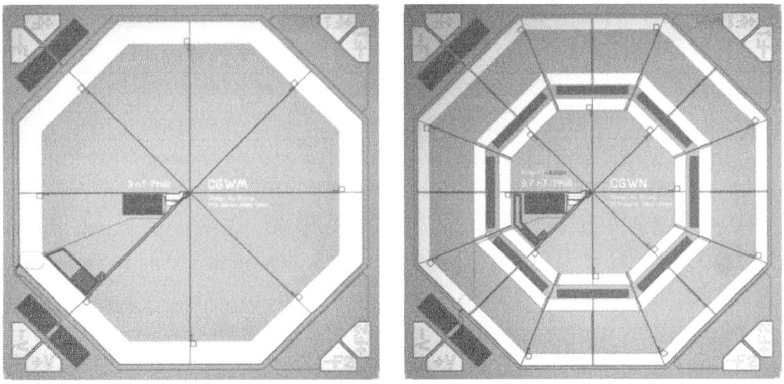

(a) (b)

Figure 8.13 Examples of integrated multiloop devices. (a) Magnetometer (PTB type WM) with a field sensitivity of 3 nT/Φ_0, (b) concentric gradiometer (PTB type WN) intended for noise thermometry. The displayed area is 3 mm × 3 mm.

Figure 8.13 shows two examples of multiloop SQUIDs, a magnetometer and a concentric gradiometer. The magnetometer uses eight parallel-connected loops, yielding an effective SQUID inductance of about 120 pH. In spite of its small size (2.8 mm outer dimension, 6.5 mm^2 area), the device achieves an effective area of 0.69 mm^2 corresponding to a field sensitivity of 3 nT/Φ_0. The typical white noise level is 3 fT/$\sqrt{\text{Hz}}$. The gradiometer in Fig. 8.13b was optimized for noise thermometry (cf. Section 8.4.3). It involves a total of 8 inner loops and 16 outer loops. Each inner loop is connected in series with two parallel-connected outer loops. The resulting eight coil combinations are connected in parallel, resulting in a low effective SQUID inductance of about 130 pH. The nominal field sensitivity of the inner loops is 9.7 nT/Φ_0.

8.3 SQUID Readout

In this Section, we describe the two most common concepts for operating a SQUID, flux modulation and direct readout. Digital SQUIDs are not discussed because they are rarely used in practice. Bias current reversal schemes for the suppression of critical current fluctuations are also omitted because modern Nb–AlO_x–Nb tunnel junctions typically do not require them. Other types of all-refractory junctions may show large levels of excess low-frequency noise from critical current fluctuations, for example NbN junctions with MgO barrier [55] or Nb junctions with barriers from amorphous silicon [56] or HfTi [57]. SQUIDs involving these junctions often require bias reversal schemes for low-frequency applications. Also, high-T_c SQUIDs are almost always operated with bias current reversal. Detailed reviews on SQUID readout including the various bias reversal schemes are found in literature [4, 58].

8.3.1 Flux-Locked Loop Basics

In principle, a SQUID can be operated in a small-signal mode around the optimum working point W which is typically located near the steepest part of the V–Φ characteristic (the inflection point). As illustrated in Fig. 8.14a, a small change in the applied flux $\delta\Phi$ will produce a proportional change in the voltage $\delta V = V_\Phi \delta\Phi$ ($V_\Phi = \partial V/\partial\Phi$ is the transfer coefficient at the working point). However, the proportionality between voltage and flux is maintained only for very small $\delta\Phi$, and the output becomes strongly distorted if the applied flux exceeds the linear flux range Φ_{lin}, which is typically a few percent of a flux quantum only. As SQUIDs are commonly applied to measure weak signals, this small dynamic range might just be sufficient. However, in practice there are usually much larger disturbing signals (e.g., the 50 Hz or 60 Hz power line interference) superimposed to the measurement signal, which makes a small-signal readout usually impossible unless the SQUID is very well shielded. Further disadvantages of the small-signal readout are that the transfer coefficient V_Φ depends on the bias settings of the SQUID, and that the SQUID noise increases if the applied flux shifts the working point too far away from optimum.

The dynamic range can be considerably increased by negative feedback. The basic circuit of the so-called flux-locked loop (FLL) is depicted in Fig. 8.14b. The SQUID is biased at the working point W as in the small-signal readout. The deviation of the SQUID voltage V from that at the working point V_b is amplified, integrated, and fed back into the SQUID via a feedback resistor R_F and a

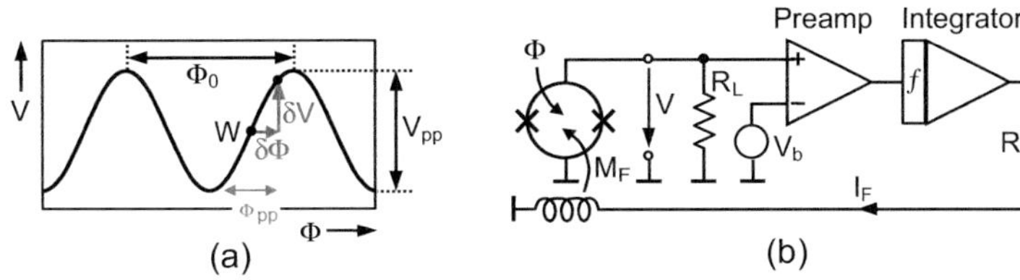

Figure 8.14 Fundamentals of SQUID readout: (a) V–Φ characteristic and (b) basic flux-locked loop (FLL) circuit. The SQUID is drawn as a circle with two crosses indicating the resistively-shunted Josephson junctions, and the bias current source is omitted for clarity.

feedback coil that is magnetically coupled to the SQUID via a mutual inductance M_{F}. Commonly, feedback resistances in the kΩ range are used, making the impedance of the feedback coil negligible in the frequency range of interest. For infinite integrator gain, the flux in the SQUID is kept constant by the negative feedback and the voltage V_{F} across the feedback resistor depends linearly on the applied flux. In this case, the transfer coefficient of the flux-locked SQUID

$$G_{\mathrm{FLL}} = \partial V_{\mathrm{F}}/\partial\Phi = -R_{\mathrm{F}}/M_{\mathrm{F}} \tag{8.18}$$

becomes independent of the working point. The noise does not degrade with applied flux because the SQUID is always kept at the chosen working point. However, the integrator gain decreases with frequency, and deviations occur at high frequencies due to the reduced open-loop gain G_{OL}. For the analysis of the FLL dynamics we assume the most common case of an integrator with a single pole in the frequency response (other types of integrators are described in [3, 4]). In this case, the room temperature electronics (preamplifier plus integrator) has an overall gain $|\partial V_{\mathrm{F}}/\partial V| = f_{\mathrm{GBW}}/f$ and is fully characterized by the gain-bandwidth product f_{GBW}. The SQUID can be considered as a current-to-voltage converter with a transresistance

$$A_{\mathrm{TR}} = \partial V/\partial I_{\mathrm{F}} = V_{\Phi} M_{\mathrm{F}}/(1 + R_{\mathrm{dyn}}/R_{\mathrm{L}}), \tag{8.19}$$

where R_{dyn} is the dynamic resistance of the SQUID at the working point and R_{L} is the input resistance of the readout electronics. Often the term in parentheses can be neglected because $R_{\mathrm{dyn}} \ll R_{\mathrm{L}}$. For wideband systems, however, the transmission lines between the

SQUID and the room temperature amplifier should be terminated (or at least be resistively shunted). The electronics in [59] has $R_L = 50\ \Omega$ (realized by negative feedback for minimum noise) which is comparable to the R_{dyn} of PTB's 16-element SSAs. In this case, the complete Eq. 8.19 has to be used.

To analyze the FLL dynamics, we first assume that the feedback loop is opened (for example by disconnecting the feedback resistor from the integrator output). Knowing A_{TR}, the overall gain of the open feedback loop can be calculated as

$$|G_{OL}| = f_1/f \text{ with } f_1 = f_{GBW} A_{TR}/R_F. \tag{8.20}$$

The open-loop gain $|G_{OL}|$ scales inversely proportional to frequency. It falls to unity at the unity-gain frequency f_1. Now we assume that the feedback loop is closed to obtain FLL operation. The idealized FLL in Fig. 8.14b exhibits a first-order low-pass response with a 3 dB bandwidth $f_{3dB} = f_1$. The 3 dB bandwidth is the frequency at which the amplitude falls to $1/\sqrt{2}$ or –3 dB. Note that Eq. 8.20 was derived for the common case of a current-biased SQUID. With voltage bias, a similar analysis can be performed by describing the SQUID as a current-to-current converter and the feedback electronics as a current-to-voltage converter with a transresistance proportional to $1/f$. The expression for f_1 will differ from Eq. 8.20, but once f_1 is fixed, the dynamic behavior of the FLL is given independent of the way the SQUID is biased. Furthermore, the noise is generally not influenced by the SQUID bias mode. Therefore, all considerations hereinafter will apply for both bias modes.

In the basic FLL circuit in Fig. 8.14b, the 3 dB bandwidth can be made arbitrarily large by increasing f_1. In practice, however, an upper limit is imposed by phase lag in the transmission lines and the readout electronics. A simple but efficient model describes the combined parasitic effects in the FLL by an effective dead time t_d [3, 4]. It was shown that with finite dead time the unity-gain frequency is limited to

$$f_{1,max} = 0.08/t_d \approx f_{3dB,max}/2.25. \tag{8.21}$$

For larger values of f_1, the FLL exhibits a peak in the frequency response or even becomes unstable (oscillation at $\approx 0.25/t_d$). According to Eq. 8.21, the maximum FLL bandwidth $f_{3dB,max}$ is a factor of 2.25 larger than $f_{1,\max}$. A relative increase in $f_{3\mathrm{dB},\max}/f_{1,\max}$ of up to a factor of 3 is common for wideband systems with feedback (not only for SQUIDs). Note that f_1 determines the open-loop gain and hence the linearization effect, not the FLL bandwidth $f_{3\mathrm{B}}$. Unfortunately, commonly $f_{3\mathrm{B}}$ is quoted rather than f_1 which gives a too optimistic view of the dynamic performance. For example, if one likes to have an open-loop gain of >4, the highest signal frequency would be $f_1/4 = f_{3\mathrm{dB}}/9$. Thus, in this example the maximum signal frequency is about one order of magnitude smaller than the FLL bandwidth $f_{3\mathrm{dB}}$. Generally, for signal frequencies close to $f_{3\mathrm{dB}}$, FLL operation is not recommended because the feedback loop does not reduce the nonlinear distortion due to phase lag, but rather increases it. In such cases, the high-frequency signal of interest is preferably measured in a small-signal readout, and a "slow" FLL could be used to suppress environmental interference at frequencies much below that of the signal.

For a typical separation of 1 m between SQUID and room temperature electronics, the dead time in the cables amounts to $t_\mathrm{d} \approx 10$ ns. The resulting limits according to Eq. 8.21 are $f_{1,\max} \approx 8$ MHz and $f_{3\mathrm{dB},\max} \approx 18$ MHz, respectively. State-of-the-art readout electronics allow FLL dynamics close to these limits [59]. In 2006, a prototype FLL with SiGe transistors was operated in liquid helium nearby a 16-element SSA [38]. Due to the short distance and the wideband setup, a very small dead time $t_\mathrm{d} \approx 0.65$ ns was achieved. The measured small-signal bandwidth of 350 MHz was even higher than $f_{3\mathrm{dB},\max} \approx 280$ MHz predicted from Eq. 8.21, suggesting that the ratio $f_{3\mathrm{dB},\max}/f_{1,\max}$ was probably about 2.8 instead of 2.25. Two years later, the cold semiconductor feedback loop was substituted by a large series-parallel array of 640 SQUIDs acting as a current amplifier [60]. A low-frequency open-loop gain of about 20 and a unity-gain frequency $f_1 > 200$ MHz were reported. In both experiments, coaxial lines between 4.2 K and room temperature were mandatory. Although these experiments clearly confirm the dead time model, such high bandwidth is commonly not needed. In most cases, the commercial variant of the electronics in [59] provides sufficient bandwidth ($f_{3\mathrm{dB},\max} \approx 20$ MHz), even when connecting the SQUID to the readout electronics via 1 m long twisted wires.

In the above discussion it was implied that f_{GBW} can be selected sufficiently high to reach $f_{3\mathrm{dB,max}}$. For wideband systems, very high values of f_{GBW} may be required depending on the SQUID. For example, the electronics in [59] allows one to select f_{GBW} up to 7.2 GHz. The full bandwidth can be achieved at acceptable feedback resistances in the range of 10 kΩ even for SQUIDs with a small A_{TR}. Decreasing the feedback resistance to boost the bandwidth lowers the output signal amplitude and increases the demands on the data acquisition system following the analog output of the FLL.

Another important parameter is the slew rate, i.e., the maximum temporal change in the feedback flux $|\partial\Phi_{\mathrm{F}}/\partial t|_{\mathrm{max}}$. In practice, the slew rate is often more critical than the bandwidth of the FLL. It is commonly measured by applying a sinusoidal signal flux and increasing the amplitude until the FLL becomes unstable or the output saturates. Here, we discuss the slew rate at high signal frequencies, i.e., in the regime where it is not limited by the static feedback range. For the basic FLL with one-pole integrator one obtains

$$\dot{\Phi}_{\mathrm{F,max}} = \pi\Phi_{\mathrm{pp}} f_1 \lesssim \Phi_0 f_1 \text{ with } \Phi_{\mathrm{pp}} = V_{\mathrm{pp}}/|V_\Phi| \lesssim \Phi_0/\pi. \quad (8.22)$$

The maximum slew rate with one-pole integrator is frequency-independent. It is proportional to the unity-gain frequency f_1 and the peak–peak flux Φ_{pp} calculated from the peak–peak voltage V_{pp} according to the right side of Eq. 8.22. One sees that a high slew rate requires a high intrinsic linearity. For a sinusoidal V–Φ characteristic (which is often a useful first approximation) one obtains the practical upper limit $\Phi_{\mathrm{pp}} = \Phi_0/\pi$ and a resulting slew rate $\Phi_0 f_1$, i.e., in the best case one obtains about 1 Φ_0/μs per megahertz of bandwidth. Note that the usable linear range Φ_{lin} for the small-signal readout is much smaller than Φ_{pp} defining the "intrinsic linearity" for calculating the slew rate. Driving a SQUID to $\pm\,\Phi_{\mathrm{pp}}$ would cause very high dynamic distortions. In fact, the error flux in the SQUID becomes large when approaching the slew rate limit, but the net effect with feedback is strongly reduced if the open-loop gain is high at the chosen signal frequency. The sinusoidal characteristic depicted in Fig. 8.14 has symmetric voltage swings $\pm V_{\mathrm{pp}}/2$ around the working point W. For asymmetric characteristics, V_{pp} in Eq. 8.22 has to be replaced by twice the smaller voltage swing.

Beside linearization of the transfer function, another important task of the readout electronics is to amplify the weak signal from the SQUID without adding (too much) noise. The preamplifier in the readout electronics is characterized by a noise voltage $V_{\mathrm{N,amp}}$ and a noise current $I_{\mathrm{N,amp}}$ at its input. The corresponding power spectral densities are $S_{\mathrm{V,amp}}$ and $S_{\mathrm{I,amp}}$, respectively. It is commonly assumed that voltage noise and current noise are uncorrelated, which is not exactly true but a reasonable assumption to simplify noise analysis. The amplifier's noise voltage is superimposed to the measured SQUID voltage which increases the total noise. The amplifier's noise current flows through the SQUID, thereby generating a voltage drop via the dynamic resistance R_{dyn} of the SQUID at its working point. To describe the effect of preamplifier current noise, it is often convenient to use the current sensitivity

$$M_{\mathrm{dyn}} = R_{\mathrm{dyn}}/V_{\Phi} \approx \pm(1\ldots 2)L. \tag{8.23}$$

M_{dyn} depends less strongly on the working point and on parasitic effects in the SQUID (e.g., due to resonances in the input coil) than R_{dyn}. Roughly speaking, M_{dyn} scales with the SQUID inductance L. One obtains values of $|M_{\mathrm{dyn}}|$ between L and $2L$ for low-T_{c} SQUIDs covering a wide range of SQUID inductances between 7 pH and 400 pH [3]. The total noise of the SQUID including preamplifier noise is given by

$$S_{\Phi,t} = S_{\Phi} + S_{\mathrm{V,amp}}/V_{\Phi}^2 + S_{\mathrm{I,amp}}M_{\mathrm{dyn}}^2. \tag{8.24}$$

In Eq. 8.24 the effect of amplifier voltage and current noise is expressed as effective flux noise contributions via V_{Φ} and M_{dyn}. It is convenient to measure V_{Φ} and M_{dyn} in the FLL mode by superimposing small test signals to the SQUID bias voltage and current, respectively, and calculating the resulting flux change from the FLL output voltage change. This way, V_{Φ} and M_{dyn} are determined under the conditions of the noise measurement (that is always performed in the FLL mode) and the preamplifier noise contributions can be accurately determined. The transfer coefficients V_{Φ} and M_{dyn} are also applicable to more complex circuits, for example a two-stage SQUID setup. In practice, one considers the SQUID as a "black box" and measures V_{Φ} and M_{dyn} without regarding the actual type of circuit.

From Eq. 8.24 the requirements for the preamplifier can be deduced. Assuming a typical Nb-based SQUID at 4.2 K with $V_\Phi = 100\ \mu\text{V}/\Phi_0$ and $1/M_{\text{dyn}} = 10\ \mu\text{A}/\Phi_0$, the preamplifier voltage and current noise levels should be $<0.1\ \text{nV}/\sqrt{\text{Hz}}$ and $<10\ \text{pA}/\sqrt{\text{Hz}}$ for keeping the amplifier contributions below the typical SQUID noise of $1\ \mu\Phi_0/\sqrt{\text{Hz}}$. The current noise requirement is easily met, but the amplifier's voltage noise is a severe issue. For arrays with N_S SQUIDs in series, the flux noise density S_Φ of the SSA scales with $1/N_\text{S}$, V_Φ is proportional to N_S and M_{dyn} is independent of N_S. As a result, the amplifier voltage noise term in Eq. 8.24 falls with N_S^2 but the current noise term remains constant and becomes the limiting contribution to $S_{\Phi,t}$ for large value of N_S. We see that amplifier noise can substantially degrade the overall noise performance. In the following sections we will describe the two most common methods to minimize the amplifier's noise contribution.

8.3.2 Flux Modulation Readout

Amplifier noise effects are conveniently discussed on the basis of the noise temperature [58]. For a resistive source, the total effect of amplifier noise is converted into an equivalent increase in the source resistor's temperature. For given voltage and current noise levels, the amplifier noise temperature depends on the source resistance. It has a minimum T_{min} for an optimum source resistance R_{opt} for which the contributions from amplifier voltage and current noise are equal: $R_{\text{opt}} = (S_{\text{V,Amp}}/S_{\text{I,amp}})^{1/2}$. For lowest noise, the amplifier should be designed such that R_{opt} is matched to the dynamic resistance of the SQUID R_{dyn}. The resulting amplifier noise temperature should be sufficiently low compared to the noise temperature of the SQUID, which is about four times its operation temperature according to simulations [13, 14].

The achievable T_{min} and R_{opt} depend of the type of transistors used in the input stage of the amplifier. Bipolar transistors allow low values $R_{\text{opt}} \approx 50\ \Omega$, but have a relatively a high T_{min} between about 30 K and 100 K (about 200 K are possible for $R_{\text{opt}} \approx 10\ \Omega$). These noise temperatures are acceptable for the readout of high-T_c SQUIDs, but for low-T_c SQUIDs special measures have to be adopted to boost the SQUID output (Section 8.3.3). In contrast, amplifiers based on junction-field effect transistors (JFETs) allow very low noise temperatures down to $T_{\mathrm{min}} \approx 1$ K at source resistances above about 1 kΩ. Therefore, the noise of JFET-based amplifiers is sufficiently low for SQUID readout, but there is a large mismatch between R_{opt} and typical values of R_{dyn}.

The straightforward method for impedance matching is the utilization of a cold transformer in a flux-modulated readout scheme [61, 62]. As shown in Fig. 8.15, a square-wave modulation flux $\Phi_{\mathrm{mod}}(t)$ is applied to the SQUID to toggle between two working points W$^+$ and W$^-$ placed at adjacent slopes of the V–Φ characteristic. Without applied flux ($\delta\Phi = 0$), zero voltage across the SQUID is obtained. Applying a positive flux $\delta\Phi > 0$ results in a square-wave SQUID voltage $V(t)$ which is out-of-phase to the modulation flux $\Phi_{\mathrm{mod}}(t)$, as illustrated in Fig. 8.15a. Accordingly, a negative flux $\delta\Phi < 0$ leads to a SQUID voltage in phase with the modulation flux. Thus, the applied flux can be sensed by synchronously detecting the SQUID voltage at the modulation frequency. This is commonly done with a synchronous switch (or a mixer in the case of wideband systems) after amplifying the SQUID output with a cold transformer followed by the room temperature preamplifier. The output of the lock-in detector is integrated and send back as a current into a feedback coil in order to counterbalance the flux applied to the SQUID. As for the basic FLL circuit in Fig. 8.14, the output voltage V_{F} represents the linearized output signal.

The resistance "seen" by the room temperature amplifier increases with the square of the transformer's turn ratio. For a suitably chosen turn ratio, the low SQUID impedance is noise-matched to the amplifier, resulting in minimum overall noise. A small resistance may be placed between the SQUID and the transformer primary to avoid that the SQUID is shorted by the transformer (the SQUID is operated with voltage bias rather than current bias).

The flux-modulation technique was introduced soon after the invention of the SQUID [61] and was the only practical readout method until the early 1990s when direct readout schemes emerged. The cylindrical dc SQUID of Clarke et al. [62] (which was a first milestone in the development of reliable thin-film devices) involved a cold inductor–capacitor resonant circuit. However, this is presently uncommon due to the reduced bandwidth compared to transformer

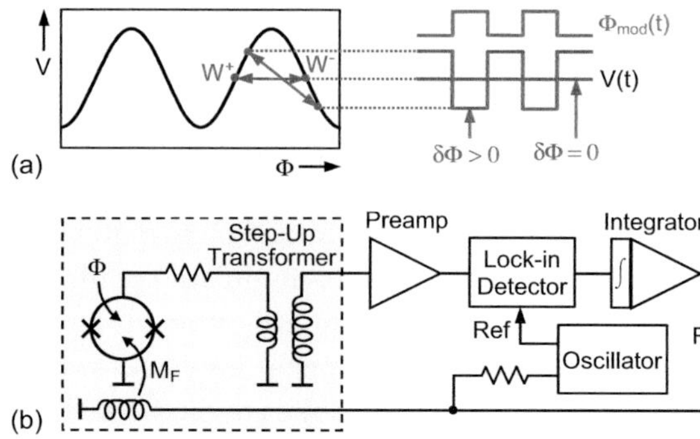

Figure 8.15 SQUID readout with flux modulation: (a) V–Φ characteristic and (b) FLL circuit. A square-wave modulation flux Φ_{mod} toggles the SQUID periodically between working points W^+ and W^- with positive and negative transfer coefficient V_Φ. Components inside the dashed box are at cryogenic temperature. The dc source for biasing the SQUID is omitted for clarity.

coupling. Flux-modulation is an efficient way to read out SQUIDs with smooth, well-behaved V–Φ characteristics. In practice, the noise might increase if the characteristics are strongly asymmetric due to parasitic resonances in the input coil structure. Typical modulation frequencies range between 100 kHz and 500 kHz where FET amplifiers have excellent noise performance. At higher frequencies, the current noise rises due to parasitic capacitance in the transistors, and the noise temperature correspondingly degrades. Square-wave modulation is ideal in terms of noise because the SQUID is always biased at points with best noise. However, in particular at high modulation frequencies, switching spikes can increase the noise due to down-mixing. Sinusoidal modulation circumvents these problems but increases the noise because the SQUID dynamically passes through points with reduced and even no sensitivity.

The main restriction of flux-modulation readout is a limited FLL bandwidth. Obviously, the maximum FLL bandwidth is lower than the modulation frequency, which also results in a reduced slew rate. Early systems used modulation at 100 kHz. In 1984, a first wideband system with 500 kHz square-wave modulation was reported involving two transformers (one cooled and the other at room temperature) [63]. In the mid-1990s, wideband

SQUID electronics with 16 MHz flux modulation were developed using a resonant superconducting thin-film transformer [64] or a nonresonant terminated transmission-line transformer [65]. A closed-loop bandwidth exceeding 2.5 MHz and a slew rate greater than 1 $\Phi_0/\mu s$ at frequencies up to 1 MHz were reported in [64], which are roughly consistent with the dead time $t_d \approx 100$ ns deduced from the measured phase response. An even higher modulation frequency of 33 MHz was reported for a high-T_c system involving two transformers and 56 cm long 50 Ω cables between 77 K and 300 K [66]. This high modulation frequency enabled an FLL bandwidth of 10 MHz and slew rates of up to about 10 $\Phi_0/\mu s$ which are record values for flux modulated systems. However, modulation frequencies substantially above 1 MHz are inconvenient in practice due to increased complexity and high demands on the wiring between the cryogenic part and the room temperature electronics. Therefore, flux-modulated SQUIDs are commonly operated at modulation frequencies well below 1 MHz.

8.3.3 Direct Readout

In the early 1990s, direct readout schemes without flux modulation were developed, stimulated by the need to simplify the electronics of biomagnetic multichannel systems [51, 53, 67, 68]. Nowadays, direct readout is widely used, in particular when applying SQUIDs as preamplifiers for superconducting detectors. Over the past decades, various concepts for amplifier noise reduction were introduced by different research groups. This has led to a quite confusing diversity of acronyms and notations. Recently, a general approach for understanding and analyzing direct readout schemes for SQUIDs was published [69]. It was pointed out that all existing methods for suppression of room temperature amplifier noise are based on feeding the SQUID voltage and/or current back into the SQUID loop. Voltage and current feedback were introduced in the early 1990s under the names *additional positive feedback* (APF) [53] and *bias current feedback* (BCF) [51], respectively. It was further shown in [69] that direct SQUID readout schemes can be conveniently analyzed by considering the SQUID and the amplifier separately. This approach allows an intuitive understanding of the various readout concepts reported in literature, and leads to simple mathematical expressions for the expected overall behavior.

It was stressed in [69] that the noise suppression does not depend on the way the SQUID is biased (constant current or voltage). The bias mode can be selected independently from the noise optimization according to the requirements in dynamic range and linearity. Current bias is more straightforward, but voltage bias yields a better intrinsic linearity of the SQUID and thus a better slew rate at given bandwidth. For wideband systems, the cable between the SQUID and the amplifier should be terminated (or at least be resistively shunted). This means that wideband systems commonly apply neither ideal current bias nor ideal voltage bias, but rather a mixture of both.

Preamplifier voltage noise commonly makes direct readout of single SQUIDs impossible. To circumvent this problem, voltage feedback was introduced in 1990 under the name APF and applied to a current-biased SQUID magnetometer [53]. Subsequently, it was utilized for a voltage-biased SQUID [67] and later named *noise cancellation* (NC) scheme [68]. The basic voltage feedback circuit is depicted in Fig. 8.16a. It consists of a resistor R_A and a coil L_A in series, both connected in parallel to the SQUID. The coil L_A is magnetically coupled to the SQUID via a mutual inductance M_A. Assuming that the SQUID is biased at a working point W at the positive slope of the V–Φ characteristic, a positive change in flux $\delta\Phi$ will cause a positive voltage change δV. The resulting increase of the current in the coil L_A induces an additional, positive flux in the SQUID via M_A which increases the SQUID voltage further and leads to an enhanced transfer coefficient V_Φ. Similarly, the transfer coefficient is lowered when W is located at the negative slope of the V–Φ characteristic. Therefore, the V–Φ characteristic becomes strongly asymmetric as schematically shown in Fig. 8.16a. The I–Φ characteristic, however, is not affected by voltage feedback since it is measured at constant SQUID voltage and thus at constant current through the feedback coil L_A.

With voltage feedback, the transfer coefficient V_Φ can be sufficiently boosted to make the amplifier voltage noise contribution smaller than the intrinsic SQUID noise. The peak–peak voltage swing V_pp is somewhat reduced because R_A resistively loads the SQUID.

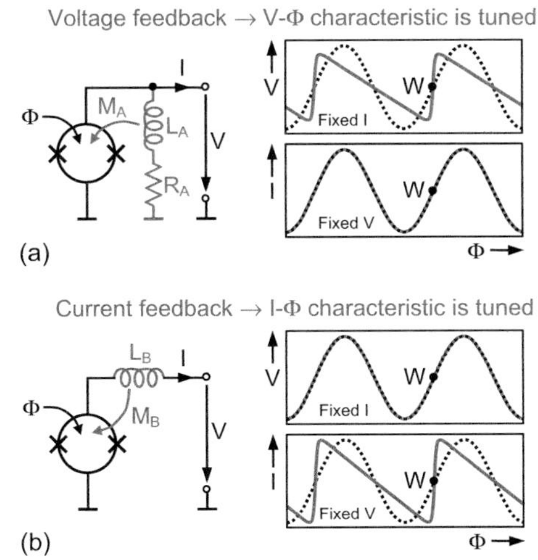

Figure 8.16 Basic circuits for (a) voltage feedback and (b) current feedback. The SQUID characteristics without feedback (dotted lines) and with feedback (solid lines) are schematically depicted on the right side. With voltage feedback the swing in the V–Φ characteristic is reduced because R_A resistively loads the SQUID, whereas with current feedback the full swing is preserved. The working points W yield positive feedback in both cases (a) and (b). The flux offsets are chosen such that the working points without and with feedback coincide.

Thermal noise in R_A causes some extra flux noise in the SQUID which is acceptable if R_A has the same operation temperature as the SQUID (typically R_A is integrated on the SQUID chip). It was shown that for adequate dimensioning ($R_A \approx 2.5R$) the increase in the noise energy ε due to voltage feedback is about half the theoretical value according to Eq. 8.5 [3].

As generally the case, positive feedback leads to a bandwidth reduction. This is not a severe limitation in practice since the feedback circuit is commonly integrated on the sensor chip and has a sufficiently high bandwidth. For a current-biased SQUID, the reduction in the intrinsic linearity Φ_{pp} leads to a degradation in slew rate. However, this is often acceptable because the bandwidth with direct readout is rather high. With voltage bias the slew rate is not degraded because the I–Φ characteristic is not affected. If the voltage feedback is made too strong, hysteresis occurs in the V–Φ

characteristic. Due to wideband noise "smearing" out the hysteresis, the onset of hysteresis might not be recognized in the experimental setup and a too high feedback gain might be chosen. This can result in excess noise from random switching, although the measured V–Φ characteristic may look smooth and nonhysteretic.

There are other methods to obtain a large voltage transfer coefficient, for example using SQUIDs with weakly shunted Josephson junctions operated near the hysteresis limit [59, 70] or SQUIDs with unshunted junctions based on relaxation oscillations [71]. Although the V–Φ characteristics remain symmetric in these cases, the intrinsic linearity Φ_{pp} is reduced compared to a standard SQUID. Only SSAs offer large transfer coefficients combined with highest possible linearity. If it is unavoidable to increase the transfer coefficient at the detriment of linearity, the feedback gain should not be chosen unnecessarily high but just high enough to obtain an acceptable amplifier noise contribution.

For direct readout of single SQUIDs, preamplifier current noise is generally not critical except in the low-frequency regime or at very high frequencies where the current noise of semiconductor amplifiers typically increases. Current noise was completely disregarded in the early phase of direct SQUID readout [53, 67]. However, a few years later it became an issue at PTB due to relatively high-ohmic SQUID magnetometers and a high $1/f$ noise level of the amplifiers used at this time (Linear Technology LT1028). As a countermeasure, current feedback was introduced under the name BCF [51]. The corresponding circuit is depicted in Fig. 8.16b. A coil L_{B} (which is connected in series to the SQUID) is magnetically coupled to the SQUID loop via a mutual inductance M_{B}. For the analysis we assume positive feedback at the chosen working point W. If the current changes by a small amount δI, the resulting voltage change via the magnetic coupling $V_{\Phi} M_{\mathrm{B}} \delta I$ will be added to the voltage change $-R_{\mathrm{dyn}} \delta I$ resulting from the SQUID's dynamic resistance. Converting the total voltage change into flux one obtains $\delta\Phi = (M_{\mathrm{B}} - M_{\mathrm{dyn}})\delta I$. The total flux change becomes zero for $M_{\mathrm{B}} = M_{\mathrm{dyn}}$, which means zero dynamic resistance of the SQUID with current feedback and complete suppression of amplifier current noise. In other words, the transfer coefficient of the I–Φ characteristic $I_{\Phi} = \partial I/\partial\Phi$ becomes infinite.

We conclude that both voltage and current feedback lead to strongly asymmetric characteristics with substantially boosted transfer coefficient V_Φ or I_Φ at the slope with positive feedback. Note that in Fig. 8.16 the direction of the current I is reversed as in [69] contrary to the normal convention. With reversed current direction optimum amplifier noise suppression always occurs for positive feedback so that the similarity of voltage and current feedback can be demonstrated more clearly.

Complementary to voltage feedback, current feedback modifies the I–Φ characteristic but does not affect the V–Φ characteristic. In contrast to voltage feedback, there is no reduction in the current swing and no excess flux noise. However, wideband noise may distort the SQUID characteristic similar to the case with voltage feedback (in particular for SQUIDs with high inductance). It is advisable to connect a resistor R_B in parallel to the feedback coil L_B in order to limit the feedback bandwidth [51]. The extra flux noise due to thermal noise in R_B is typically insignificant. Feeding the bias current asymmetrically into the SQUID has a similar effect as current feedback. In this case, half of the SQUID inductance L acts as an "intrinsic" feedback coil with an effective mutual inductance $M_B = \pm L/2$ [72]. Most SQUID sensors developed at PTB are equipped with this technique to reduce the dynamic resistance at "no expense."

Current feedback is also well suited for a two-stage SQUID to increase the flux gain between the first and the second stage. This was first demonstrated in [38] under the somewhat misleading acronym APF (which was at this time not considered as a synonym for voltage feedback but rather used to indicate that the additional feedback was positive). PTB's two-stage sensor are normally equipped with current feedback (see Fig. 8.10). Using negative instead of positive current feedback, the linearity of a SQUID or SSA can be increased to enable operation without room temperature FLL. This technique was introduced independently by two research groups under the names *output current feedback* (OCF) [60] and *current-sampling feedback* [73].

Voltage and current feedback can be combined for maximum suppression of amplifier noise. This was first demonstrated in 1993 for an integrated multiloop magnetometer with current bias [51]. This particular device involved a special design where the feedback resistor R_A is merged into the junction shunt resistors and only one feedback coil is used to realize the functions of L_A and L_B. A combination of voltage and current feedback was recently applied to a voltage biased SQUID and published under the name *SQUID bootstrap circuit* (SBC) [74]. A description of the different implementations is given in [69].

With direct readout a high bandwidth can easily be obtained because there is no limiting modulation frequency and the loop delay can be made very small. The first system with voltage feedback, published in 1990 [53], achieved a bandwidth of 0.5 MHz. Five years later, the bandwidth was increased to 5 MHz [75], comparable to the fastest systems with flux modulation at this time. Another factor of 3 improvement to 15 MHz was achieved in the following two years [76]. Since then, there was only slight improvement to 20 MHz bandwidth [59] because the overall FLL performance is limited by the loop delay in the wiring between the cryogenic part and the room temperature electronics. Several 100 MHz bandwidth were demonstrated with "cold" FLL electronics [38] or on-chip feedback by large SQUID arrays [60], but these techniques have not become established. In contrast, direct readout electronics with a bandwidth $\gtrsim$5 MHz are nowadays commercially available and widely used. The noise spectra shown as examples in this chapter were measured with the commercial variant of the electronics in [59] (XXF-1 from Magnicon GmbH, Hamburg) and include the noise contribution from the preamplifier.

8.4 SQUID Applications

8.4.1 Introductory Discussion

As shown in the previous sections, the SQUID is a very sensitive magnetic flux detector. A common phrase is that "it is well suited for the measurement of magnetic flux or any quantity that can be converted into magnetic flux." Due to this feature and the low SQUID noise, the device has found a large number of applications since its invention half a century ago, ranging from ultrasensitive laboratory setups involving a single SQUID to multichannel instruments with hundreds of SQUIDs. Comprehensive reviews of the various SQUID applications are found in literature (e.g., [3, 5, 6]). To give an impression of the versatility of the SQUID, we briefly summarize the main application fields. Some more detailed examples of SQUID work at PTB are presented in the following sections.

In medicine, SQUIDs are used for non-invasive diagnostics of organs (most common brain and heart) based on measuring their magnetic field (biomagnetism) or the response to applied magnetic fields (susceptometry, low-field magnetic resonance). Geophysical exploration is another large field (see Section 9.3.4 in [6]). The field inhomogeneity at the surface of the Earth is mapped with SQUIDs to prospect for oil or minerals. Magnetic exploration has also been applied to detect archaeological objects or unexploded ordnance and mines. In transient electromagnetics (TEM), a large current through a transmitter coil is switched on and off, and the resulting secondary magnetic field is sensed to get information about the resistivity of the subsoil. In astronomy, SQUIDs are employed in large telescopes as amplifiers for superconducting detectors (e.g., [77]). In nondestructive evaluation of materials, flaws in the tested material (for instance aircraft wheels and fuselage) can be detected by measuring the inhomogeneity of the distribution of a current-induced magnetic field [78]. The widely used magnetic property measurement systems (MPMSs) allow the measurement of remnant field, magnetization, or magnetic susceptibility of small samples with a SQUID magnetometer as a function of applied field or temperature. Scanning SQUID microscopes can be utilized to measure the local magnetic field just above the surface of a sample.

In metrology, SQUID-based noise thermometers are established tools for the measurement of the absolute temperature, and cryogenic current comparators (involving a SQUID as a null detector) are commonly applied for the measurement of resistance or electric current ratios with highest accuracy. SQUIDs are also applied in nuclear magnetic resonance (NMR) spectroscopy at low and ultralow fields [79–85], in quantum computing [86] and in nanotechnology [12]. They have been used to measure extremely small mechanical displacements for gravitational wave detection [3, 87] or in free-fall experiments to test Einstein's equivalence principle at an improved level of sensitivity [88]. Finally, some important physical experiments involve SQUIDs such as the space experiment Gravity Probe B for testing two fundamental predictions of Einstein's theory of general relativity [89], the search for a permanent electric dipole moment of the neutron [90] or for axion dark matter (where a microstrip SQUID amplifier [91, 92] acts as a nearly-quantum-limited rf detector), the investigation of Hawking radiation [93], or the observation of the dynamical Casimir effect [94].

The most obvious application of SQUIDs is to replace other types of magnetometers in order to improve existing setups. However, in contrast to other magnetic field sensors, the SQUID has to be cooled to very low temperatures. In some applications (for example magnetic property measurement systems), cryocoolers may be employed to obtain a user-friendly system and wide acceptance. However, in demanding magnetometry applications such as biomagnetism or geophysical exploration, cryocoolers typically produce too much vibrational noise, and cooling with liquid helium (or liquid nitrogen in the case of high-T_c SQUIDs) is required for best noise performance. However, this substantially reduces the acceptance by the users. Of course, if the signal source has to be operated at cryogenic temperature anyway, the SQUID is a natural choice as long as it can be sufficiently well coupled to the source. There are many of such applications in low-temperature physics and low-temperature particle detection (see Section 8.4.4).

If the SQUID has to compete with existing sensors (e.g., in geophysics), a substantial benefit in performance is required to compensate for the above-mentioned complications from cooling. When considering upgrading a sensor by a SQUID in an existing setup or using a SQUID in an experiment for the first time, a detailed noise analysis of the setup is recommended to obtain a realistic (or better a conservative) estimate of the final performance. There should remain enough "headroom" in performance compared to the competing (commonly room-temperature) technologies. One should bear in mind that, although the flux noise of the SQUID in a well shielded environment is excellent, the signal-to-noise ratio in the true experiment might be degraded due to problems in coupling the SQUID to the signal source and/or excess noise picked up by the SQUID.

When relying on data from literature for predicting the overall performance of a new SQUID application, one should consider that published information is sometimes too optimistic or even misleading. For example, in a recent paper [95], the authors presented high-T_c magnetometers based on SSAs and claimed that these devices are "ideal candidates to replace single-SQUIDs in many applications." They compared the best *flux noise* of their SSAs (0.25 $\mu\Phi_0/\sqrt{\text{Hz}}$) with that of a typical single Nb-based device (1 $\mu\Phi_0/\sqrt{\text{Hz}}$). Besides the fact that the total flux noise of an SSA is always lower than that of a single SQUID, a correct comparison has to be done on the basis of the quantity to be measured, for a magnetometer the flux density B rather than the flux Φ. Taking a field sensitivity of about 4 μT/Φ_0 from Fig. 2 in [95], a flux density noise $\sqrt{S_B} \approx 1$ pT/$\sqrt{\text{Hz}}$ is obtained for the high-T_c SSA. Single Nb-based integrated magnetometers optimized for field sensing have demonstrated a factor of 1000 lower noise levels of about 1 fT/$\sqrt{\text{Hz}}$ already in 1993 [33]. Even the bare low-T_c SQUID in Fig. 8.6 has a white noise level comparable to the high-T_c SSA in [95], but superior performance in the $1/f$ regime.

8.4.2 Biomagnetism

In biomagnetism, the magnetic field produced by movements of ions inside the human body is measured outside, and the localization of the source (approximated by a current dipole in the simplest and most common assumption) is deduced from the measured magnetic field distribution by solving the inverse problem. Spatial resolutions in the range of several millimeters and temporal resolutions in the millisecond range are obtained. Biomagnetism is a contact-free, completely non-invasive technique and has no influence on the subject. The field distribution is not distorted by the human tissue, which is an advantage over the measurement of electric surface potentials with electrodes (electrocardiogram, electroencephalogram). Due to their low noise levels, SQUIDs are well suited and widely used for biomagnetic measurements. Comprehensive reviews on biomagnetic SQUID applications are found in literature [5, 6, 96–101].

Since the heart generates the strongest magnetic signal among the human organs, biomagnetic research started with magnetocardiography (MCG). About five years after the invention of the SQUID, a first magnetocardiogram was measured with a point-contact rf SQUID to which a second-order wire-wound pickup coil was coupled [102, 103]. Since then, SQUIDs were applied to record the magnetic field of numerous other body organs, for example from the brain (magnetoencephalography, MEG), the fetal heart (fMCG) and brain (fMEG), the eye (magnetooculogram, MOG), the peripheral nerves (magnetoneurogram, MNG), the liver (liver susceptometry), the stomach (magnetogastrogram, MGG), the small intestine (magnetoenterogram, MENG), the skeletal muscles (magnetomyogram, MMG), and the lungs.

Today, large biomagnetic systems that are designed for clinical applications have become common. They typically involve hundreds of SQUID channels. Basically, two geometrical configurations are used: helmet-type sensor arrays optimized for MEG and nearly flat sensor arrays designed for MCG (which are also well suited for most other human organs of interest). It was estimated in Chapter 11 of [5] that by the end of 2004 the number of SQUIDs installed in MEG and MCG applications alone was over 20000. The most intensive biomagnetic application, MEG, is gradually moving from research laboratories to clinical practice. It is currently routinely used to map sources of epileptic activities in pre-surgical evaluation of epilepsy [100]. A current research focus is to combine MEG with ultralow-field magnetic resonance imaging (MRI), i.e., to use the SQUID sensor array also for recording MRI data. Combining those two methods could improve source localization accuracy by restricting the available source space by anatomical knowledge. Another research goal of ultralow-field MRI is the direct measurement of stimulated brain currents by MRI. Here, the shift in the local resonance frequency caused by the magnetic field of the brain currents is measured directly and thus the inverse problem is avoided.

In the past decade, low-noise optical magnetometers came up that offer the potential to replace SQUID sensors in biomagnetic applications (see, e.g., [104, 105]). This would remove the obstacle of cryogenic cooling and remove the limiting noise contribution from the dewar. However, one should not be too euphoric in thinking that this will make the breakthrough of biomagnetic methods and bring them rapidly into clinics. Firstly, there is a big difference between demonstrating a sufficiently low noise level with a few sensors in a laboratory environment and implementing a complete and reliable measurement system with hundreds of channels for clinical use. One should not forget that the first demonstrations of biomagnetic measurements were done soon after the invention of the SQUID, but that it took decades to get robust biomagnetic multichannel systems. Secondly, the shielding issue still remains and it has to be demonstrated that the established noise cancellation methods are equally well suited for new magnetometer technology. For highest performance, a shielded room is typically required which is expensive and occupies a large space. Thirdly, to establish biomagnetism for clinical use, the inverse problem (i.e., the localization of the source from its magnetic field distribution outside the body) has to be solved efficiently and accurately with adequate source models. This field has advanced in the recent decades, but there is still a lot to do.

We conclude that, even if the SQUID is replaced by a non-cryogenic sensor, biomagnetic multichannel systems remain quite expensive and that the main barrier for a widespread clinical use is the clear demonstration of sufficient diagnostic benefits over existing methods. The latter aspect also holds for less expensive small-scale systems intended for the use in doctor's offices or bedside applications in clinics (for example in MCG where the competing technology of electrocardiography is very old and well established). Nevertheless, biomagnetism is a very important field in research, and reliable and low-noise Nb-based dc SQUIDs are still the most widely used sensors in commercial systems.

Biomagnetic research has a long tradition at PTB. In 1980, a first heavily shielded room was built, the Berlin magnetically shielded room (BMSR). This room consists of six layers of mu-metal plus one eddy-current shield of 15 mm thick Cu, and has a total weight of 25 tons. The shielding factor is 10^4 at 0.1 Hz and increases to $>10^5$ above 2 Hz. Twenty years later, an even larger and more heavily shielded room with further improved shielding factor (two to three orders of magnitude with active shielding) was completed, the BMSR-2. This room consists of seven mu-metal layers plus one eddy-current shield of Al [106]. It is surrounded by an extra rf shield. The obtained shielding factors of both rooms are high enough to enable the use of magnetometers. For this reason, between 1980 and 2000 the SQUID research at PTB was focused on magnetometers rather than the commonly used (mostly wire-wound) gradiometers.

Several biomagnetic multichannel systems involving integrated multiloop magnetometers were manufactured at PTB, a 37-channel system in 1991, an 83-channel system in 1994, and a 304-channel system in 2003. The 83-channel magnetometer was operated in a moderately shielded room in a Berlin clinic. Due to the rapid progress in sensor technology in the early 1990s, the 83-channel system achieved a record noise level of typically 2.5 $fT/\sqrt{Hz}$, which was about a factor of two lower than that of the best commercial systems at this time [107]. This low noise level made it possible to detect the very small signals from the peripheral nervous system (about 5 fT when studying the nervous propagation in the spinal cord). For comparison, the brain signals are about 100 fT to 1 pT and the heart signal is about 100 pT, respectively.

Figure 8.17 depicts two magnetocardiograms as example of biomagnetic signals. They were recorded in the BMSR to compare the noise levels of high-T_c and low-T_c magnetometers under normal operation conditions. The high-T_c device shows a larger signal due to the smaller cold-warm distance of the dewar. Although the noise level of this magnetometer (18 $fT/\sqrt{Hz}$ at 100 Hz) was substantially higher that of the low-T_c device, both magnetocardiograms are quite clear and exhibit a high signal-to-noise ratio. The sensor noise is sufficiently low to enable much more demanding biomagnetic applications like recording brain signals with high quality or the very weak signals of the peripheral nervous system. We see that with state-of-the-art SQUID sensors the intrinsic sensor noise is commonly not the main limiting factor. Even high-T_c devices would yield acceptable signal-to-noise ratios, but unfortunately their reliability is still a critical issue.

One of the most demanding applications of SQUID magnetometers is ultra-low-field MRI, where pulsed fields of up to about 100 mT are applied and signals down to the lower femtotesla range are subsequently measured. Due to these large pulsed fields,

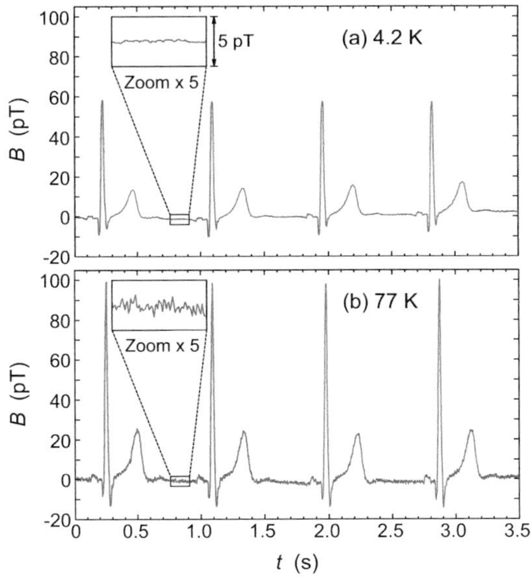

Figure 8.17 Magnetocardiograms of one of the authors (DD) recorded in the mid-1990s for characterizing the noise levels of different types of SQUID magnetometers. (a) Low-T_c multiloop magnetometer at 4.2 K, (b) high-T_c magnetometer at 77 K (flip-chip device in [108]). The measurements were performed in the BMSR. Raw data are shown without averaging (measurement bandwidth 0.016–200 Hz). A $\times$ 5 magnification window shows the base noise level during a "quiet" phase in the magnetocardiogram.

PTB's integrated multiloop magnetometer can no longer be used. For this reason, PTB is currently developing a field-tolerant vector magnetometer based on SQUID current sensors with wire-wound pickup coils [109]. This system will be operated in the BMSR-2.

A prototype module with 18 magnetometer channels is depicted in Fig. 8.18. There are coils with two different sizes, located at two levels separated by a baseline of 90 mm. The bottom coils form the sensor array, while the top ones are mainly intended as references to optionally realize software gradiometers if this improves the signal-to-noise ratio. The effective diameters of the pickup coils are 17.1 mm and 74.5 mm, respectively. The quoted diameter of the large coils corresponds to an equivalent circle with the same area as the polygonal loop. The different coil sizes aim at improving

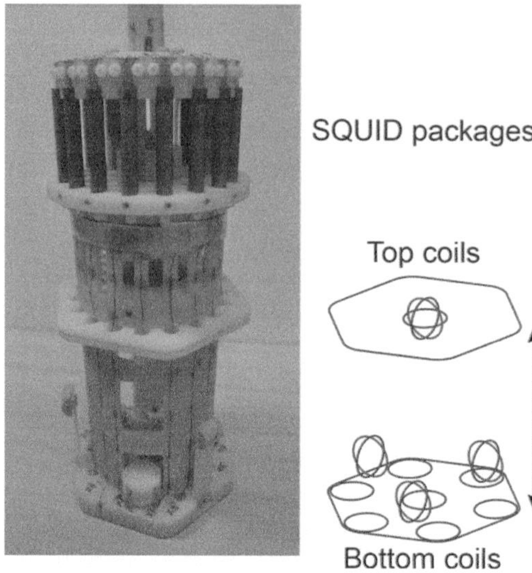

Figure 8.18 Photograph of the prototype module of PTB's vector magnetometer involving 18 magnetometer channels. On the right side, the arrangement of the 16 small and two large pickup coils is sketched.

the signal-to-noise ratio depending on the depth of the source with respect to the coil diameter. In the complete system, seven modules with a total of 126 channels can be arranged in a hexagonal configuration, covering an area with an equivalent diameter of about 210 mm. The pickup coils of the channels measuring the horizontal field components are arranged in the module such that they are also forming regular hexagonal grids in the complete system [109].

The SQUID current sensors are single-stage variants of the device depicted in Fig. 8.10. They are enclosed by superconducting shields. To minimize the effect of field distortion due to the shields, the SQUID packages are located relatively far away from the pickup coils. The distance was deduced from numerical simulations. The input coil inductances were selected to match the inductance of the small and large pickup coils, respectively. The use of integrated voltage and current feedback (APF and BCF) enables direct readout. The feedback current is passed into the feedback transformer (contact pads $\pm$FIN in Fig. 8.10) to keep the current in the input coil constant which minimizes crosstalk between channels. The

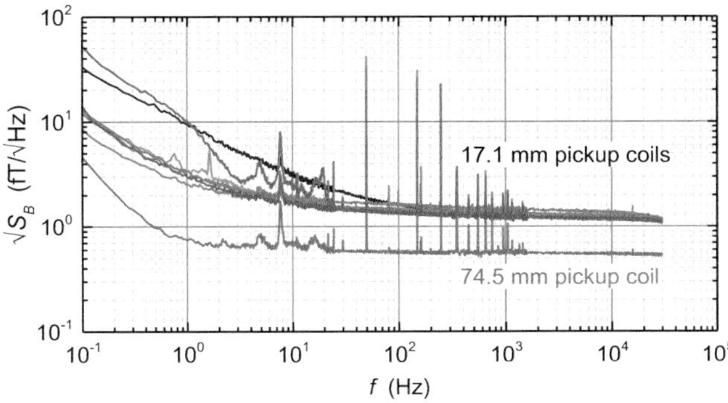

Figure 8.19 Noise spectra of the eight bottom *z*-channels (sensitive to the vertical component B_Z) of the prototype module of PTB's vector magnetometer operated in the BMSR-2 [109]. A software first-order gradiometer configuration was realized by subtracting the output of the top large magnetometer as a reference from all *z*-channels. One of the channels has an atypically high low-frequency noise, another one is probably distorted by a magnetic contamination in the fiberglass dewar, resulting in excess low-frequency and vibrational noise.

integrated current limiters protect the input coils against large currents in magnetic resonance experiments.

Figure 8.19 shows noise spectra obtained with the prototype module in the BMSR-2. Software first-order gradiometers were built (with the upper large magnetometer used as a reference) to remove the environmental magnetic noise contribution of about 0.6 fT/$\sqrt{\text{Hz}}$ at 100 Hz and vibrational noise distorting the measurement in the frequency range between 3 Hz and 30 Hz. The environmental noise contribution was increased most likely because the sensor array was placed nearer to the walls of the shielded room than usual. The white noise levels are dominated by the dewar's contribution of about 1.1 fT/$\sqrt{\text{Hz}}$ and 0.5 fT/$\sqrt{\text{Hz}}$ for the small and large coils, respectively. It is well known that the dewar noise contribution decreases for larger coils due to the mismatch between the coil size and the spatial distribution of the thermal noise currents in the superinsulation foils of the dewar [110].

The intrinsic magnetometer noise levels in Fig. 8.19 are typically 0.7 fT/$\sqrt{\text{Hz}}$ and 0.07 fT/$\sqrt{\text{Hz}}$ for the two coil sizes. The prototype module was operated with the commercial variant of the electronics in [59], allowing the lowest possible magnetometer noise. For a single module, the relatively large size and high power consumption of this electronics are acceptable. The final system will be equipped with a newly designed and much more compact readout electronics being battery powered thanks to a substantially reduced power consumption. This will remove the strong power line interference visible in Fig. 8.19 (an artifact of the experimental setup due to ground loops in the wiring). However, the final sensor noise will be somewhat higher. Assuming a typical $V_\Phi = 1\ \mathrm{mV}/\Phi_0$, the intrinsic noise levels are expected to be typically 1 $\mathrm{fT}/\sqrt{\mathrm{Hz}}$ and 0.1 $\mathrm{fT}/\sqrt{\mathrm{Hz}}$ for the two coil sizes. Nevertheless, the increase in the intrinsic sensor noise will have only minor effect on the overall system performance due to the dewar's limiting noise contribution.

8.4.3 Metrology

In metrology, highest accuracy is always the most important issue. Therefore, costs play a minor role and the need of cryogenic temperatures is commonly no reason for exclusion if the improvement in performance justifies the extra efforts. We briefly discuss two metrology applications of SQUIDs in which PTB is involved for many years: thermometry and metrology of electrical units.

SQUIDs can be used to design powerful thermometers for temperature measurements in the low and ultralow temperature range. For a practical thermometer, a good thermal contact to the location where the temperature has to be measured and negligible heating effects by the thermometer itself are crucial, in particular at millikelvin temperatures. Resistance thermometers are based on measuring the temperature-dependent resistivity of the sensor. They have to be calibrated and need an excitation signal for operation which can cause heating problems. External excitation can be omitted and consequently heating effects be minimized if the thermal noise in a conducting sensor is measured rather than the resistivity. Such noise thermometers relate the measured noise voltage or current to the thermodynamic temperature via the well-known Nyquist formula. Due to its low intrinsic noise, the SQUID is well suited for noise thermometry (for a review see Chapter 9.4 in [6]).

There are different types of SQUID-based noise thermometers. In 1971, Kamper and Zimmerman introduced the resistive SQUID (RSQUID) where the noise-generating resistor is an integral part of the SQUID loop [111]. This device was implemented as a bulk rf SQUID with adjustable point contact. It is working as a voltage-to-frequency converter, i.e., it transforms the voltage across the resistor into a frequency via the ac Josephson effect. The noise voltage, which is a measure of the temperature, is obtained from the frequency fluctuations. Unfortunately, the inconveniently long measurement times and additional problems with parasitic noise sources restricted the use of the RSQUID noise thermometers to a very few metrology laboratories. Nevertheless, noise thermometry utilizing this type of sophisticated mechanical SQUID played an important role in extending the International Temperature Scale of 1990 (ITS-90) to lower temperatures and establishing the Provisional Low Temperature Scale 2000 (PLTS-2000). The ITS-90 ranges down to 0.65 K, while the PLTS-2000 covers the range between 0.9 mK and 1 K.

In the late 1990s it was attempted to combine the RSQUID concept with thin-film dc SQUIDs technology in order to significantly simplify the thermometer operation [112]. From the technological point of view, the most challenging issue was the integration of a low-value resistor (a few 10 $\mu\Omega$) into the SQUID loop. Although the fabrication, experimental setup, and operation of the new generation of dc RSQUIDs were much more simple and reliable compared to the traditional bulk devices, the required measurement times were still quite long and the use in the typical temperature range of dilution refrigerators was impaired by inherent thermal problems. Because of these reasons, integrated RSQUIDs have neither gained significant recognition from potential users nor reached marketability.

Another more practical device, the current-sensing noise thermometer (CSNT), was introduced by Giffard et al. in the early 1970s [113, 114]. In this type of thermometer, the noise resistor is connected to the input coil of a SQUID current sensor. In early devices, bulk Nb rf SQUIDs with point contacts were used. They were later replaced by reliable Nb-based dc SQUIDs [115]. The temperature is determined from the noise spectrum measured at the output of the SQUID. For constant input coil sensitivity and temperature-independent resistance the measured power spectral density is exactly proportional to temperature. In contrast to the RSQUID, the CSNT is a semi-primary thermometer, i.e., has to be calibrated at a known temperature. On the other hand, the CSNT allows much shorter measurement times due to the high sensitivity of modern dc SQUIDs. For example, a relative uncertainty of about 1% at 100 mK can be achieved in a very short measurement time of 100 ms [116]. An important issue is to minimize contact resistances between the normal conducting noise resistor and the superconducting input coil.

Yet another thermometer, the magnetic-field fluctuation thermometer (MFFT) was introduced about ten years ago [117]. Instead of galvanically connecting a noise resistor of bulk Cu with a SQUID current sensor, the magnetic field of the noise currents in the Cu block is measured with a SQUID magnetometer. This avoids electrical contacts to the noise resistor (temperature sensor) and eliminates potential problems with contact resistances. As for the CSNT, the temperature is deduced from the measured noise spectrum, and calibration at one known temperature is required [118]. Recently, a prototype of a calculable MFFT was developed aiming at operating the device in a primary mode without the need for calibration [119]. This device involved an integrated conductivity measurement of the Cu temperature sensor and applied a correlation-based SQUID readout with two independent magnetic field sensors [120].

In the original MFFT approach [117], a wire-wound pickup coil was used. At PTB, the integrated concentric multiloop gradiometer depicted in Fig. 8.13b was developed as an improved magnetic field sensor for the MFFT. Figure 8.20 shows a commercially available version of the MFFT. The high-purity copper temperature sensor has a volume of 1.2 cm^3. The multiloop gradiometer chip is directly glued onto the Cu body. The chip thickness is reduced to about 0.1 mm to enhance the SQUID's sensitivity to the nearby thermal noise currents in the Cu block underneath the chip. The module is easily mounted, for instance to the mixing chamber plate of a dilution refrigerator, with a threaded stud at the end of the Cu body.

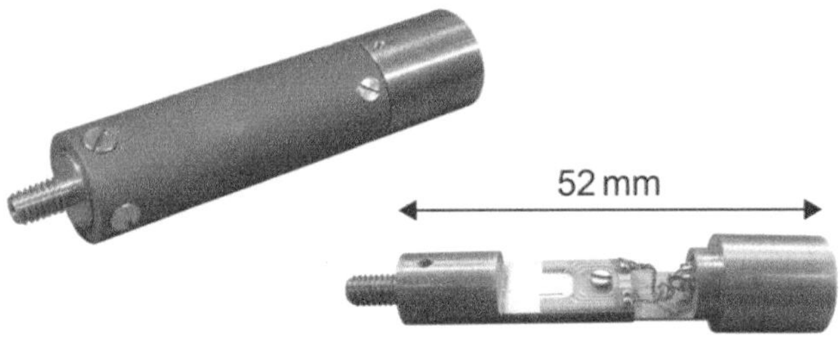

Figure 8.20 Sensor module of a commercial MFFT. The multiloop gradiometer in Fig. 8.13b is used to measure the thermal noise currents in the temperature sensor, a solid block of high-purity Cu. Picture courtesy of Magnicon GmbH, Hamburg.

Another important SQUID application is the cryogenic current comparator (CCC), which was invented by Harvey in 1972 [121]. A CCC is basically a superconducting transformer, which keeps the ratio of two currents constant with ultimate accuracy (the relative error is typically below 10^{-10}). It commonly involves wire-wound coils with a large number of turns (typically several thousands in total) to obtain a high sensitivity and to enable large current ratios.

A convenient technical realization is the so-called binary CCC involving a set of windings with turn numbers scaling in powers of two. Figure 8.21 shows a 12-bit CCC where the individual windings range from $2^0 = 1$ to $2^{11} = 2048$ turns [122]. Completed by some additional windings, 18 windings with a total of 4647 turns are accessible at the room temperature side of the probe stick. The high accuracy of the CCC is achieved by enclosing the coils in a superconducting shield that overlaps itself like a snake swallowing its tail. Using Pb foil for the shield, the multilayer shield can be implemented by successively soldering a layer, isolating it with heat-resistant Kapton tape (yellow part in the torus in Fig. 8.21), soldering the next layer, isolating it, and soldering the final layer. A SQUID magnetometer mounted in the hole of the torus detects the net ampere-turns balance and keeps it at a constant level via a feedback loop. As a result, the current ratio is exactly equal to the inverse of the turns ratio. For effective suppression of external magnetic fields (six orders of magnitude are typically required), the

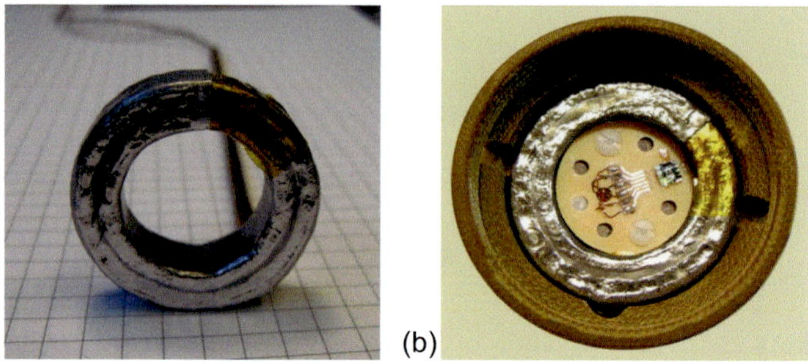

Figure 8.21 Photographs of a 12-bit CCC torus (a) before and (b) after mounting it into the probe stick. The inner and outer diameters of the torus are 18.4 mm and 31.4 mm, respectively. In (b) the circular carrier with the 3 mm $\times$ 3 mm SQUID chip is visible. The single-turn pickup coil is placed on the rear side of the carrier in a groove at the outer edge (reproduced with permission from Götz et al. [122]). The CCC support is made of laminated fabric.

sensitive components CCC and SQUID are placed inside a two-layer shield consisting of an inner Nb and an outer Cryoperm layer.

CCCs are used at national metrology institutes for resistance calibration [122–124]. In this application, the two resistors R_1 and R_2 to be compared and the CCC primary/secondary windings N_1 and N_2 are incorporated into a bridge circuit. Two currents I_1 and I_2 are passed through the resistors to produce nearly identical voltage drops across the resistors (typically 0.5 V). The difference between the voltage drops is measured by a nanovoltmeter. It is used to determine the resistance ratio R_1/R_2 from the current ratio I_2/I_1 with an accuracy down to about 10^{-9}. To suppress the effect of thermal drifts, I_1 and I_2 are periodically reversed. At PTB, a repetition frequency of 0.05 Hz is typically chosen. Therefore, in CCC applications a good low-frequency noise of the SQUID is crucial.

The discovery of the quantum Hall effect (QHE) by von Klitzing in 1980 revolutionized resistance metrology and was honored with the 1985 Nobel Prize in physics. Soon after his discovery, the QHE was introduced as primary resistance standard with quantum accuracy [125]. Due to the low temperatures and high magnetic fields involved (typically $T \approx 1$ K and $B \approx 10$ T), the QHE is normally not directly used for resistance calibration. Rather, national metrology institutes have a set of highly stable and accurate standard resistors which are occasionally calibrated against the quantum Hall resistance, typically a few times a year depending on the drift in the resistors. Due to their high accuracy, CCC-based resistance bridges have become a vital tool in this traceability chain of standard resistors, and the SQUID is a small but essential component. Today, CCC-based resistance bridges are commercially available as complete, ready-to-use solution for high-end resistance calibration.

The CCC is also involved in the redefinition of the unit ampere. The present definition (based on the force between two current-carrying wires) is difficult to realize with high accuracy in practice. The new definition is based on the elementary charge of the electron $e \approx 1.602 \times 10^{-19}$ A·s. A known electric current can be generated with high accuracy by clocking the flow of single electrons through so-called single-electron pumps. Unfortunately, the resulting current is small. Assuming a typical clock rate of 1 GHz, a current of 10^9 $e/\text{s} \approx$ 160 pA is obtained. It was suggested to use a CCC with $\gtrsim$10000 turns as an accurate current amplifier [126]. Due to the large number of turns, a high current sensitivity and a resulting low noise level down to a few fA/$\sqrt{\text{Hz}}$ can be achieved, which is necessary to measure the small currents from single-electron pumps with sufficiently low statistical uncertainty.

Figure 8.22 shows a current noise spectrum obtained with PTB's new 14-bit CCC, which has the best low-frequency noise performance reported for a CCC so far [127]. Connecting seven CCC windings in series to form an "input coil" of 17,152 turns, one obtains an overall mutual inductance $M_\text{i} \approx 3.2$ μH and a current sensitivity of 0.65 nA/Φ_0. This is a factor of 370 more sensitive than the SQUID current sensor depicted in Fig. 8.10. The current noise level is correspondingly low, for example 2 fA/$\sqrt{\text{Hz}}$ at 0.4 Hz. The coupling constant $k \approx 0.17$ between the input inductance $L_\text{i} \approx 4.5$ H and the SQUID inductance $L \approx 80$ pH is substantially lower than for a well-coupled SQUID current sensor. However, for the application as a direct current amplifier this does not matter.

The SQUID is a strongly nonlinear device. It was recently argued that mixing-down effects in the SQUID can lead to systematic errors in CCC-based amplifiers when measuring small currents [127]. To

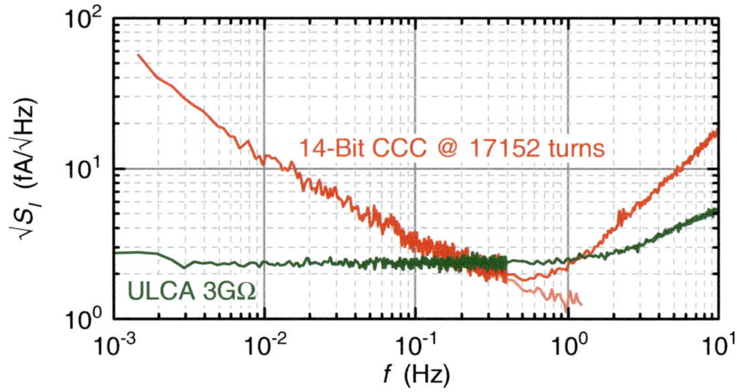

Figure 8.22 Low-frequency noise spectra of (a) PTB's 14-bit CCC and (b) the ultrastable low-noise current amplifier (ULCA). The current noise of the CCC was obtained from the measured flux noise for an input coil with 17152 turns having a current sensitivity of 0.65 nA/Φ_0. The light red line shows the CCC noise corrected for the thermal noise in the damping circuit connected across the 8192-turn winding (series shunt of $R_d = 2$ kΩ and $C_d = 0.9$ μF) [127].

avoid this potential problem, a novel approach was introduced, the ultrastable low-noise current amplifier (ULCA) [128]. The ULCA is a noncryogenic, high-accuracy current amplifier based on specially designed operational amplifiers and resistor networks. Its transfer coefficient is extremely stable versus time, temperature and amplitude within the full dynamic range. Due to a high feedback resistance of 3 GΩ in the first amplifier stage, the ULCA achieves a very low current noise level of 2.4 fA/$\sqrt{\text{Hz}}$. Below about 0.2 Hz, the ULCA's noise is even lower than that of the 14-bit CCC (see Fig. 8.22). However, the ULCA needs calibration, and this is done with a very small uncertainty of $<10^{-7}$ by using the CCC. Thus, the new approach combines the best of two technologies: the CCC calibrates the ULCA at high currents of about 10 nA where mixing-down effects in the SQUID are sufficiently low, and the ULCA is applied to measure the small currents from single-electron pumps with excellent low-frequency noise and maximal user-friendliness. This example shows how superconducting and room temperature electronics can be combined to obtain best possible solution for a measurement task.

8.4.4 Readout of Superconducting Detectors

SQUID sensors have enabled the development of two categories of superconducting low-temperature detectors for radiation and particles, namely power-sensitive detectors (bolometers) and energy-dispersive detectors (calorimeters) that are based on transition edge sensors (TESs) or magnetically coupled calorimeters (MCCs). These thermal detectors are highly sensitive and versatile. They can achieve noise-equivalent power levels or energy resolutions that are significantly lower than conventional semiconductor-based radiation detectors. As an example, silicon drift detectors (SDDs) that are widely used as detectors in energy dispersive X-ray spectrometry provide an energy resolution of typically 120–150 eV at 5.9 keV. TES- and MCC-based calorimeters enable at the same energy a spectral resolution that is up to two orders of magnitude better than that of SDDs. SQUIDs have significantly contributed to this success. They are essential for TES/MCC applications and there are practically no alternatives to using SQUIDs for the readout of these low-temperature detectors.

In a TES the steep resistive transition of a superconducting material at its critical temperature T_c is utilized to form a highly sensitive thermistor. The TES output signal is a change in the current through the thermistor. MCCs include metallic magnetic calorimeters (MMCs) and magnetic penetration thermistors (MPTs). In an MMC the temperature dependence of the magnetization of a paramagnetic metallic material exposed to a magnetic field is used to detect a temperature rise upon absorption of a photon or energetic particle. Here, the measured signal is the change in magnetization δM of the detector material. The MPT operational principle is very similar, except that the paramagnetic metallic material of the MMC is replaced by a type-I superconducting material. Well below its T_c this superconductor becomes a near-perfect diamagnet. The magnetic field the superconductor is exposed to is slightly below its critical magnetic field at the MPT operation temperature. An increase in temperature caused by an absorption event then induces a magnetic transition of the superconducting detector material. For reviews on the operational principles and properties we refer to [129] for

TES-based detectors and to [130, 131] for MMC- and MPT-based detectors.

Over the last two to three decades, the development of TES- and MCC-based detectors for radiation and particles has been mainly driven by efforts to improve the sensitivity of precision measurement instruments for cosmology, astronomy and fundamental physics (see references in [132]). Examples of other applications of TESs and MCCs are passive THz security cameras [133], photon counters in the near-infrared and visible range for quantum optics [134, 135], high resolution X-ray spectrometers for materials analysis [136] or photo-biophysics [137], and gamma-ray spectrometers for radionuclide metrology [138, 139] or nuclear materials analysis [139, 140]. These examples illustrate the very broad energy range of existing and emerging applications of these two detector types. TESs, in particular, can be used for a wide range of energies, from bolometric detection of sub-millimeter waves to calorimetric detection of X- and gamma-rays. The operation temperature of TES bolometers and calorimeters for these very different applications ranges from around 1 K down to about 50 mK. MCCs typically address high-energy applications, namely spectrometry of soft and hard X-rays with the highest energy resolution demonstrated, so far. They are mostly operated in the temperature range below 50 mK.

SQUIDs are the natural choice to read out these detector types: they can provide sufficiently high sensitivity, exhibit low power dissipation and are principally compatible with the low operating temperatures of TESs and MCCs. In the following we discuss the main similarities and differences of SQUID readout of single TESs and MCCs and give an example of a SQUID sensor developed at PTB that is suitable for both.

TESs and MCCs set somewhat different requirements with regards to their readout with SQUID sensors. The simplified schemes in Fig. 8.23 depict a TES and an MCC from the readout perspective. A TES in operation can be regarded as a current source with an internal resistance R_{TES} at the working point. The various implementations of TES detectors cover a wide range for R_{TES}, below 1 mΩ up to a few ohms. The TES signal current δI_{TES} upon absorption of radiation or particles can differ, as well. It can range

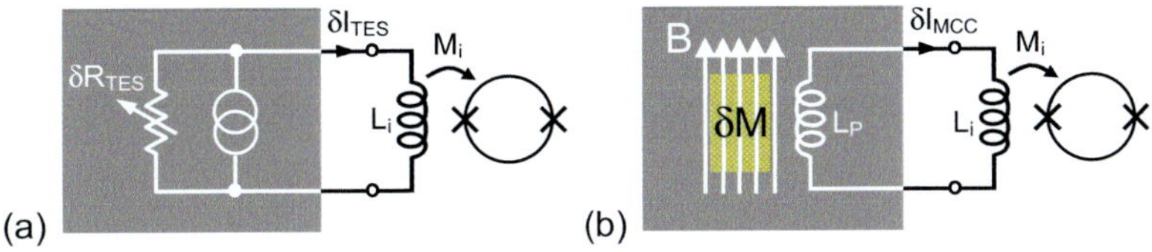

Figure 8.23 Simplified scheme of the SQUID readout configuration for two superconducting detectors: (a) TES and (b) MCC. The TES is symbolized by a current source with an internal variable resistance. A change in resistance δR_{TES} upon absorption of radiation or particles is measured by a SQUID current sensor via the resulting change in current δI_{TES}. In the case of the MCC, the detector material (yellow) is exposed to a static magnetic field B and magnetically coupled to a superconducting pickup coil with inductance L_{p}. This coil forms a superconducting flux transfer loop with the SQUID input coil. An absorption event causes a change in the magnetization of the detector material. The corresponding current change δI_{MCC} in the flux transfer loop is measured with the SQUID sensor.

from tens of nA to tens of μA. TESs show current noise levels typically in the range of 10 to 100 pA/$\sqrt{\mathrm{Hz}}$. For the TES readout the SQUID sensor is a "genuine" current sensor. That is to say the relevant parameter with regard to the sensitivity of this readout configuration is the current noise $\sqrt{S_{\mathrm{I}}}$ referred to the SQUID input. Obviously, $\sqrt{S_{\mathrm{I}}}$ needs to be sufficiently below the TES current noise level in order to avoid degradation of the TES signal-to-noise ratio.

As discussed in Section 8.2.3, it is possible to achieve a necessary $\sqrt{S_{\mathrm{I}}}$ for a SQUID with a given flux noise level by choosing the appropriate number of turns of the input coil resulting in the value of the SQUID input inductance L_{i} according to Eq. 8.14. The choice of L_{i} can be made to a certain extent independent of the TES parameters R_{TES} and δI_{TES} and the TES current noise. One has to keep in mind, however, that the TES/SQUID input circuit forms an R–L circuit that acts as a low-pass filter. L_{i} is often the dominant inductive component of this R–L circuit. Naturally, the bandwidth of this filter needs to sufficiently exceed the TES signal bandwidth of interest. Furthermore, electrothermal stability of the TES operation requires the filter bandwidth to be at least a factor of about 6 larger than the TES bandwidth. As long as this condition is fulfilled, the input inductance is not an important parameter for the choice of a SQUID current sensor to read out a TES. Given the wide range of

TES designs and corresponding thermal and electrical parameters, SQUID current sensors with input inductances from a few nH to hundreds of nH are employed for their readout.

MCCs differ from TESs in this respect. Here, the detector material is inductively coupled to a superconducting pickup coil with inductance L_P. The pickup coil and the SQUID input coil form a fully superconducting circuit. This configuration is conceptually identical to the SQUID magnetometer formed by a SQUID current sensor to the input of which a superconducting pickup coil is connected. Therefore, as discussed in Section 8.2.4, flux coupling to the SQUID loop is maximal for $L_i \approx L_P$ for a given L_P and δM. This means that for the MCC readout configuration the SQUID input inductance is more stringently determined by the detector design and dimensioning.

Existing MCCs require relatively low SQUID input inductances, typically a few nH to about 20 nH. In conjunction with the requirement $L_i \approx L_P$, the relevant parameter with regards to the SQUID sensitivity in MCC readout configurations is the coupled energy sensitivity $\varepsilon_c = L_i S_I/2$. Although not shown in Fig. 8.23, parasitic inductances in the flux coupling loop would degrade the flux coupling and are to be avoided. As a consequence, the SQUID sensor is often located in close vicinity to the MCC. Here, thermal loading from the SQUID sensor to the MCC detector can be an issue. A low SQUID power dissipation as well as a solid thermal anchoring of the SQUID sensor (chip) are then necessary. MCCs are nonresistive and nondissipative detectors. However, as thermal detectors they exhibit thermodynamic energy fluctuations between the detector and the thermal bath as well as between subsystems of the detector (e.g., between spins and electrons in the case of MMCs). These thermal noise contributions fundamentally determine the MCC intrinsic noise. Given the lower temperature of operation the readout of MCCs typically requires lower SQUID noise as compared to TES readout.

At PTB, a range of SQUID sensors have been developed specifically for TES and MCC readout. Beyond sufficient sensitivity and dynamic performance, these sensors also address system aspects, for example easy operability in millikelvin refrigerators in which TESs or MCCs are mostly operated. As an example, Fig. 8.24a shows a TES/SQUID module that combines two TES detectors made from

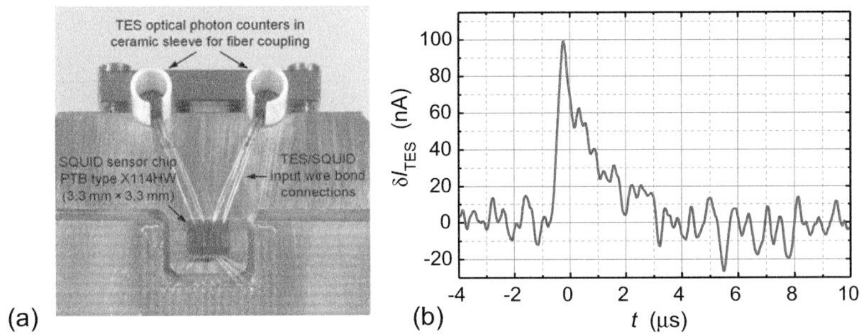

Figure 8.24 (a) TES/SQUID module with two TES-based, number-resolving photon counters for the near-infrared and optical range. The detectors are coupled to the inputs of two SQUID current sensors (PTB type X114HW). (b) Measured detector current change in response to the absorption of a single photon with a wavelength of 810 nm.

superconducting tungsten with $T_c \approx 150$ mK and PTB SQUID current sensors for their readout. The module can be directly placed onto the experimental platform of a millikelvin refrigerator and is usually operated at temperatures below 0.1 K. The TES detectors have been developed at the National Institute of Standards and Technology (NIST) as highly sensitive, number-resolving counters of single to a few optical photons [134]. The detector design and its coupling to an optical fiber (leading to room temperature) allow a detection efficiency of close to 100% [141, 142].

The two TESs depicted in Fig. 8.24 are directly connected to the input coils ($L_i \approx 2$ nH) of two SQUID current sensors that are on one chip. Each current sensor is a two-stage SQUID configuration: a single front-end SQUID (to which the input coil is connected) is read out by a 14-element SSA integrated on the current sensor chip. Their current noise level $\sqrt{S_I}$ is below 2 pA/$\sqrt{\text{Hz}}$ for frequencies larger than 1 kHz. Note that the SQUID sensor chip is located directly on the gold-plated copper module body. Doing so enables easy thermal anchoring of the SQUID chip. However, the magnetic field noise arising from the module body below (in the MFFT this noise is used as the temperature signal) would degrade the current sensor noise if both the front-end SQUID as well as the SSA would not be of compact gradiometric design (cf. Fig. 8.7) The diagram in Fig. 8.24b shows the measured TES current in response to the absorption of a single photon with a wavelength of 810 nm. This compact TES/SQUID configuration has proven very suitable for measurements where single photons in the near-infrared and visible range need to be detected with high efficiency. For instance, it has been recently used to experimentally prove the effect of quantum entanglement (here of two photons) as predicted by quantum theory [143].

The comparably small input inductance and low current noise level of the SQUID sensors of PTB type X114HW (or type X114W without integrated bias resistors for TES operation) correspond to a coupled energy sensitivity of about 6 h. This very low noise level make these devices also very suitable for the readout of MCCs. Figure 8.25 shows a photo of an MMC detector chip containing four detectors coupled to current sensors. The MMC detectors shown are being developed for the high-resolution measurement of the energy spectrum following the electron capture process of ^{163}Ho [144]. The interconnection between the detectors and the SQUIDs are made by short Al wires that are superconducting at the detector operation temperature of below 50 mK. This results in a low parasitic inductance (less than 1 nH) in the superconducting flux transfer loop. The very low operation temperature of the configuration requires both the MMC chip and the SQUID sensor chips to be well thermally anchored. This is achieved by means of Au wires that provide a link to the refrigerator base temperature with low thermal impedance.

A specific aspect of the SQUID readout of TESs and MCCs is multiplexing [139, 145]. Many of the above-mentioned TES and MCC applications involve hundreds or even thousands of detector channels. In such large-format detector setups constraints on wiring and circuit complexity and on the cooling power of the refrigerator often prohibit the readout of every detector channel by a dedicated SQUID sensor channel. The basic idea of multiplexing is to combine signals of more than one detector into one common readout channel. In order to identify the detectors, their signals need to be individually encoded, for instance by modulating the detector signals themselves, the coupling to the SQUID, or the SQUID output signal.

Multiplexing techniques have been primarily developed to operate large format TES detector arrays. The multiplexing schemes

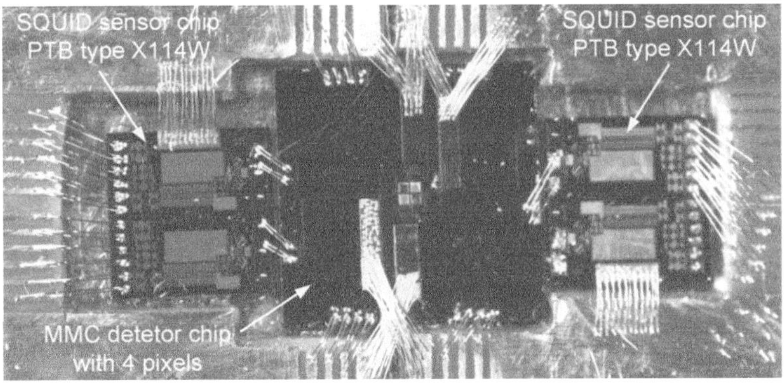

Figure 8.25 Chip with four MMC detector pixels coupled to SQUID current sensors (PTB-type X114W). The interconnection between the detectors and the SQUIDs are made by short superconducting Al wires. Both the MMC chip and the SQUID sensor chips are thermally anchored via Au wires (photo courtesy of S. Kempf, University of Heidelberg).

practically always cause degradation of the input-referred noise or coupled energy sensitivity compared to a given nonmultiplexed SQUID sensor. As mentioned above, TESs typically set lower SQUID noise requirement compared to MCCs. This noise margin makes multiplexed readout of TESs simpler. A multiplexing scheme that may prove suitable also for the readout of large format arrays of MCCs is microwave SQUID multiplexing [146, 147]. Here, nonhysteretic rf SQUIDs are coupled to superconducting microwave resonators with different resonance frequencies. Magnetic flux coupled to an rf SQUIDs via its input coil shifts the resonant frequency of the corresponding resonator. The rf SQUID does not act as a sensor that provides amplification of the detector signal, but rather as a flux-dependent inductance.

The three schemes developed for multiplexed readout of TES arrays are frequency-division (FD), time-division (TD) and code-division (CD) multiplexing [139]. In FD multiplexing [148] the TES signals are modulated at different frequencies via a particular ac biasing scheme. The amplitude modulated signals are then combined into the inputs of conventional SQUID current sensors (basically the same type can be used as for the readout of single, conventionally operated TESs). Hence, FD multiplexing of TESs does not require specific SQUID multiplexer circuits as in TD and CD multiplexing and is, therefore, not further discussed here.

The use of ac bias for the TESs is not necessary in TD and CD multiplexing. In these schemes the multiplexing is achieved by modulating the output signals (TD) of the readout SQUIDs or their input coupling (CD). In TD multiplexing, the modulation is done with boxcar functions (unipolar, low-duty cycle square waves) that switch on and off individual SQUID channels. An example for a SQUID TD multiplexer that illustrates this multiplexing scheme is shown in Fig. 8.26. This circuit represents a 4:1 TD SQUID multiplexer developed at PTB [149]. It is formed by a series connection of four identical SQUID current sensors, specifically the 16-element SSA discussed in Section 8.2.2. The TESs are connected to the input coils of the current sensors. A switch is placed in parallel with each SQUID current sensor. The switch state can be changed from superconducting (S) to normal conducting (N). A SQUID current sensor is disabled or enabled when its corresponding switch is in S or N state, respectively. The multiplexer function is obtained by addressing the switches so as to activate exactly one SQUID at a given time. This way only the output signal of that particular current sensor channel is present at the voltage output of the multiplexer. Therefore, the TESs are measured in a sequence defined by the boxcar functions that are used to address the switches. A particular aspect of this multiplexer is that the switches are implemented by arrays of low-inductance SQUIDs. This makes it possible to address the switches inductively and without additional power dissipation.

As already discussed, the channels of an N:1 TD multiplexer need to be switched on and off. A given TES is, hence, read out only a fraction $1/N$ of the time. As a result, the effective rms current noise of a SQUID channel in a TD multiplexer is degraded proportional to $\sqrt{N}$. However, despite this so-called noise penalty TD multiplexing is comparably simple and the most widely used scheme for multiplexed TES readout.

The CD multiplexing scheme avoids noise degradation [150, 151]. Here, the input coupling to a SQUID current sensor is modulated. The modulation is done with orthogonal Walsh functions (bipolar square waves with 50% duty cycle at different frequencies).

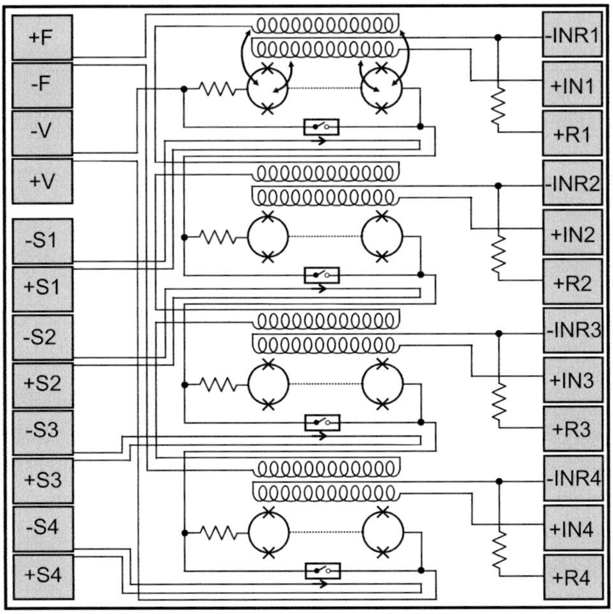

Figure 8.26 Circuit scheme of a 4:1 time-domain SQUID multiplexer (PTB-type X416FLM). The four SQUID current sensors are 16-element SSAs. Flux-actuated SQUID-based superconducting-to-normalconducting switches [149] are connected in parallel to the SSAs. Low-value resistors in series to the SSAs avoid closed superconducting loops when the switches are in the superconducting state. Connections to the room temperature readout electronics are via ±V (SSA voltage), ±F (feedback), and ±S1 to ±S4 (addressing currents).

These modulation functions encode the TES signals individually without the need for the above-mentioned TES ac bias in FD multiplexing. The SQUID current sensor itself is active at all times. Hence, there is no noise degradation as in TD multiplexing. Flux-actuated SQUID-based switches are an elegant way to implement Walsh function modulation of the signals coupled to the SQUID sensor.

SQUID multiplexer circuits have been employed for the readout of large-format TES arrays in a variety of science instruments. A prominent example is the SCUBA-2 instrument at the James Clerk Maxwell Telescope [152]. This camera in the sub-millimeter range uses more than ten thousand TES bolometer pixels that are read out by a complex three-stage TD SQUID multiplexer [153]. The multiplexing ratio is 40:1. The practical implementation of CD SQUID multiplexers is progressing and multiplexing ratios similar to TD multiplexers are foreseeable [154].

8.5 Conclusions

We are looking back to half a century of SQUID history. The initial devices were made from machined bulk Nb with point contact junctions realized by screws. A big step forward was the advent of reliable thin-film tunnel junction processes in the 1980s, predominantly the Nb–AlO_x–Nb trilayer technology. In the beginning, the development of SQUIDs was mainly driven by biomagnetism. The need for large biomagnetic multichannel systems initiated a rapid development of SQUID concepts and readout schemes, in particular in the 1990s. At the same period another important field emerged, the use of SQUIDs as preamplifiers for superconducting detectors.

Today, SQUID technology has become mature. Modern Nb-based devices are extremely sensitive, versatile and robust, but their main restriction is the need for cryogenic temperatures. Nevertheless, SQUIDs are widely used in biomagnetism, astronomy, geomagnetism, material sciences, and metrology. Many interesting experiments in fundamental research became possible thanks to the SQUID's ultimate sensitivity. SQUID current sensors are used as preamplifiers for large detector arrays, for example SCUBA-2 with more than ten thousand TES bolometer pixels [152]. Numerous large biomagnetic multichannel systems with sophisticated noise cancellation techniques are operated worldwide, partially in clinical environment. These systems are commercially available as well as small-scale laboratory SQUIDs, where the user can individually design the pickup circuit for his specific application. Although being a small component in the whole system, the SQUID decisively determines the overall system performance. An example is the magnetic property measurement system, probably the best-seller in SQUID-based instruments. Being equipped with a cryocooler, the user has not to care about cryogenic cooling. The user-friendliness of SQUID-based instruments has also been greatly improved in the past decades. Nowadays, systems are inevitably fully computer controlled and operable even for non-scientists.

Acknowledgements

The authors thank Marianne Fleischer-Bartsch for noise measurements and micrographs of the described SQUID sensors, Jan-Hendrik Storm and Rainer Körber for sharing unpublished material on the robust multichannel magnetometer currently under development at PTB (Figs. 8.18 and 8.19), Eckart Pesel for measuring the inductance of the 14-bit CCC, and Martin Burghoff for valuable comments on the biomagnetism part of the manuscript.

References

1. Ryhänen, T., Seppä, H., Ilmoniemi, R., and Knuutila, K. (1989). SQUID magnetometers for low-frequency applications. *J. Low Temp. Phys.* **76**, 287–386.
2. Barone, A., ed. (1992). *Principles and Applications of Superconducting Quantum Interference Devices* (World Scientific Publishing, Singapore).
3. Weinstock, H., ed. (1996). *SQUID Sensors: Fundamentals, Fabrication and Applications*, NATO ASI Series E: Applied Sciences, Vol. 329 (Kluwer Academic Publishers, Dordrecht).
4. Clarke, J., and Braginski, A. I., eds. (2004). *The SQUID Handbook*, Vol. I: Fundamentals and Technology of SQUIDs and SQUID Systems (Wiley-VCH, Weinheim).
5. Clarke, J., and Braginski, A. I., eds. (2006). *The SQUID Handbook*, Vol. II: Applications of SQUIDs and SQUID Systems (Wiley-VCH, Weinheim).
6. Seidel, P., ed. (2015). *Applied Superconductivity*, Vol. 2, Chapter 9: Superconducting Quantum Interference (SQUIDs) (Wiley-VCH, Weinheim), pp. 949–1110.
7. Jaklevic, R. C., Lambe, J., Silver, A. H., and Mercereau, J. E. (1964). Quantum interference effects in Josephson tunneling. *Phys. Rev. Lett.* **12**, 159–160.

8. Zimmerman, J. E., and Silver, A. H. (1966). Macroscopic quantum interference effects through superconducting point contacts. *Phys. Rev.* **141**, 367–375.
9. Stewart, W. C. (1968). Current-voltage characteristics of Josephson junctions. *Appl. Phys. Lett.* **12**, 277–280.
10. McCumber, D. E. (1968). Effect of ac impedance on dc voltage-current characteristics of superconductor weak-link junctions. *J. Appl. Phys.* **39**, 3113–3118.
11. Anders, S., Schmelz, M., Fritzsch, L., Stolz, R., Zakosarenko, V., Schönau, T., and Meyer, H.-G. (2009). Sub-micrometer-sized, cross-type Nb-AlO_x-Nb tunnel junctions with low parasitic capacitance. *Supercond. Sci. Technol.* **22**, 064012 (4 pp).
12. Granata, C., and Vettoliere, A. (2016). Nano superconducting quantum interference device: a powerful tool for nanoscale investigations. *Phys. Rep.* **614**, 1–69.
13. Tesche, C. D., and Clarke, J. (1977). dc SQUID: noise and optimization. *J. Low Temp. Phys.* **29**, 301–331.
14. Bruines, J. J. P., de Waal, V. J., and Mooij, J. E. (1982). Comment on: "dc SQUID: Noise and optimization" by Tesche and Clarke. *J. Low Temp. Phys.* **46**, 383–386.
15. de Waal, V. J., Schrijner, P., and Llurba, R. (1982). Simulation and optimization of a dc SQUID with finite capacitance. *J. Low Temp. Phys.* **54**, 215–232.
16. Ryhänen, T., Seppä, H., and Cantor, R. (1992). Effect of parasitic capacitance and inductance on the dynamics and noise of dc superconducting quantum interference devices. *J. Appl. Phys.* **71**, 6150–6166.
17. Voss, R. F. (1981). Noise characteristics of an ideal shunted Josephson junction. *J. Low Temp. Phys.* **42**, 151–163.
18. Tesche, C. D., and Clarke, J. (1979). dc SQUID: current noise. *J. Low Temp. Phys.* **37**, 397–403.
19. Martinis, J. M., and Clarke, J. (1986). Current noise measured in the dc SQUID. *J. Low Temp. Phys.* **65**, 459–468.
20. Drung, D. (1995). Theory for the multiloop dc superconducting quantum interference device magnetometer and experimental verification. *J. Appl. Phys.* **77**, 4088–4098.
21. Enpuku, K., Yoshida, K., and Kohjiro, S. (1986). Noise characteristics of a dc SQUID with a resistively shunted inductance. II. Optimum damping. *J. Appl. Phys.* **60**, 4218–4223.

22. Knuutila, J., Ahonen, A., and Tesche, C. (1987). Effects on dc SQUID characteristics of damping of input coil resonances. *J. Low Temp. Phys.* **68**, 269–284.

23. Knuutila J., Kajola, M., Seppä, H., Mutikainen, R., and Salmi, J. (1988). Design, optimization, and construction of a dc SQUID with complete flux transformer circuits. *J. Low Temp. Phys.* **71**, 369–392.

24. Drung, D., Beyer, J., Storm, J.-H., Peters, M., and Schurig, T. (2011). Investigation of low-frequency excess flux noise in dc SQUIDs at mK temperatures. *IEEE Trans. Appl. Supercond.* **21**, 340–344.

25. Wellstood, F. C., Urbina, C., and Clarke, J. (1994). Hot-electron effect in metals. *Phys. Rev. B* **49**, 5942–5955.

26. Wellstood, F. C., Urbina, C., and Clarke, J. (1987). Low-frequency noise in dc superconducting quantum interference devices below 1 K. *Appl. Phys. Lett.* **50**, 772–774.

27. Kempf, S., Ferring, A., Fleischmann, A., and Enss, C. (2015). Direct-current superconducting quantum interference devices for the read-out of metallic magnetic calorimeters. *Supercond. Sci. Technol.* **28**, 045008 (12 pp).

28. Ketchen, M. B., Stawiasz, K. G., Pearson, D. J., Brunner, T. A., Hu, C.-K., Jaso, M. A., Manny, M. P., Parsons A. A., and Stein K. J. (1992). Sub-μm linewidth input coils for low T_c integrated thin-film dc superconducting quantum interference devices. *Appl. Phys. Lett.* **61**, 336–338.

29. Jaycox, J. M., and Ketchen, M. B. (1981). Planar coupling scheme for ultra low noise dc SQUIDs. *IEEE Trans. Magn.* **17**, 400–403.

30. Ketchen, M. B. (1987). Integrated thin-film dc SQUID sensors. *IEEE Trans. Magn.* **23**, 1650–1657.

31. Ketchen, M. B., Gallagher, W. J., Kleinsasser, A. W., and Murphy, S. (1985). dc SQUID flux focuser, *SQUID '85, Superconducting Quantum Interference Devices and their Applications*, eds. Hahlbohm, H. D., and Lübbig, H. (Walter de Gruyter, Berlin), pp. 865–871.

32. Drung, D., Beyer, J., Matz, H., Lindström, M., Ludwig, F., and Schurig, T. (1998). Magnetic coupling, effective area, and inductance of SQUID sensors: New theoretical description and comparison with experiment, presented at *Applied Superconductivity Conference ASC1998*, Palm Desert, CA, USA (unpublished).

33. Drung, D., and Koch, H. (1994). An integrated dc SQUID magnetometer with variable additional positive feedback. *Supercond. Sci. Technol.* **7**, 242–245.

34. Jansman, A. B. M. (1999). High-T_c dc SQUIDs for use in a background field, *Dissertation* (University of Twente, Low Temperature Division, The Netherlands).
35. Yoshida, K., Hossain, M. S., Kisu, T., Enpuku, K., and Yamafuji, K. (1992). Modeling of kinetic-inductance coplanar stripline with NbN thin films. *Jpn. J. Appl. Phys.* **31**, 3844–3850.
36. Grover, F. W. (1962). *Inductance Calculations: Working Formulas and Tables* (Dover, New York).
37. Welty, R. P., and Martinis, J. M. (1991). A series array of dc SQUIDs. *IEEE Trans. Magn.* **27**, 2924–2926.
38. Drung, D., Assmann, C., Beyer, J., Kirste, A., Peters, M., Ruede, F., and Schurig, T. (2007). Highly sensitive and easy-to-use SQUID sensors. *IEEE Trans. Appl. Supercond.* **17**, 699–704.
39. Stawiasz, K. G., and Ketchen, M. B. (1993). Noise measurements of series SQUID arrays. *IEEE Trans. Appl. Supercond.* **3**, 1808–1811.
40. Stan, G., Field, S. B., and Martinis, J. M. (2004). Critical field for complete vortex expulsion from narrow superconducting strips. *Phys. Rev. Lett.* **92**, 097003 (4 pp).
41. Carelli, P., Castellano, M. G., Flacco, K., Leoni, R., and Torrioli, G. (1997). An absolute magnetometer based on dc superconducting quantum interference devices. *Europhys. Lett.* **39**, 569–574.
42. Häussler, Ch., Oppenländer, J., and Schopohl, N. (2001). Nonperiodic flux to voltage conversion of series arrays of dc superconducting quantum interference devices. *J. Appl. Phys.* **89**, 1875–1879.
43. Beyer, J., and Drung, D. (2008). A SQUID series array dc current sensor. *Supercond. Sci. Technol.* **21**, 095012 (6 pp).
44. Dettmann, F., Richter, W., Albrecht, G., and Zahn, W. (1979). A monolithic thin film dc-SQUID. *Phys. Status Solidi A* **51**, K185–K188.
45. Cromar, M. W., and Carelli, P. (1981). Low-noise tunnel junction dc SQUIDs. *Appl. Phys. Lett.* **38**, 723–725.
46. Ketchen, M. B., and Jaycox, J. M. (1982). Ultra-low-noise tunnel junction dc SQUID with a tightly coupled planar input coil. *Appl. Phys. Lett.* **40**, 736–738.
47. Carelli, P., Castellano, M. G., Torrioli, G., and Leoni, R. (1998). Low noise multiwasher superconducting interferometer. *Appl. Phys. Lett.* **72**, 115–117.

48. Muhlfelder, B., Johnson, W., and Cromar, M. W. (1983). Double transformer coupling to a very low noise SQUID. *IEEE Trans. Magn.* **19**, 303–307.

49. ter Brake, H. J. M., Fleuren, F. H., Ulfman, J. A., and Flokstra, J. (1986). Elimination of flux-transformer crosstalk in multichannel SQUID magnetometers. *Cryogenics* **26**, 667–670.

50. Hilbert, C., Clarke, J., Sleator, T., and Hahn, E. L. (1985). Nuclear quadrupole resonance detected at 30 MHz with a dc superconducting quantum interference device. *Appl. Phys. Lett.* **47**, 637–639.

51. Drung, D., and Koch, H. (1993). An electronic second-order gradiometer for biomagnetic applications in clinical shielded rooms. *IEEE Trans. Appl. Supercond.* **3**, 2594–2597.

52. Zimmerman, J. E. (1971). Sensitivity enhancement of superconducting quantum interference devices through the use of fractional-turn loops. *J. Appl. Phys.* **42**, 4483–4487.

53. Drung, D., Cantor, R., Peters, M., Scheer, H. J., and Koch, H. (1990). Low-noise high-speed dc superconducting quantum interference device magnetometer with simplified feedback electronics. *Appl. Phys. Lett.* **57**, 406–408.

54. Schmelz, M., Stolz, R., Zakosarenko, V., Schönau, T., Anders, S., Fritzsch, L., Mück, M., and Meyer, H.-G. (2011). Field-stable SQUID magnetometer with sub-fT $Hz^{-1/2}$ resolution based on sub-micrometer cross-type Josephson tunnel junctions. *Supercond. Sci. Technol.* **24**, 065009 (5 pp).

55. Dössel, O., David, B., Fuchs, M., Kullmann, W. H., and Lüdeke, K. M. (1991). A modular low noise 7-channel SQUID-magnetometer. *IEEE Trans. Magn.* **27**, 2797–2800.

56. Daalmans, G. M., Bär, L., Bömmel, F. R., Kress, R., and Uhl, D. (1991). Ultra low noise all niobium dc-SQUIDs. *IEEE Trans. Magn.* **27**, 2997–3000.

57. Bechstein, S., Ruede, F., Drung, D., Storm, J.-H., Kieler, O. F., Kohlmann, J., Weimann, T., and Schurig, T. (2015). HfTi-nanoSQUID gradiometers with high linearity. *Appl. Phys. Lett.* **106**, 072601 (4 pp).

58. Drung, D. (2003). High-T_c and low-T_c dc SQUID electronics. *Supercond. Sci. Technol.* **16**, 1320–1336.

59. Drung, D., Hinnrichs, C., and Barthelmess, H.-J. (2006). Low-noise ultra-high-speed dc SQUID readout electronics. *Supercond. Sci. Technol.* **19**, S235–S241.

60. Drung, D., Beyer, J., Peters, M., Storm, J.-H., and Schurig, T. (2009). Novel SQUID current sensors with high linearity at high frequencies. *IEEE Trans. Appl. Supercond.* **19**, 772–777.
61. Forgacs, R. L., and Warnick, A. (1967). Digital-analog magnetometer utilizing superconducting sensor. *Rev. Sci. Instrum.* **38**, 214–220.
62. Clarke, J., Goubau, W. M., and Ketchen, M. B. (1976). Tunnel junction dc SQUID: fabrication, operation, and performance. *J. Low Temp. Phys.* **25**, 99–144.
63. Wellstood, F., Heiden, C., and Clarke, J. (1984). Integrated dc SQUID magnetometer with a high slew rate. *Rev. Sci. Instrum.* **55**, 952–957.
64. Koch, R. H., Rozen, J. R., Wöltgens, P., Picunko, T., Goss, W. J., Gambrel, D., Lathrop, D., Wiegert, R., and Overway, D. (1996). High performance superconducting quantum interference device feedback electronics. *Rev. Sci. Instrum.* **67**, 2968–2976.
65. Penny, R. D., Lathrop, D. K., Thorson, B. D., Whitecotton, B. R., Koch, R. H., and Rosen, J. R. (1997). Wideband front end for high-frequency SQUID electronics. *IEEE Trans. Appl. Supercond.* **7**, 2323–2326.
66. Matlashov, A., Espy, M., Kraus, Jr., R. H., Ganther, Jr., K. R., and Snapp, L. D. (2001). Electronic gradiometer using HTc SQUIDs with fast feedback electronics. *IEEE Trans. Appl. Supercond.* **11**, 876–879.
67. Seppä, H., Ahonen, A., Knuutila, J., Simola, J., and Vilkman, V. (1991). dc-SQUID electronics based on adaptive positive feedback: experiments. *IEEE Trans. Magn.* **27**, 2488–2490.
68. Kiviranta, M., and Seppä, H. (1995). dc SQUID electronics based on the noise cancellation scheme. *IEEE Trans. Appl. Supercond.* **5**, 2146–2148.
69. Drung, D. (2010). Simplified analysis of direct SQUID readout schemes. *Supercond. Sci. Technol.* **23**, 065006 (7 pp).
70. Polushkin, V., Glowacka, D., Hart, R., and Lumley, J. (1999). Effect of an input coil microwave resonance on dynamics and noise properties of a dc superconducting quantum interference device operating close to the hysteretic mode. *Rev. Sci. Instrum.* **70**, 1713–1718.
71. Adelerhof, D. J., Nijstad, H., Flokstra, J., and Rogalla, H. (1994). (Double) relaxation oscillation SQUIDs with high flux-to-voltage transfer: simulations and experiments. *J. Appl. Phys.* **76**, 3875–3886.
72. Uehara, G., Matsuda, N., Kazami, K., Takada, Y., and Kado, H. (1993). Asymmetric bias injection technique for Drung-type superconducting quantum interference devices. *Jpn. J. Appl. Phys.* **32**, L1735–L1738.

73. Kiviranta, M. (2008). SQUID linearization by current-sampling feedback. *Supercond. Sci. Technol.* **21** 045009 (6 pp).

74. Xie, X., Zhang, Y., Wang, H., Wang, Y., Mück, M., Dong, H., Krause, H.-J., Braginski, A. I., Offenhäusser, A., and Jiang, M. (2010). Voltage biased superconducting quantum interference device bootstrap circuit. *Supercond. Sci. Technol.* **23**, 065016 (4 pp).

75. Drung, D., Matz, H., and Koch, H. (1995). A 5-MHz bandwidth SQUID magnetometer with additional positive feedback. *Rev. Sci. Instrum.* **66**, 3008–3015.

76. Drung, D. (1997). Improved dc SQUID read-out electronics with low $1/f$ noise preamplifier. *Rev. Sci. Instrum.* **68**, 4066–4074.

77. Ruhl, J. E., et al. (2004). The South Pole telescope. *Proc. SPIE* **5498**, 11–29.

78. Jenks, W. G., Sadeghi, S. S. H., and Wikswo Jr., J. P. (1997). SQUIDs for nondestructive evaluation. *J. Phys. D: Appl. Phys.* **30**, 293–323.

79. Greenberg, Ya. S. (1998). Application of superconducting quantum interference devices to nuclear magnetic resonance. *Rev. Mod. Phys.* **70**, 175–222; Erratum *Rev. Mod. Phys.* **72**, 329 (2000).

80. McDermott, R., Trabesinger, A. H., Mück, M., Hahn, E. L., Pines, A., and Clarke, J. (2002). Liquid-state NMR and scalar couplings in microtesla magnetic fields. *Science* **295**, 2247–2249.

81. Körber, R., Casey, A., Shibahara, A., Piscitelli, M., Cowan, B. P., Lusher, C. P., Saunders, J., Drung, D., and Schurig, T. (2007). Nuclear magnetic resonance on room temperature samples in nanotesla fields using a two-stage dc superconducting quantum interference device sensor. *Appl. Phys. Lett.* **91**, 142501 (3 pp).

82. Levitin, L. V., Bennett, R. G., Casey, A., Cowan, B. P., Lusher, C. P., Saunders, J., Drung, D., and Schurig, T. (2007). A nuclear magnetic resonance spectrometer for operation around 1MHz with a sub-10-mK noise temperature, based on a two-stage dc superconducting quantum interference device sensor. *Appl. Phys. Lett.* **91**, 262507 (3 pp).

83. Espy, M., et al. (2010). Ultra-low-field MRI for the detection of liquid explosives. *Supercond. Sci. Technol.* **23**, 034023 (8 pp).

84. Hartwig, S., Voigt, J., Scheer, H.-J., Albrecht, H.-H., Burghoff, M., and Trahms, L. (2011). Nuclear magnetic relaxation in water revisited. *J. Chem. Phys.* **135**, 054201 (4 pp).

85. Kraus, Jr., R., Espy, M., Magnelind, P., and Volegov, P. (2014). *Ultra-Low Field Nuclear Magnetic Resonance: A New MRI Regime* (Oxford University Press, New York).

86. Ruggiero, B., Delsing, P., Granata, C., Pashkin, Y. A., and Silvestrini, P., eds. (2006). *Quantum Computing in Solid State Systems* (Springer Science & Business Media, New York).

87. Podt, M., Gottardi, L., de Waard, A., Frossati, G., and Flokstra, J. (2003). A spherical gravitational wave detector readout by nearly quantum limited SQUIDs. *Supercond. Sci. Technol.* **16**, 1531–1535.

88. Vodel, W., Koch, H., Nietzsche, S., Glyscinski, J. V. Z., Neubert, R., Pilz, M., Dittus, H., Lochmann, S., Mehls, C., and Lockowandt, D. (1999). Application of high performance LTS SQUID systems in gravitational experiments. *IEEE Trans. Appl. Supercond.* **9**, 4119–4122.

89. Everitt, C. W. F., et al. (2011). Gravity Probe B: Final results of a space experiment to test general relativity. *Phys. Rev. Lett.* **106**, 221101 (5 pp).

90. Gemmel, C., et al. (2010). Ultra-sensitive magnetometry based on free precession of nuclear spins. *Eur. Phys. J. D* **57**, 303–320.

91. Mück, M., André, M.-O., Clarke, J., Gail, J., and Heiden, C. (1998). Radio-frequency amplifier based on a niobium dc superconducting quantum interference device with microstrip input coupling. *Appl. Phys. Lett.* **72**, 2885–2887.

92. Mück, M., Kycia, J. B., and Clarke, J. (2001). Superconducting quantum interference device as a near-quantum-limited amplifier at 0.5 GHz. *Appl. Phys. Lett.* **78**, 967–969.

93. Nation, P. D., Blencowe, M. P., Rimberg, A. J., and Buks, E. (2009). Analogue Hawking radiation in a dc-SQUID array transmission line. *Phys. Rev. Lett.* **103**, 087004 (4 pp).

94. Wilson, C. M., Johansson, G., Pourkabirian, A., Simoen, M., Johansson, J. R., Duty, T., Nori, F., and Delsing, P. (2011). Observation of the dynamical Casimir effect in a superconducting circuit. *Nature* **479**, 376–379.

95. Chesca, B., John, D., and Mellor, C. J. (2015). Flux-coherent series SQUID array magnetometers operating above 77 K with superior white flux noise than single-SQUIDs at 4.2 K. *Appl. Phys. Lett.* **107**, 162602.

96. Hämäläinen, M., Hari, R., Ilmoniemi, R. J., Knuutila, J., and Lounasmaa, O. V. (1993). Magnetoencephalography—theory, instrumentation, and applications to noninvasive studies of the working human brain. *Rev. Mod. Phys.* **65**, 413–497.

97. Del Gratta, C., Pizzella, V., Tecchio, F., and Romani, G. L. (2001). Magnetoencephalography - a noninvasive brain imaging method with 1 ms time resolution. *Rep. Prog. Phys.* **64**, 1759–1814.

98. Koch, H. (2001). SQUID magnetocardiography: status and perspectives. *IEEE Trans. Appl. Supercond.* **11**, 49–59.

99. Andrä, W., and Nowak, H., eds. (2007). *Magnetism in Medicine*, 2nd Ed. (Wiley-VCH, Weinheim).

100. Supek, S., and Aine, C. J. (2014). *Magnetoencephalography: From Signals to Dynamic Cortical Networks* (Springer-Verlag Berlin Heidelberg).

101. Hobbie, R. K., and Roth, B. J. (2015). *Intermediate Physics for Medicine and Biology*, Chapter 8: Biomagnetism (Springer International Publishing, Switzerland), pp. 213–238.

102. Silver, A. H. (2006). How the SQUID was born. *Supercond. Sci. Technol.* **19**, S173–S178.

103. Cohen, D., Edelsack, E. A., and Zimmerman, J. E. (1970). Magnetocardiograms taken inside a shielded room with a superconducting point-contact magnetometer. *Appl. Phys. Lett.* **16**, 278–280.

104. Budker, D., and Romalis, M. (2007). Optical magnetometry. *Nat. Phys.* **3**, 227–234.

105. Shah, V., Knappe, S., Schwindt, P. D. D., and Kitching, J. (2007). Subpicotesla atomic magnetometry with a microfabricated vapour cell. *Nat. Photonics* **1**, 649–652.

106. Bork, J., Hahlbohm, H.-D., Klein, R., and Schnabel, A. (2001). The 8-layered magnetically shielded room of the PTB: design and construction, *Biomag 2000: Proceedings of the 12th International Conference on Biomagnetism*, eds. Nenonen, J., Ilmoniemi, R. J., and Katila, T. (Helsinki University of Technology, Espoo, Finland), pp. 970–973.

107. Drung, D. (1995). The PTB 83-SQUID system for biomagnetic applications in a clinic. *IEEE Trans. Appl. Supercond.* **5**, 2112–2117.

108. Drung, D., Dantsker, E., Ludwig, F., Koch, H., Kleiner, R., Clarke, John, Krey, S., Reimer, D., David, B., and Doessel, O. (1996). Low noise $YBa_2Cu_3O_{7-x}$ SQUID magnetometers operated with additional positive feedback. *Appl. Phys. Lett.* **68**, 1856–1858.

109. Storm, J.-H., Drung, D., Burghoff, M., and Körber, R. (2016). A modular, extendible and field-tolerant multichannel vector magnetometer based on current sensor SQUIDs. *Supercond. Sci. Technol.* **29**, 094001 (9 pp).

110. Nenonen, J., Montonen, J., and Katila, T. (1996). Thermal noise in biomagnetic measurements. *Rev. Sci. Instrum.* **67**, 2397–2405.

111. Kamper, R. A., and Zimmerman, J. E. (1971). Noise thermometry with the Josephson effect. *J. Appl. Phys.* **42**, 132–136.

112. Menkel, S., Drung, D., Greenberg, Ya. S., and Schurig, T. (2000). Integrated thin-film dc RSQUIDs for noise thermometry. *J. Low Temp. Phys.* **120**, 381–400.

113. Giffard, R. P., Webb, R. A., and Wheatley, J. C. (1972). Principles and methods of low-frequency electric and magnetic measurements using an rf-biased point contact superconducting device. *J. Low Temp. Phys.* **6**, 533–610.

114. Webb, R. A., Giffard, R. P., and Wheatley, J. C. (1973). Noise thermometry at ultralow temperatures. *J. Low Temp. Phys.* **13**, 383–429.

115. Lusher, C.P., Li, J., Maidanov, V. A., Digby, M. E., Dyball, H., Casey, A., Nyéki, J., Dmitriev, V. V., Cowan, B. P., and Saunders, J. (2001). Current sensing noise thermometry using a low T_c dc SQUID preamplifier. *Meas. Sci. Technol.* **12**, 1–15.

116. Casey, A., et al. (2014). Current sensing noise thermometry: a fast practical solution to low temperature measurement. *J. Low Temp. Phys.* **175**, 764–775.

117. Netsch, A., Hassinger, E., Enss, C., and Fleischmann, A. (2006). Novel, non-contact noise thermometer for Milli-kelvin temperatures. *AIP Conf. Proc.* **850**, 1593–1594.

118. Beyer, J., Schmidt, M., Engert, J., AliValiollahi, S., and Barthelmess, H.-J. (2013). Reference measurements of SQUID-based magnetic-field fluctuation thermometers. *Supercond. Sci. Technol.* **26**, 065010 (11 pp).

119. Kirste, A., Regin, M., Engert, J., Drung, D., and Schurig, T. (2014). A calculable and correlation-based magnetic field fluctuation thermometer. *J. Phys. Conf. Ser.* **568**, 032012 (7 pp).

120. Rothfuß, D., Reiser, A., Fleischmann, A., and Enss, C. (2013) Noise thermometry at ultra low temperatures. *Appl. Phys. Lett.* **103**, 052605 (4 pp).

121. Harvey, I. K. (1972). A precise low temperature dc ratio transformer. *Rev. Sci. Instrum.* **43**, 1626–1629.

122. Götz, M., Drung, D., Pesel, E., Barthelmess, H.-J., Hinnrichs, C., Aßmann, C., Peters, M., Scherer, H., Schumacher, B., and Schurig, T. (2009). Improved cryogenic current comparator setup with digital current sources. *IEEE Trans. Instrum. Meas.* **58**, 1176–1182.

123. Drung, D., Götz, M., Pesel, E., Storm, J.-H., Aßmann, C., Peters, M., and Schurig, T. (2009). Improving the stability of cryogenic current comparator setups. *Supercond. Sci. Technol.* **22**, 114004 (8 pp).

124. Williams, J. M., Janssen, T. J. B. M., Rietveld, G., and Houtzager, E. (2010). An automated cryogenic current comparator resistance ratio bridge for routine resistance measurements. *Metrologia* **47**, 167–174.

125. Jeckelmann, B., and Jeanneret, B. (2001). The quantum Hall effect as an electrical resistance standard. *Rep. Prog. Phys.* **64**, 1603–1655.

126. Gay, F., Piquemal, F., and Genevès, G. (2000). Ultralow noise current amplifier based on a cryogenic current comparator. *Rev. Sci. Instrum.* **71**, 4592–4595.

127. Drung, D., Götz, M., Pesel, E., and Scherer, H. (2015). Improving the traceable measurement and generation of small direct currents. *IEEE Trans. Instrum. Meas.* **64**, 3021–3030.

128. Drung, D., Krause, C., Becker, U., Scherer, H., and Ahlers, F. J. (2015). Ultrastable low-noise current amplifier: a novel device for measuring small electric currents with high accuracy. *Rev. Sci. Instrum.* **86**, 024703 (10 pp).

129. Irwin, K. D., and Hilton, G. C. (2005). Transition-edge sensors, *Cryogenic Particle Detection*, Topics in Applied Physics, Vol. 99, Enss, C., ed. (Springer, Berlin), pp. 63–150.

130. Fleischmann, A., Enss, C., and Seidel, G. M. (2005). Metallic magnetic calorimeters, *Cryogenic Particle Detection*, Topics in Applied Physics, Vol. 99, Enss, C., ed. (Springer, Berlin), pp. 151–216.

131. Bandler, S. R., Irwin, K. D., Kelly, D., Nagler, P. N., Porst, J. P., Rotzinger, H., Sadleir, J. E., Seidel, G. M., Smith, S. J., and Stevenson, T. R. (2012). Magnetically coupled microcalorimeters. *J. Low Temp. Phys.* **167**, 254–268.

132. Enss, C., ed. (2005). *Cryogenic Particle Detection*, Topics in Applied Physics, Vol. 99 (Springer, Berlin).

133. May, T., Heinz, E, Peiselt, K., Zieger, G., Born, D., Zakosarenko, V., Brömel, A., Anders, S., and Meyer, H.-G. (2013). Next generation of a sub-millimetre wave security camera utilising superconducting detectors. *J. Instrum.* **8**, P01014 (11 pp).

134. Rosenberg, D., Lita, A. E., Miller, A. J., Nam, S., and Schwall, R. E. (2005). Performance of photon-number resolving transition-edge sensors with integrated 1550 nm resonant cavities. *IEEE Trans. Appl. Supercond.* **15**, 575–578.

135. Fukuda, D., Fujii, F., Yoshizawa, A., Tsuchida, H., Damayanthi, R. M. T., Takahashi, H., Inoue, S., and Ohkubo, M. (2008). High speed photon-number resolving detector with titanium transition edge sensor. *J. Low Temp. Phys.* **151**, 100–105.

136. Wollman, D. A., Irwin, K. D., Hilton, G. C., Dulcie, L. L., Newbury, D. E., and Martinis, J. M. (1997). High-resolution, energy-dispersive microcalorimeter spectrometer for X-ray microanalysis. *J. Microsc.* **188**, 196–223.

137. Uhlig, J., et al. (2013). Table-top ultrafast X-ray microcalorimeter spectrometry molecular structure. *Phys. Rev. Lett.* **110**, 138302 (5 pp).

138. Loidl, M., Leblanc, E., Bouchard, J., Branger, T., Coron, N., Leblanc, J., de Marcillac, P., Rotzinger, H., Daniyarov, T., Linck, M., Fleischmann, A., and Enss, C. (2004). High energy resolution X-ray, gamma and electron spectroscopy with cryogenic detectors. *Appl. Radiat. Isot.* **60**, 363–368.

139. Ullom, J. N., and Bennett, D. A. (2015). Review of superconducting transition-edge sensors for X-ray and gamma-ray spectroscopy. *Supercond. Sci. Technol.* **28**, 084003 (36 pp).

140. Horansky, R. D., Ullom, J. N., Beall, J. A., Hilton, G. C., Irwin, K. D., Dry, D. E., Hastings, E. P., Lamont, S. P., Rudy, C. R., and Rabin, W. R. (2008). Superconducting calorimetric alpha particle sensors for nuclear nonproliferation applications. *Appl. Phys. Lett.* **93**, 123504 (3 pp).

141. Lita, A. E., Miller, A. J., and Nam, S. W. (2008). Counting near-infrared single-photons with 95% efficiency. *Opt. Express* **16**, 3032–3040.

142. Miller, A. J., Lita, A. E., Calkins, B., Vayshenker, I., Gruber, S. M., and Nam, S. W. (2011). Compact cryogenic self-aligning fiber-to-detector coupling with losses below one percent. *Opt Express* **19**, 9102–9110.

143. Giustina, M., et al. (2015). Significant-loophole-free test of Bell's theorem with entangled photons. *Phys. Rev. Lett.* **115**, 250401 (7 pp).

144. Gastaldo, L., et al. (2014). The electron capture ^{163}Ho experiment ECHo. *J. Low Temp. Phys.* **176**, 876–884.

145. Irwin, K. D. (2002). SQUID multiplexers for transition-edge sensors. *Physica C* **368**, 203–210.

146. Irwin, K. D., and Lehnert, K. W. (2004). Microwave SQUID multiplexer. *Appl. Phys. Lett.* **85**, 2107–2109.

147. Kempf, S., Gastaldo, L., Fleischmann, A., and Enss, C. (2014). Microwave SQUID multiplexer for the readout of metallic magnetic calorimeters. *J. Low. Temp. Phys.* **175**, 850–860.

148. Yoon, J., Clarke, J., Gildemeister, J. M., Lee, A. T., Myers, M. J., Richards, P. L., and Skidmore, J. T. (2001). Single superconducting quantum interference device multiplexer for arrays of low-temperature sensors. *Appl. Phys. Lett.* **78**, 371–373.

149. Beyer, J., and Drung, D. (2008). A SQUID multiplexer with superconducting-to-normalconducting switches. *Supercond. Sci. Technol.* **21**, 105022 (5 pp).

150. Niemack, M. D., Beyer, J., Cho, H. M., Doriese, W. B., Hilton, G. C., Irwin, K. D., Reintsema, C. D., Schmidt, D. R., Ullom, J. N., and Vale, L. R. (2010). Code-division SQUID multiplexing. *Appl. Phys. Lett.* **96**, 163509 (3 pp).

151. Irwin, K. D., Niemack, M. D., Beyer, J., Cho, H. M., Doriese, W. B., Hilton, G. C., Reintsema, C. D., Schmidt, D. R., Ullom, J. N., and Vale, L. R. (2010). Code-division multiplexing of superconducting transition-edge sensor arrays. *Supercond. Sci. Technol.* **23**, 034004 (7 pp).

152. Holland, W. S., et al. (2013). SCUBA-2: the 10000 pixel bolometer camera on the James Clerk Maxwell Telescope. *Monthly Notices of the Royal Astronomical Society* **430**, 2513–2533.

153. Kent, D. I., Audley, M. D., Beall, J. A., Beyer, J., Deiker, S. D., Doriese, W. B., Duncan, W., Hilton, G. C., Holland, W., Reintsema, C. D., Ullom, J. N., Vale, L. R., and Zu, Y. (2004). In-focal-plane SQUID multiplexer. *Nuc. Instr. Meth. A* **520**, 544–547.

154. Irwin, K., Cho, H., Doriese, W., Fowler, J., Hilton, G., Niemack, M., Reintsema, C., Schmidt, D., Ullom, J., and Vale, L. (2012). Advanced code-division multiplexers for superconducting detector arrays. *J. Low Temp. Phys.* **167**, 588–594.

Chapter 9

Application in Adiabatic Quantum Annealing

Siyuan Han
Department of Physics and Astronomy, University of Kansas, Lawrence, Kansas 66045, USA
han@ku.edu

One of the emerging applications of Josephson tunnel junctions is quantum computation [1]. Quantum computation can solve mathematical problems that are intractable to conventional computers. For example, the amount of classical computing power required to factorize an N-bit integer into two prime numbers, upon which the most widely used RAS cryptosystem [2] is based, grows exponentially with N and thus is intractable to the conventional computers. Furthermore, the need to have new paradigms of computation is made more urgent as the 50-year-old Moore's law is expected to the end set by the fundamental laws of physics as the size of transistors—the basic units of digital information processing—approaches a few nanometers. This chapter intends to give a concise review on the application of Nb Josephson junctions (JJs) in quantum annealing (QA) [3]. It should be noted that many practical technological and engineering problems confronted in the development of a large-scale superconducting quantum annealing

Josephson Junctions: History, Devices, and Applications
Edited by Edward Wolf, Gerald Arnold, Michael Gurvitch, and John Zasadzinski

ISBN 978-981-4745-47-5 (Hardcover), 978-1-315-36452-0 (eBook)
www.panstanford.com

processor (QAP) [4, 5] are also of great concern to those interested in implementing gate-model quantum computing (GMQC) processors [1] using superconducting technologies.

The chapter begins with a brief introduction to diffident strategies for implementing quantum computation (QC) with an emphasis on quantum annealing. Section 9.2 presents the basic physics of the simplest superconducting flux qubit: RF-SQUID. Section 9.3, 9.4, and 9.5 give a brief description on the design and realization of scalable flux qubits, couplers, and control and measurement circuitry using Nb Josephson junctions, respectively. Section 9.6 describes a scalable architecture upon which quantum annealing processors consisting up to 1024 superconducting flux qubits made of more than 10^4 Nb Josephson junctions. Section 9.7 discusses evidence on whether D-Wave 2X, which is the latest quantum annealing processor available, works as expected. Section 9.8 is on the future prospect of the quantum annealing processors made of Nb flux qubits. The final section is a short summary. It should be pointed out that the scope of the materials presented and references cited is by no means comprehensive and complete.

9.1 Introduction

Conventional (i.e., classical) computers store and process digital information in the form of bit, which is either 0 or 1 but cannot be both 0 and 1 at the same time. Unlike the classical bit, a quantum bit (qubit) can be 0 and 1 simultaneously because of the principle of quantum state superposition. Combined with quantum entanglement, which originated from nonclassical correlations between qubits and unitary transformations that lead to reversible state evolution, quantum computers built with qubits offer the potential to be exponentially faster than conventional computers.

In 1994 Peter Shor discovered that the time required to find the primes of an integer on a quantum computer scales as polynomial as the size of the problem (i.e., the number of bits of the integer) [6]. In 1995 Lov Grover showed that using a quantum computer one can reduce the amount of time needed to search through an unstructured database of N items from $O(N)$ to $O(N^{1/2})$ [7, 8]. The discoveries of Shor's and Grover's quantum algorithms has since attracted widespread interest in the physical realization of quantum computers.

Another area in which quantum resources can be explored to turn impossible into probable is quantum simulation. Because the dimension of the Hilbert space of an N-particle quantum system increases exponentially as N grows, it is extremely hard to study the static and dynamic properties of such systems analytically and practically impossible numerically using conventional computers. For example, for a system consisting of interacting N qubits (i.e., quantum two-level systems such as electron spins) the Hilbert space is 2^N-dimensional, which cannot be tackled even with the world's most powerful supercomputers as N becomes greater than a few hundreds. In 1982 Richard Feynman raised an interesting idea of using quantum computers [9] to simulate the physics of such quantum systems. Feynman showed that while simulators based on classical computing would encounter an exponential slowdown, the proposed quantum simulators would not. In 1996, Seth Lloyd showed that a universal quantum simulator can be programmed to simulate any quantum system with polynomial amount of time [10].

At the present time there are two main approaches to quantum information processing. The first and more extensively studied is the gate-model, also called circuit-model, quantum computing. GMQC works very much like the conventional digital computers. Information is coded in qubits and is processed by applying a sequence of preprogrammed 1- and 2-qubit unitary transformations to produce solutions. It has been shown that any unitary transformations of multi-qubit state can be realized by a series of 1- and 2-qubit gates (i.e., unitary transformations of states), so that it is sufficient to operate quantum computers on 1- and 2-qubit gates at a time. It is in this sense that 1- and 2-qubit gates are said to be universal for quantum computation [11]. The greatest challenge to GMQC is decoherence caused by the unavoidable interaction between environment and the qubits which ultimately renders a coherent quantum superposition of qubits into a mixture of classical bits.

The characteristic time describing the process of losing quantum coherence is called coherence time τ_c while the time required to execute a typical quantum gate is called gate time τ_G. It is generally accepted that scalable quantum computing requires qubits with quantum quality factor τ_c/τ_G greater than the 10^4 threshold so that errors caused by decoherence can be corrected efficiently using various error correction schemes [12–17].

The second approach is adiabatic quantum computing (AQC), which is also known as ground state quantum computing. AQC can be considered as the quantum counterpart of classical analog computing. It has been shown that AQC is equivalent to gate-model quantum computation [4, 18, 19]. Thus, in principle any problems that can be solved efficiently by GMQC can also be solved efficiently by AQC. In the adiabatic approach, the Hamiltonian of the quantum simulator and its wave function $|\psi\rangle$ undergo adiabatic evolution in such a way that while the transformations of $|\psi\rangle$ represent some meaningful computation, this state also remains the instantaneous ground state of H_S throughout the process. This is achieved by starting the evolution from a sufficiently simple initial Hamiltonian, the ground state of which can be reached directly and readily, e.g., by energy relaxation, and evolving into a final Hamiltonian, which provides solution to some complex computation problem. One of the main advantage of doing computation this way is that the energy gap separating the ground and excited states of the Hamiltonian H_S ensures some measure of protection against decoherence. In particular, $|\psi\rangle$ maintains its coherence properties in time far beyond what would be the single-qubit decoherence time which limits the gate-model quantum computation and thus is expected to be more robust against noise and control errors [20, 21]. The cost one pays for this advantage of AQC is the need for more complex Hamiltonians required to implement universal computations in the ground-state mode.

While the task of building a universal adiabatic quantum computer is daunting, it is within the capability of present-day technology to build a special type of quantum simulator which is based on adiabatic quantum annealing (AQA). Annealing is a process for a system initially out of equilibrium to reach equilibrium (the state with minimum energy allowed by physics). The process is said to be classical (quantum) if it is driven primarily by thermal (quantum) fluctuations. Annealing can be used to solve hard problems. For example, the ground state of a particle in a two-dimensional (2D) potential landscape with numerous local potential minima separated by potential energy barriers can be found by starting the system at a temperature $T \sim \Delta U/k_\mathrm{B}$, where ΔU is the typical barrier height and k_B is the Boltzmann constant, then slowly let T goes to zero. It is obvious that for systems with deep local potential minima classical simulated annealing (SA) is not effective because once the particle falls into one of these potential minima the probability of escaping from it via thermal activation over the top of the barrier is quite low. In contrast, quantum annealing, which allows the particle to escape from local minima by tunneling, could be much more efficient for solving optimization problem [3].

A prototypical system used to demonstrate AQA is Ising spin glass (ISG), which describes the behavior of N interacting spin 1/2 particles in a magnetic field [22]. The Hamiltonian of a general Ising spin glass can be written as

$$H_{\mathrm{isg}} = -\sum_{i=1}^{N} \left(\varepsilon_\mathrm{i}\sigma_\mathrm{i}^z + \Delta_\mathrm{i}\sigma_i^x\right) + \sum_{i\neq j} J_{\mathrm{ij}}\sigma_\mathrm{i}^z\sigma_\mathrm{j}^z \tag{9.1}$$

where, σ_i^x and σ_i^z are the Pauli-X and Z operators of the ith spin, ε_i and Δ_i are proportional to the strength of the longitudinal and transverse magnetic fields on the ith spin, respectively. J_{ij} is the coupling between the ith and the jth spins. Because each spin can have two configurations, there are a total of 2^N possible configurations for a system of N Ising spins. In general, analytical solution of ISG is not attainable and thus numerical methods, such as simulated classical and quantum annealing protocols, have to be used. In 2002 Santoro et al. showed that QA is more effective than SA in finding the lowest energy configuration of the Ising spin glasses [5].

Hamiltonian (9.1) can be reorganized into the following form:

$$H_{\mathrm{isg}} = \left\{ -\sum_{i=1}^{N} \varepsilon_\mathrm{i}\sigma_\mathrm{i}^z - \sum_{i\neq j} J_{\mathrm{ij}}\sigma_\mathrm{i}^z\sigma_\mathrm{j}^z \right\} + \sum_{i=1}^{N} \Delta_\mathrm{i}\sigma_\mathrm{i}^x \equiv H_\mathrm{P} + \sum_{i=1}^{N} \Delta_\mathrm{i}\sigma_\mathrm{i}^x$$

where, H_P is the Hamiltonian of the classical Ising model whose ground state corresponding to the solution of the problem is to be solved. QA can be performed by slowly varying the time-dependent system Hamiltonian

$$H(t) = A(t)H_P + B(t)\sum_{i=1}^{N} \Delta_i \sigma_i^x \tag{9.2}$$

The ground state configuration of H_P can be found by starting the system from a quantum superposition of all possible states (candidate states) with equal weights. This initial state of the Ising spin glass can be obtained by setting all ε_i and J_{ij} to zero and Δ_i to maximum at $t = 0$, which is equivalent to have $A(0) = 0$ and $B(0) = 1$. By slowly decreasing $\Delta_i(t)$ to zero while at the same time gradually increasing ε_i and J_{ij} to values specified by the final problem Hamiltonian H_P (i.e., to $A = 1$ and $B = 0$). The process causes the probability amplitudes of all candidate states keep evolving according to the time-dependent local fields and coupling causing quantum tunneling between states. If the rate of changing $\Delta_i(t)$ is sufficiently slow (adiabatic), the system stays close to the ground state of the instantaneous Hamiltonian. When $\Delta_i(t)$ finally becomes zero, the system is expected to reach the ground state of H_P, which corresponds to the solution to the original optimization problem. In 1999 Brooke et al. reported an experimental demonstration of the quantum annealing for random magnets [23]. In December 2015, a team of researchers at Google announced that using a D-Wave 2X quantum annealer [13], consisting of 1024 superconducting flux qubits based on Nb Josephson tunnel junctions, they are able to outperform a single processor core by 10^8 times that runs either quantum Monte Carlo (QMC) or classical simulated annealing (SA) algorithm for solving instances of nearly 1000-binary-variable global optimization problem crafted to have tall and narrow energy barriers between local minima [24]. Furthermore, it has been proposed that a quantum annealer with $2N^2$ qubits can be used to factor $2N$-bit integers. Thus, a 2048-qubit D-Wave-type quantum annealer will be able to factor 32-bit integers. It is reasonable to expect that practically useful quantum annealers will come to fruition ahead of its GMQC counterpart.

9.2 Superconducting Flux Qubit

The basic building block of quantum computers and simulators, either digital or analog, is qubit—any objects that have two distinguishable quantum states well separated from other states. It is generally recognized that a physical system should meet the so-called DiVincenzo criteria to be a good candidate of qubits, of which the most important ones are long coherence time and large qubit quality factor, ability to prepare initial states, to control, and to measure the states of individual qubits. Another important requirement is that one must be able to scale up the size of the quantum circuits to contain tens of thousands or even millions of physical qubits.

Any quantum systems having two states that can be effectively separated from the other states can be used as qubits. Examples of physical qubits include nature-made systems such as the polarization states of optical photons, electron and nuclear spins, energy eigenstates of ions and neutral atoms, etc., and engineered systems such as superconducting devices based on Josephson junctions and semiconducting quantum dots. The nature-made qubits are usually microscopic objects with identical property. These microscopic objects in general interact weakly with the environment, which leads to long coherence times. However, the weak coupling to the external world also makes it difficult to control and measure the states of individual qubits. In addition, there does not seem to have a clear path to scale up quantum circuits made from nature-made qubits such as nuclear spins, ions, and atoms. Because engineered qubits are solid state devices that can be fabricated by integrated circuit technology it is straightforward to scale up circuits made of engineered solid state qubits. It is also much easier to control and measure the states of individual engineered qubits because their coupling to environment can be designed and controlled. The flip side of the strong qubit-environment coupling also induces strong decoherence, which might be more harmful to the gate model than ground state quantum computing [25].

There are three broad categories of superconducting qubits depending on whether the electric charge, the phase difference across a Josephson junction, or the magnetic flux enclose by a superconducting loop, respectively, is the relevant quantum mechanical observable [26]. Among them the flux qubit is the best to implement ISG Hamiltonian (9.1) because it is relatively straightforward to couple a flux qubit with its neighbors. Since the Nb/AlO_x/Nb trilayer Josephson tunnel junction has superior manufacturability and excellent controllability of device parameters, is durable with respect to thermal cycles and environmental elements, and compatible with advanced integrated circuit technology, it has been chosen as the top candidate for building analog quantum simulators. Furthermore, because the fundamental building blocks of the most advanced quantum annealers demonstrated so far are variations of superconducting flux qubits with one Josephson junction [27], a.k.a. the radio frequency superconducting quantum interference devices (RF-SQUID), this chapter will focus on the physics and application of RF-SQUID flux qubits.

An RF-SQUID consists of a superconducting ring of inductance L interrupted by a Josephson junction, as shown schematically in Fig. 9.1a. The junction is characterized by its critical current I_c, shunt capacitance C, and shunt resistance R. For the simple sinusoidal phase-current relationship given by the DC Josephson equation (1.4), the potential energy of the SQUID is given by

$$U(\Phi) = \frac{(\Phi - \Phi_e)^2}{2L} - E_J \cos\left(\frac{2\pi\Phi}{\Phi_0}\right) \tag{9.3}$$

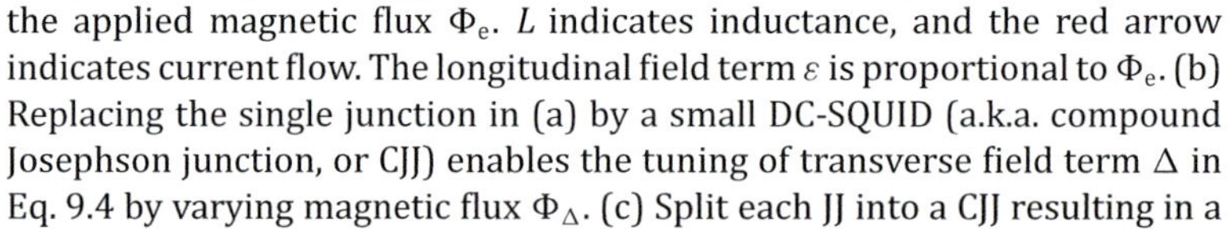

Figure 9.1 Circuit schematics of (a) RF-SQUID. The black cross represents the Josephson junction (JJ). The blue cross enclosed by a circle represent the applied magnetic flux Φ_e. L indicates inductance, and the red arrow indicates current flow. The longitudinal field term ε is proportional to Φ_e. (b) Replacing the single junction in (a) by a small DC-SQUID (a.k.a. compound Josephson junction, or CJJ) enables the tuning of transverse field term Δ in Eq. 9.4 by varying magnetic flux Φ_Δ. (c) Split each JJ into a CJJ resulting in a so-called CCJJ (compound-compound-JJ) RF-SQUID. By adjusting the fluxes inside each CJJ (unlabeled for clarity), the critical current of the two CJJs can be made practically identical. The novel CCJJ flux qubit eliminates $\varepsilon - \Delta$ crosstalk discussed in the text.

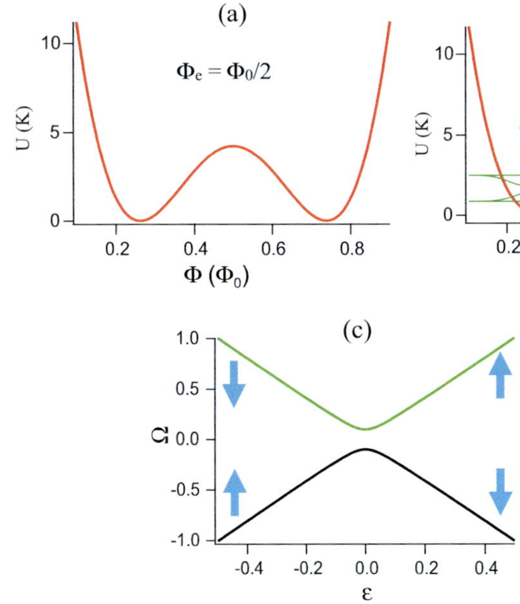

Figure 9.2 (a) Symmetric RF-SQUID potential at $\Phi_e = \Phi_0/2$. (b) Quantized energy levels of RF-SQUID. Only the lowest four levels with corresponding wavefunctions in the Φ space are shown. The supercurrent circulating the SQUID loop is counterclockwise (clockwise) when the "flux particle" is in the left (right) well. The state with counterclockwise (clockwise) circulation current can be mapped to the spin-up (-down) states of the σ^z operator represented by the up (down) arrow. (c) Energy of the ground state (black line) and the first excited state (green line) of the SQUID flux qubit as a function of external magnetic flux.

This potential energy is the sum of the magnetic energy stored in the inductor and the Josephson coupling energy of the tunnel junction, where $E_J = \hbar I_c/2e$ is the Josephson coupling energy, while Φ and Φ_e are the total and external magnetic flux threading the SQUID ring, respectively.

For $\Phi_e = \Phi_0/2$ the SQUID potential (9.3) is symmetric as shown in Fig. 9.2a. The barrier height depends on L and the value of the dimensionless parameter $\beta_L = 2\pi L I_c/\Phi_0$. Changing Φ_e slightly away from $\Phi_0/2$ creates an energy difference ε between the left and the right potential minima. For $\Delta\Phi_e = (\Phi_e - \Phi_0/2) \ll \Phi_0$ the energy bias is $\varepsilon = 2I_p\Delta\Phi_e$, where I_p is the superconducting circulation current (a.k.a. persistent current) in the SQUID loop.

The Hamiltonian of an RF-SQUID is

$$H = \frac{Q^2}{2C} + \frac{(\Phi - \Phi_{ex})^2}{2L} - E_J \cos\left(\frac{2\pi\Phi}{\Phi_0}\right), \tag{9.4}$$

where Q is the charge on junction's shunt capacitance satisfying $[\Phi, Q] = i\hbar$. According to (9.3) the eigen-energy spectrum of RF-SQUIDs is quantized. The validity of RF-SQUID Hamiltonian was confirmed by Rouse et al. [28] in 1995 by matching the experimentally measured positions of the resonant tunneling peaks with that of obtained by diagonalizing H. Additional experiments by Han et al. and Friedman et al. demonstrated a number of unique quantum phenomena, ranging from population inversion to coherent superposition of states, in Nb Josephson tunnel junction based RF-SQUIDs [29, 30]. The result of these experiments showed that RF-SQUIDs are quantum objects and can be treated as artificial atoms. These early works laid the foundation for the use of RF-SQUIDs as flux qubits for scalable AQA.

For energy bias ε smaller than the intra-well-level spacing, the two lowest-lying levels are very well isolated from the other states due to the strong anharmonicity of the SQUID potential and thus can be utilized as the computational states of the flux qubit. The truncated Hamiltonian in the subspace spanned by these two states can be written conveniently as

$$H_q = -\frac{1}{2}\left(\varepsilon\sigma^z + \Delta\sigma^x\right), \tag{9.5}$$

where Δ is the level spacing of the ground and excited states at $\Delta\Phi_e = 0$. This Hamiltonian is identical to that of a spin 1/2 particle in a longitudinal field ε, which can be adjusted by varying the external flux Φ_e applied to the RF-SQUID loop, and a constant transverse field Δ. The eigenenergies of the flux qubit as a function of ε is given by $\Omega_\pm = \pm\sqrt{\varepsilon^2 + \Delta^2}$ as depicted in Fig. 9.2c. However, in order to implement the QA protocol Δ also needs to be adjustable, which can be realized by replacing the single junction by a double-junction DC-SQUID of inductance much smaller than L. The resulting device is equivalent to a RF-SQUID whose effective critical current can be tuned in situ by changing the external magnetic flux applied to the DC-SQUID loop [31], as shown in Fig. 9.2b. The critical current tunable low-inductance DC-SQUID is also referred to a compound Josephson junction and has been used widely in applications where it is necessary to have devices with tunable Josephson coupling energy such as the frequency-tunable superconducting flux, transmon, and Xmon qubits [32, 33]. Notice that for qubits with typical parameters, separation between adjacent energy levels in the same potential well is about 10 GHz. It is thus necessary to cool qubits to below about 30 mK to ensure the system remains in its ground state.

9.3 Robust and Scalable Flux Qubit

For the precise control of single qubit parameters ε and Δ, it is highly desirable to have minimal crosstalk between them. For example, ideally the current targeted for changing ε should not cause changes in Δ of the qubit. In principle, linear crosstalk can be completely eliminated by properly compensating the effect of crosstalk with the help of carefully measured inductance matrix M that relates the source currents to the target fluxes. In practice this approach is not efficient because the time and hardware resource required grows quadratically with the number of qubits N. In theory, at the single qubit level, perfect "orthogonality" between ε and Δ can be achieved by exploring the symmetry property of the Δ-tunable flux qubit assuming the two junctions could be made identical [34]. The problem is that the device parameters of Josephson junctions, such as critical current, are always not as uniform as one desires due to fabrication variability. Therefore, even if one is able to design circuits with diagonalized inductance matrices M, there would still be an effective crosstalk between ε and Δ that has originated from, for instance, spreads in junctions' critical currents.

In 2010 Harris et al. designed and tested a novel type of flux qubit aimed at solving the problem of effective ε–Δ crosstalk [27] arising from fabrication variations in Josephson junction critical currents. The new design uses four Josephson junctions, which could have a few percent critical current variation, in place of the two junctions in the Δ-tunable flux qubit. It essentially replaced each

Josephson junction in the original Δ-tunable flux qubit by a smaller two-junction DC-SQUID (a.k.a. compound Josephson junction or CJJ), so that the critical currents of the two CJJs can be made equal by setting properly the fluxes applied to the CJJs, thus reducing ε–Δ crosstalk to negligible levels.

The design is shown to be *robust* in that the undesirable effects from realistic levels of fabrication variations can be readily compensated in situ. In addition, the design is deemed to be *scalable* in two respects. First, the signals needed to compensate for fabrication variations are static flux biases, as opposed to custom-tuned time-dependent flux biases. This is a significant advantage as the qubit can then be controlled by static on-chip magnetic memory devices that are programmed using a scalable control signal architecture based on single flux quantum (SFQ) superconducting logic circuitry [35]. Such a scheme makes economical use of what will inevitably be a limited number of on-chip bias lines in any practical processor. Second, the design allows one to compensate for fabrication variations both within single qubits and between multiple qubits, thus providing a means of homogenizing qubit parameters across a multiqubit device. It is emphasized that this latter issue is of critical importance in the development of useful large-scale quantum information processors that could potentially involve thousands of qubits.

9.4 Coupler

One approach to implement the spin–spin interaction energy J_{ij} between spins to run AQA involves setting up a network of inductively coupled flux qubits. A simple way to couple two flux qubits is to have them both inductively coupled to a flux transformer (a.k.a. coupler), which is just a superconducting loop. The problem of this simple coupler design is that the coupling strength is fixed while the AQA protocol described above requires the adjustment of J_{ij} in situ. In 2005, Maassen van den Brink et al. proposed the use of an RF-SQUID to implement tunable J_{ij} [36]. Experiments on systems of coupled flux qubits verified that such couplers did perform as expected [37]. However, additional work not reported in the literature revealed two serious drawbacks: first, the tuning mechanism involves threading flux through the RF-SQUID (coupler) loop, thus inducing a large persistent current that, in turn, biases the qubits. This is a significant problem if the qubit biases need to be controlled to high precision atop what can be a very large nonlinear crosstalk imparted by the coupler. Second, $J_{\mathrm{ij}} = 0$ can only be achieved if $L_{\mathrm{C}}/L_{\mathrm{JC}} < 1$ where L_{C}, $L_{\mathrm{JC}} L_{\mathrm{JC}} = \Phi_0/2\pi I_0$ and I_0 are the RF-SQUID coupler's geometric inductance, Josephson inductance, and critical current, respectively. On the other hand, in order to achieve appreciable nonzero coupling, it proved necessary to design couplers with $L_{\mathrm{C}}/L_{\mathrm{JC}} > 0.9$. Such devices were acutely sensitive to fabrication variations, where higher than expected critical current density of the $\mathrm{Nb/AlO}_x\mathrm{/Nb}$ trilayer could make $J_{\mathrm{ij}} = 0$ unattainable. Thus the simple RF-SQUID coupler design proved troublesome in practice. Note that the dc-SQUID [36, 38] and the unipolar [39] couplers suffer from similar deficiencies. The challenge was then to design a tunable J_{ij} that invokes minimal persistent current in the coupler loop and is robust against fabrication variations.

To address the challenge Harris et al. [40] proposed and demonstrated a new form of coupler, based on the CJJ RF-SQUID (Fig. 9.1b). The difference between a CJJ RF-SQUID coupler and a CJJ RF-SQUID flux qubit is that the geometric inductance-to-Josephson inductance ratio of the coupler is less than, while that of the qubit is greater than, unity. Figure 9.3 shows the circuit schematics of two CJJ RF-SQUID flux qubit coupled by a CJJ RF-SQUID coupler.

This coupler provides both sign- and magnitude-tunable mutual inductance in a manner that invokes minimal nonlinear crosstalk from the coupler tuning parameter to the qubits. Furthermore, this crosstalk can be reduced to negligible levels with improved fabrication uniformity and subtle improvements in device layout. Modulation of the qubit inductance via changes in the coupler settings has been characterized and shown to be predictable using an effective one-dimensional model of the coupler potential. The effective mutual inductance between the qubit is

$$M_{\mathrm{eff}} = M_{\mathrm{co},1} M_{\mathrm{co},2} \chi^{(1)}, \tag{9.6}$$

where $\chi^{(1)}$ is the first-order (linear) susceptibility of the coupler. Since the derivation of the dependence of $\chi^{(1)}$ is quite lengthy, we

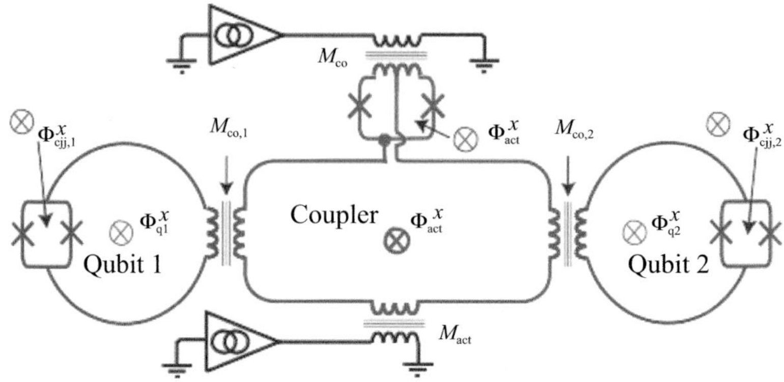

Figure 9.3 Schematic of a CJJ RF-SQUID coupler interacting with two CJJ rf-SQUID qubits. The coupler is controlled via bias currents that are coupled to the device through mutual inductances M_{co} and M_{act}, respectively. These give rise to the fluxes Φ^x_{co} and Φ^x_{act}. The transverse and longitudinal fields Δ and ε of the qubits are controlled via fluxes $\Phi^x_{cjj\alpha}$ and $\Phi^x_{q\alpha}$ ($\alpha = 1, 2$). The qubits interact with the coupler via mutual inductances $M_{co,\alpha}$.

will not repeat it here. Those who are interested in the detail are referred to Ref. [40].

9.5 Control and Measurement Circuit

The ε- and Δ-adjustable flux qubits together with the sign and magnitude tunable coupler provides a good starting point to the development of scalable superconducting quantum simulators. However, the operation of a quantum simulator also requires the ability to precisely control ε_j, Δ_j, and J_{ij} (i, $j = 1, 2, \ldots, N$) and to read out the states of all flux qubits at the end of the annealing process. Since ε_j, Δ_j, and J_{ij} are all current-driven, each of them can be controlled by a room temperature source through a bias line running from ~300 K to the qubit circuit cooled to about 20 mK. This simple approach requires at least two wires for each qubit and one wire for each coupler, which is okay for circuits having no more than a few tens of qubits, but quickly becomes unmanageable for a large number of qubits because of the limited cooling power of the dilution refrigerator. An elegant solution is to use SFQ on-chip programmable integrated superconducting digital circuits (e.g., XY-addressable arrays, digital-to-analog converters, demultiplexers), so that each line going into the fridge can be used to control multiple qubits and/or couplers and a few hundred control lines are sufficient to control many thousands of qubits as demonstrated by Johnson et al. [41].

One of the critical problems of developing scalable quantum annealers is to read out the states of all qubits truthfully. The state of a flux qubit is represented by the direction of superconducting circulation current in the RF-SQUID ring. The usual way of measuring small magnetic signal using DC-SQUID running in either the flux-locked loop mode or small signal mode would not work, because for both methods the DC-SQUID is operated in the finite voltage branch of the current–voltage characteristics (IVC), which not only generate on-chip heat but also microwaves with frequency in the range of 10–10^2 GHz, enough to excite the qubits out of the ground state, due to the AC Josephson effect (Eq. 1.2). An alternative way is to measure the state of the qubit using an underdamped hysteretic DC-SQUID inductively coupled to the qubit. Because the critical current of the DC-SQUID depends on the magnetic flux threading its loop, the state of the qubit can be inferred by measuring the critical current of the DC-SQUID. However, the sudden jump from the zero voltage to the finite voltage state generates a large transient pulse. The jolt is so severe that it can randomly flip the states of other qubits, thus destroying the ground state of the ISG system reached at the end of adiabatic quantum evolution. To solve this so-called readout destruction problem, Berkley et al. developed a scalable XY-addressable readout system for superconducting adiabatic quantum annealer [42]. The readout system is designed specifically for reading out the states of the superconducting flux qubits in an adiabatic quantum annealer. For an N-qubit adiabatic quantum annealer, this readout system uses N hysteretic DC-SQUIDs and N hysteretic ($L_J/L \gg 1$) RF-SQUIDs operated as quantum flux parametrons (QFPs). The qubits are coupled to QFPs, which then couple to the DC-SQUIDs. This readout architecture not only solves the problem of readout destruction but also has significantly higher flux sensitivity than DC-SQUIDs directly coupled to flux qubits. The readout system has been tested and demonstrated successfully and was subsequently deployed in D-Wave 1 (128-qubit), 2 (512-qubit), and 2X (1024-qubit) quantum annealers [24, 43].

9.6 Scalable Architecture

A scalable architecture is critical to the development of quantum annealers built upon superconducting flux qubits. Researchers at D-Wave invented a scalable architecture tailored to the CCJJ RF-SQUID qubits (Fig. 9.1c) and CJJ RF-SQUID couplers (Fig. 9.1b and Fig. 9.3) described above. The architecture is based on 8-qubit unit cells that form an expandable network. In each unit cell the eight qubits are divided into two 4-qubit groups. One group consists of four qubits with vertical skinny long loops (q_1, q_2, q_3, q_4) and the other group consists of four qubits with horizontal loops (q_5, q_6, q_7, q_8) as shown in Fig. 9.4a. Inter-qubit couplers are placed at the intersections of

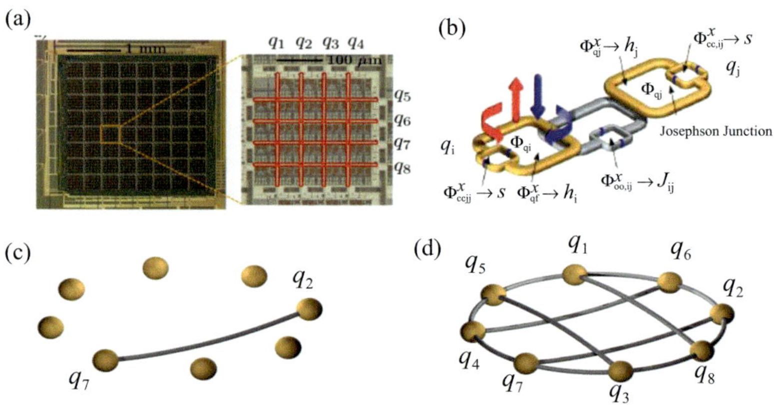

Figure 9.4 (a) Left: Photograph of a QA processor. Right: The eight-qubit unit cell. The bodies of the qubits are extended loops of Nb wiring (highlighted with red rectangles). Inter-qubit couplers are located at the intersections of the qubit bodies. (b) Schematic diagram of a pair of coupled superconducting flux qubits with external control flux biases and with flux through the body of the ith qubit. An inductive coupling between the qubits is tuned with the bias $\Phi_{co,ij}$. (c, d) The two- and eight-qubit systems investigated by Lanting et al. [44] were programmed to have the topologies shown. Qubits are represented as gold spheres, and inter-qubit couplers are represented as silver lines. Adapted from Lanting et al. [44].

the vertical and horizontal qubit bodies. Therefore, within a unit cell each vertical (horizontal) qubit is coupled to four horizontal (vertical) qubits but not to other vertical (horizontal) qubits. The topology corresponding to maximum intra-unit cell connectivity is depicted in Fig. 9.4d. In addition, each vertical (horizontal) qubit is coupled to its counterpart in the upper (left) and lower (right) nearest neighbor cells. Therefore, a unit cell can be coupled to a maximum of four nearest neighbor unit cells. With this architecture one can construct a 32-qubit quantum annealing processor (QAP) with 4 unit cells, a 128-qubit QAP with $4^2 = 16$ unit cells, etc. It is easy to show that a 2^{2k+3}-qubit QAP can be put together using a square lattice of 4^k unit cells with essentially the same technology. In this sense, the architecture is scalable. For example, the first generation 128-qubit D-Wave 1 QAP has 16 unit cells, the second generation 512-qubit D-Wave 2 purchased by Google and NASA jointly a few years ago has 64 unit cells, and the latest model available is a 2.5 generation 1024-qubit D-Wave 2X acquired recently by Google.

9.7 Does It Work?

The QAP hardware developed by D-Wave over the last 10 years is truly impressive. For example, the parameters of Nb Josephson tunnel junctions are tightly controlled, the latest generation QAP contains tens of thousands Josephson junctions, and the integration of on-chip SFQ superconducting logic circuitry. However, coherence time of superconducting qubits made from Nb junctions is typically on the order of 10^2 ns, which is much shorter than their Al junction counterparts. It is considered that for GMQC the coherence time of typical superconducting qubits should be on the order of 10^{-1} ms to implement error correction that is indispensable for scalable digital quantum processors. Although it is not entirely clear about the minimum coherence time required for running QA the fact that Nb qubits' coherence timescale is orders of magnitude shorter than the adiabatic evolution time significantly does not seem to inspire confidence. Thus, the question in the minds of many people, especially in academic circles, is whether the D-Wave type of QAP built upon Nb qubits works or not. As discussed below, this question has three levels of meaning.

At the lowest level "does it work" means "Does it work as an annealer that can actually find solutions for problems it is designed to solve?" At the intermediate level the question means "Is it quantum?" or "Does the process involves quantum physics?" The question ultimately means "Does it show quantum speedup?" Tests performed on D-Wave QAPs of various sizes have answered the first two questions positively [24, 44–47]. In particular, the work of Denchev et al. [24] showed that the 1024-qubit D-Wave QAP is not only capable of solving ISG and computationally more difficult multivariable global optimization problems but also the critical role played by finite range quantum tunneling. In addition, the experiments of Johnson et al. [45], Boixo et al. [46], and Lanting et al. [44] demonstrated quantum coherence and entanglement across the 8-qubit unit cells investigated. These experimental studies provide strong evidence that during the process of quantum annealing all of the qubits in the unit cell were coherent and entangled and that the unit cell-scale coherence and entanglement persisted even as the system reached thermal equilibrium with its environment. It is noted that the unit cell-scale coherence and entanglement were observed during a significant portion of the QA process. Furthermore, it also observed that QA is surprisingly robust against environment noise, qubits parameter variations, and perturbations in flux biases. Because the total run time of the process $t_{QA} \sim 10 - 10^2$ μs and the coherence time of the Nb flux qubits, though not measured directly by the standard time domain techniques [26], is most likely much less than 1 μs these results support the notion that ground state quantum computing (e.g., QA) in general is more robust with respect to decoherence [20].

One important question that remains open is how environment-induced decoherence affects ground state quantum computing. Theoretical [20, 48–50] and experimental [44–46] evidence indicate decoherence is less harmful to AQC than to GMQC. The robustness of ground state quantum computing can be understood intuitively with the following argument: during adiabatic evolution of adiabatic quantum computation the system stays closely to its instantaneous ground state. Energy relaxation which produces errors in GMQC might actually assist ground state quantum computing by reducing the probability of occupying excited states. Furthermore, staying in the ground state also makes the effect of dephasing much less relevant. Overall, the role of decoherence in ground state quantum computing is not well understood and more theoretical and experimental studies are needed to clarify this problem of fundamental importance.

The latest experimental result demonstrated that for problems whose potential energy landscape have many local minima separated by high and thin barriers QA has significant runtime advantage, about a factor of 10^8 in fact, over SA and QMC [24]. To an ordinary person this is a remarkable speedup. However, to experts working in quantum information science it may or may not be the evidence for "quantum speedup," which refers to the qualitative improvement in the scaling of the total run time versus the problem size of a quantum algorithm over that of the best classical algorithm. For example, the famous Shor's factoring algorithm is said to have exponential quantum speedup while Grover's search algorithm achieves polynomial quantum speedup.

According to this formal definition of "quantum speedup" D-Wave QAPs have not shown definitive evidence of quantum speedup, despite the demonstrated 10^8 performance improvement over a single core processor running either SA or QMC. So the jury is still out on whether without global coherence and entanglement of all flux qubits in a QAP over a significant portion of the annealing time quantum speedup is expected. On one hand, if quantum speedup indeed requires global scale coherence and entanglement, in contrast to the local scale ones observed in 8-qubit unit cells of D-Wave QAPs, then even the best Nb qubit technology currently available may not be sufficient to reach the required coherence timescale. On the other hand, the 10^8 so called "constant speedup" in solving the special class of problems tested indicate that at the equivalence of less than a dollar per single core level D-Wave QAPs could have a competitive performance/price edge over classical computing technology.

9.8 Future Prospects

The progress made in the last ten years on the development of quantum annealing processors based on superconducting flux qubits made of trilayer Nb Josephson junction technology is remarkable. It is expected that the 3rd generation 2048-qubit D-Wave QAP will come out soon and according to recent trend of doubling the qubit count every 12 months we may have a QAP with more than 16,000 qubits before year 2020. In addition to solving hard optimization problems this type of QAPs could also be used for pattern recognition, financial analysis, and artificial intelligence. Furthermore, it will be very interesting to see if prime factoring algorithm is implementable on QAPs. However, it is clear that acceptance and adoption of superconducting Nb QAPs depends on several critical issues.

The most fundamental issue that obviously needs more theoretical understanding and/or experimental demonstration is the importance of global coherence and/or entanglement to ground state QC. It is easy to see from the quantum annealing Hamiltonian (9.3) that at $t = 0$, the system is in the product state $|\Psi(0)\rangle = |+\rangle_1 \otimes |+\rangle_2 \ldots \otimes |+\rangle_i \ldots \otimes |+\rangle_N$, where $|+\rangle_i$ is the eigenstate of σ_i^x operator, and at the end of annealing $t = t_f$ in the ground state, which encodes the solution of the optimization problem, must be separable. Thus, neither entanglement nor quantum coherence of processor scale is required for the entire annealing process. Nevertheless, the lack of global coherence and/or entanglement during the critical stage of annealing may very well lead to no quantum speedup over classical computers. The work of Dickson et al. [51] and Lanting et al. [44] provide empirical evidence indicating that quantum coherence and entanglement in the tested 8- and 16-qubit systems persisted well beyond the single qubit coherence time. It is found even with annealing times many orders of magnitude longer than the estimated single-qubit coherence time the performance, in terms of the probability of finding the correct ground state of the problem Hamiltonian, is similar to those expected for a fully coherent system.

In practice, increasing coherence time of Nb flux qubits will improve the performance QAPs and the probability of achieve quantum speedup. Currently, the coherence time of Nb qubits is about 10^3 less than that of Al qubits. Because there does not seem to be a fundamental microscopic mechanism limiting the coherence time of Nb qubits [27] the huge performance gap between the Nb and Al superconducting qubits could be originated from material and processing issues. It is noticed that the almost all of the present-day technology used to fabricate Nb JJs and integrated circuits are based on the $Nb/AlO_x/Nb$ trilayer process invented by Gurvitch et al. (Chapter 5 and references therein). The Nb and Al films are usually deposited by DC sputtering in argon plasma of a few times of 10^{-1} Pa. After sputtering Al the chamber is evacuated to the base pressure of a few times of 10^{-6} Pa. High purity oxygen gas is then introduced to react with Al to form the tunnel barrier. In comparison, Al JJs used in all of the long coherence time superconducting qubits are fabricated in vacuum systems having base pressure on the order 10^{-9} Pa. The much lower base pressure significantly reduces the probability of contaminating the tunnel barrier and the superconducting electrode/barrier interfaces. The contaminant most likely result in microscopic two-level fluctuators inside the barrier and/or at the superconductor/oxide interface which have been identified as a major source of decoherence [52]. By making Nb trilayers in ultrahigh vacuum systems one may significantly improve the coherence time of Nb qubits.

As the qubit count of future generations of QAP and the problem size increase the run time for reach the ground state of the problem Hamiltonian by quantum annealing increase as well. Therefore, the run time of future QAP will be orders of magnitude greater than the coherence time of the best Nb qubits and error correction [53, 54] must be incorporated into the design and operation of the quantum processors. The is a tough theoretical and practical challenge that must be overcome to build scalable ground state quantum computers that can harness the power of quantum speedup.

Finally, in order to moving beyond QA to universal adiabatic quantum computing one must be able to implement Heisenberg spin interaction terms in the system Hamiltonian of Ref. [55]. Because for CJJ and CCJJ RF-SQUID qubits the realizable range of $\sigma_i^x\sigma_j^x$ interaction strength is much smaller than that of $\sigma_i^z\sigma_j^z$ new types of superconducting qubits need to be invented to implement universal AQC.

9.9 Summary

Adiabatic quantum computing is a viable alternative to gate model quantum computing. In AQC the system stays closely to the time-dependent instantaneous ground state and thus is expected to be more robust than GMQC against environment noise which ultimately destroy quantum coherence and entanglement. Quantum annealing is a special type of ground state quantum computing. Although QA is not equivalent to universal GMQC it is an effective strategy for solving difficult optimization problems such as ISG and problems with a large number of binary variables.

Quantum annealers have been developed and demonstrated using RF-SQUID-based flux qubits and couplers made of Nb junctions as the building blocks. The architectural complexity of QAPs, the need for SFQ on-chip control and measurement circuitry, and the demand for durability and highly uniform junction parameters makes the Nb Josephson junction the top candidate for the implementation of quantum annealing processors. The debut of D-Wave 1 a few years ago represents a milestone in the development of practically useful superconducting quantum annealing processors. Tests performed on a 1024-qubit D-Wave 2X programmable spin simulator revealed some very interesting results. The most noticeable is the demonstration of a factor of 10^8 "constant speedup" over optimized QMC or SA algorithms running on a single core processor for solving optimization problem instances with 945 binary variables. However, the performance gap between QA and QMC or SA did not widen indicating a lack of quantum speedup. It is interesting to find out what types of problems will show evidence of quantum speedup when solved on a QAP.

References

1. Nielsen, M. A., and Chuang, I. L. (2000). *Quantum Computation and Quantum Information* (Cambridge University Press).
2. Rivest, R. L., Shamir, A., and Adleman, L. (1978). A method for obtaining digital signatures and public-key cryptosystems. *Commun. ACM* **21**, 120–126.
3. Finnila, A. B., Gomez, M. A., Sebenik, C., Stenson, C., and Doll, J. D. (1994). Quantum annealing: a new method for minimizing multidimensional functions. *Chem. Phys. Lett.* **219**, 343–348.
4. Farhi, E., Goldstone, J., Gutmann, S., Lapan, J., Lundgren, A., and Preda, D. (2001). A quantum adiabatic evolution algorithm applied to random instances of an NP-complete problem. *Science* **292**, 472–475.
5. Santoro, G. E., Martoňák, R., Tosatti, E., and Car, R. (2002). Theory of quantum annealing of an ising spin glass. *Science* **295**, 2427–2430.
6. Shor, P. W. (1994). Algorithms for quantum computation: discrete log and factoring. In *Proceedings of the 35th Annual Symposium on Foundation of Computer Science*, Goldwasser, S., ed. (IEEE Computer Society Press, Los Alamitos, CA), pp. 124–134.
7. Grover, L. K. (1996). A fast quantum mechanical algorithm for database search. In *Proceedings of the 28th Annual ACM Symposium on the Theory of Computing* (ACM Press, New York), pp. 212–219.
8. Grover, L. K. (1997). Quantum mechanics helps in searching for a needle in a haystack. *Phys. Rev. Lett.* **79**, 325–328.
9. Feynman, R. (1982). Simulating physics with computers. *Int. J. Theor. Phys.* **21**, 467–488.
10. Lloyd, S. (1996). Universal quantum simulators. *Science* **273**, 1073–1078.
11. DiVincenzo, D. P. (1995). Two-bit gates are universal for quantum computation. *Phys. Rev. A* **51**, 1015–1022.
12. Kribs, D., Laflamme, R., and Poulin, D. (2005). Unified and generalized approach to quantum error correction. *Phys. Rev. Lett.* **94**, 180501.
13. Knill, E., Laflamme, R., and Viola, L. (2000). Theory of quantum error correction for general noise. *Phys. Rev. Lett.* **84**, 2525–2528.
14. Barnes, J. P., and Warren, W. S. (2000). Automatic quantum error correction. *Phys. Rev. Lett.* **85**, 856–859.
15. Steane, A. M. (1996). Error correcting codes in quantum theory. *Phys. Rev. Lett.* **77**, 793–797.

16. Laflamme, R., Miquel, C., Paz, J. P., and Zurek, W. H. (1996). Perfect quantum error correcting code. *Phys. Rev. Lett.* **77**, 198–201.
17. DiVincenzo, D. P., and Shor, P. W. (1996). Fault-tolerant error correction with efficient quantum codes. *Phys. Rev. Lett.* **77**, 3260–3263.
18. Mizel, A., Lidar, D. A., and Mitchell, M. (2007). Simple proof of equivalence between adiabatic quantum computation and the circuit model. *Phys. Rev. Lett.* **99**, 070502–070504.
19. Aharonov, D., van Dam, W., Kempe, J., Landau, Z., Lloyd, S., and Regev, O. (2008). Adiabatic quantum computation is equivalent to standard quantum computation. *SIAM Rev.* **50**, 755–787.
20. Childs, A. M., Farhi, E., and Preskill, J. (2001). Robustness of adiabatic quantum computation. *Phys. Rev. A* **65**, 012322.
21. Roland, J., and Cerf, N. J. (2005). Noise resistance of adiabatic quantum computation using random matrix theory. *Phys. Rev. A* **71**, 032330.
22. Ising, E. (1925). Beitrag zur Theorie des Ferromagnetismus. *Z. Phys.* **31**, 253–258.
23. Brooke, J., Bitko, D., Rosenbaum, T. F., and Aeppli, G. (1999). Quantum annealing of a disordered magnet. *Science* **284**, 779–781.
24. Denchev, V. S., Boixo, S., Isakov, S. V., Ding, N., Babbush, R., Smelyanskiy, V., Martinis, J., and Neven, H. (2015). What is the computational value of finite range tunneling?. *Phys. Rev. X* **6**, 031015.
25. Amin, M. H. S., Averin, D. V., and Nesteroff, J. A. (2009). Decoherence in adiabatic quantum computation. *Phys. Rev. A* **79**, 022107–022104.
26. Clarke, J., and Wilhelm, F. K. (2008). Superconducting quantum bits. *Nature* **453**, 1031–1042.
27. Harris, R., Johansson, J., Berkley, A. J., Johnson, M. W., Lanting, T., Han, S., Bunyk, P., Ladizinsky, E., Oh, T., Perminov, I., Tolkacheva, E., Uchaikin, S., Chapple, E. M., Enderud, C., Rich, C., Thom, M., Wang, J., Wilson, B., and Rose, G. (2010). Experimental demonstration of a robust and scalable flux qubit. *Phys. Rev. B* **81**, 134510.
28. Rouse, R., Han, S., and Lukens, J. E. (1995). Observation of resonant tunneling between macroscopically distinct quantum levels. *Phys. Rev. Lett.* **75**, 1614.
29. Han, S., Rouse, R., and Lukens, J. E. (1996). Generation of a population inversion between quantum states of a macroscopic variable. *Phys. Rev. Lett.* **76**, 3404.
30. Friedman, J. R., Patel, V., Chen, W., Tolpygo, S. K., and Lukens, J. E. (2000). Quantum superposition of distinct macroscopic states. *Nature* **406**, 43–46.

31. Han, S., Lapointe, J., and Lukens, J. E. (1989). Thermal activation in a two-dimensional potential. *Phys. Rev. Lett.* **63**, 1712.
32. Koch, J., Yu, T. M., Gambetta, J., Houck, A. A., Schuster, D. I., Majer, J., Blais, A., Devoret, M. H., Girvin, S. M., and Schoelkopf, R. J. (2007). Charge-insensitive qubit design derived from the Cooper pair box. *Phys. Rev. A* **76**, 042319.
33. Barends, R., Kelly, J., Megrant, A., Sank, D., Jeffrey, E., Chen, Y., Yin, Y., Chiaro, B., Mutus, J., Neill, C., O'Malley, P., Roushan, P., Wenner, J., White, T. C., Cleland, A. N., and Martinis, J. M. (2013). Coherent Josephson qubit suitable for scalable quantum integrated circuits. *Phys. Rev. Lett.* **111**, 080502.
34. Han, S., Lapointe, J., and Lukens, J. E. (1993). Thermally activated barrier crossings in superconducting quantum interference devices. In *Activated Barrier Crossing*, Fleming, G. R., and Hanggi, P., eds. (World Scientific, Singapore), pp. 241–267.
35. Likharev, K. K., and Semenov, V. K. (1991). RSFQ logic/memory family: a new Josephson-junction technology for sub-terahertz-clock-frequency digital systems. *IEEE Trans. Appl. Supercond.* **1**, 3–28.
36. Maassen van den Brink, A. (2005). Hamiltonian for coupled flux qubits. *Phys. Rev. B* **71**, 064503.
37. Harris, R., Berkley, A. J., Johnson, M. W., Bunyk, P., Govorkov, S., Thom, M. C., Uchaikin, S., Wilson, A. B., Chung, J., Holtham, E., Biamonte, J. D., Smirnov, A. Y., Amin, M. H. S., and Maassen van den Brink, A. (2007). Sign- and magnitude-tunable coupler for superconducting flux qubits. *Phys. Rev. Lett.* **98**, 177001.
38. Hime, T., Reichardt, P. A., Plourde, B. L. T., Robertson, T. L., Wu, C.-E., Ustinov, A. V., and Clarke, J. (2006). Solid-state qubits with current-controlled coupling. *Science* **314**, 1427–1429.
39. Castellano, M. G., Chiarello, F., Leoni, R., Simeone, D., Torrioli, G., Cosmelli, C., and Carelli, P. (2005). Variable transformer for controllable flux coupling. *Appl. Phys. Lett.* **86**, 152504.
40. Harris, R., Lanting, T., Berkley, A. J., Johansson, J., Johnson, M. W., Bunyk, P., Ladizinsky, E., Ladizinsky, N., Oh, T., and Han, S. (2009). Compound Josephson-junction coupler for flux qubits with minimal crosstalk. *Phys. Rev. B* **80**, 052506.
41. Johnson, M. W., Bunyk, P., Maibaum, F., Tolkacheva, E., Berkley, A. J., Chapple, E. M., Harris, R., Johansson, J., Lanting, T., Perminov, I., Ladizinsky, E., Oh, T., and Rose, G. (2010). A scalable control system for a superconducting adiabatic quantum optimization processor. *Supercond. Sci. Technol.* **23**, 065004.

42. Berkley, A. J., Johnson, M. W., Bunyk, P., Harris, R., Johansson, J., Lanting, T., Ladizinsky, E., Tolkacheva, E., Amin, M. H. S., and Rose, G. (2010). A scalable readout system for a superconducting adiabatic quantum optimization system. *Supercond. Sci. Technol.* **23**, 105014.
43. Rønnow, T. F., Wang, Z., Job, J., Boixo, S., Isakov, S. V., Wecker, D., Martinis, J. M., Lidar, D. A., and Troyer, M. (2014). Defining and detecting quantum speedup. *Science* **345**, 420–424.
44. Lanting, T., Przybysz, A. J., Smirnov, A. Y., Spedalieri, F. M., Amin, M. H., Berkley, A. J., Harris, R., Altomare, F., Boixo, S., Bunyk, P., Dickson, N., Enderud, C., Hilton, J. P., Hoskinson, E., Johnson, M. W., Ladizinsky, E., Ladizinsky, N., Neufeld, R., Oh, T., Perminov, I., Rich, C., Thom, M. C., Tolkacheva, E., Uchaikin, S., Wilson, A. B., and Rose, G. (2014). Entanglement in a quantum annealing processor. *Phys. Rev. X* **4**, 021041.
45. Johnson, M. W., Amin, M. H. S., Gildert, S., Lanting, T., Hamze, F., Dickson, N., Harris, R., Berkley, A. J., Johansson, J., Bunyk, P., Chapple, E. M., Enderud, C., Hilton, J. P., Karimi, K., Ladizinsky, E., Ladizinsky, N., Oh, T., Perminov, I., Rich, C., Thom, M. C., Tolkacheva, E., Truncik, C. J. S., Uchaikin, S., Wang, J., Wilson, B., and Rose, G. (2011). Quantum annealing with manufactured spins. *Nature* **473**, 194–198.
46. Boixo, S., Albash, T., Spedalieri, F. M., Chancellor, N., and Lidar, D. A. (2013). Experimental signature of programmable quantum annealing. *Nat. Commun.* **4**, 2067.
47. Boixo, S., Ronnow, T. F., Isakov, S. V., Wang, Z., Wecker, D., Lidar, D. A., Martinis, J. M., and Troyer, M. (2014). Evidence for quantum annealing with more than one hundred qubits. *Nat. Phys.* **10**, 218–224.
48. Amin, M. H. S., Love, P. J., and Truncik, C. J. S. (2008). Thermally assisted adiabatic quantum computation. *Phys. Rev. Lett.* **100**, 060503–060504.
49. Kamleitner, I., Cresser, J., and Twamley, J. (2008). Adiabatic information transport in the presence of decoherence. *Phys. Rev. A* **77**, 032331–032310.
50. Deng, Q., Averin, D. V., Amin, M. H., and Smith, P. (2013). Decoherence induced deformation of the ground state in adiabatic quantum computation. *Sci. Rep.* **3**, 1479.
51. Dickson, N. G., Johnson, M. W., Amin, M. H., Harris, R., Altomare, F., Berkley, A. J., Bunyk, P., Cai, J., Chapple, E. M., Chavez, P., Cioata, F., Cirip, T., deBuen, P., Drew-Brook, M., Enderud, C., Gildert, S., Hamze, F., Hilton, J. P., Hoskinson, E., Karimi, K., Ladizinsky, E., Ladizinsky, N., Lanting, T., Mahon, T., Neufeld, R., Oh, T., Perminov, I., Petroff, C., Przybysz, A., Rich, C., Spear, P., Tcaciuc, A., Thom, M. C., Tolkacheva, E., Uchaikin, S.,

Wang, J., Wilson, A. B., Merali, Z., and Rose, G. (2013). Thermally assisted quantum annealing of a 16-qubit problem. *Nat. Commun.* **4**, 1903.

52. Simmonds, R. W., Lang, K. M., Hite, D. A., Nam, S., Pappas, D. P., and Martinis, J. M. (2004). Decoherence in Josephson phase qubits from junction resonators. *Phys. Rev. Lett.* **93**, 077003–077004.
53. Jordan, S. P., Farhi, E., and Shor, P. W. (2006). Error-correcting codes for adiabatic quantum computation. *Phys. Rev. A* **74**, 052322–052325.
54. Lidar, D. A. (2008). Towards fault tolerant adiabatic quantum computation. *Phys. Rev. Lett.* **100**, 160506–160504.
55. Biamonte, J. D., and Love, P. J. (2008). Realizable Hamiltonians for universal adiabatic quantum computers. *Phys. Rev. A* **78**, 012352–012357.

Chapter 10

Application to Josephson Voltage Standards

J. Kohlmann
Physikalisch-Technische Bundesanstalt (PTB), Bundesallee 100, 38116 Braunschweig, Germany
johannes.kohlmann@ptb.de

As already mentioned in Chapter 2, Josephson junctions are, among others, the basic element of Josephson voltage standards. This chapter summarizes the development of modern Josephson voltage standards, a development that is closely linked to the availability of Josephson junctions made of refractory materials.

10.1 Introduction

The point of origin for Josephson voltage standards is located in Josephson's first paper [1]. On the basis of the AC Josephson effect, Brian Josephson predicted that the irradiation of external microwaves with frequency f can phase-lock the Josephson oscillating current over a certain bias current range and thus can generate constant-voltage steps V_n:

$$V_n = n \cdot (h/2e) \cdot f \qquad (10.1)$$

Josephson Junctions: History, Devices, and Applications
Edited by Edward Wolf, Gerald Arnold, Michael Gurvitch, and John Zasadzinski

ISBN 978-981-4745-47-5 (Hardcover), 978-1-315-36452-0 (eBook)
www.panstanford.com

where n denotes the integer step number ($n = 1, 2, 3 \ldots$), h is Planck's constant, and e is the elementary charge. All Josephson voltage standards are based on this equation relating voltages to frequencies, which can be finely controlled with high precision and accurately referenced to atomic clocks. However, a single Josephson junction operated on the first-order constant-voltage step generates a voltage of only 145 μV when irradiated by 70 GHz microwaves. Highly integrated junction series arrays are therefore needed to achieve practical output voltages of 1 V or even 10 V. These series arrays consist of more than 10,000 or even 100,000 junctions. Advanced fabrication technologies are therefore required for reliable fabrication processes of this huge number of junctions. In addition, the use of durable materials for the fabrication is absolutely necessary. The invention of the Nb/Al–Al_2O_3 technology by Gurvitch et al. in 1983 [2] (cf. Chapter 5) and its adaption to the fabrication of Josephson voltage standards [3] were a main breakthrough in a reliable fabrication process for robust and durable Josephson voltage standard circuits.

The area of applications for Josephson voltage standards have been changed over years. For many years, Josephson voltage standards have been used for DC applications. These conventional Josephson voltage standards based on underdamped Josephson junctions are commercially available since the mid-1990s. At this time, the increasing interest in highly precise AC voltages stimulated attempts to develop measurement tools based on Josephson junction series arrays for AC applications. Different attempts have been suggested and partly realized. The main important ones are programmable Josephson voltage standards based on binary divided series arrays, pulse-driven series arrays and digital-to-analog converters based on the dynamic logic of processing single-flux quanta (SFQ) (cf. [4]). The first two versions are described more detailed in this chapter (cf. Section 10.3), as most research activities have been focused on these two and promising results have been achieved.

This chapter describes the development of Josephson voltage standards from conventional DC to modern AC standards focused on refractory Josephson junctions based on the Nb/Al–Al_2O_3 technology and its enhancements for Josephson voltage standards.

Numerous review papers have been published in the course of time covering different topics of Josephson voltage standards; cf. for example DC standards [5–8] and its underpinning physics [9], the transition from DC to AC standards [10–15], the physics underpinning AC standards [16], and an overview of the historical developments [17].

10.2 Conventional DC Josephson Voltage Standards

The development of conventional DC Josephson voltage standards for output voltages up to 10 V was performed over a period of about 25 years. Although Brian Josephson mentioned constant-voltage steps in his first paper [1], the use of Josephson junctions for voltages standards was first suggested and demonstrated by Taylor et al. in 1967 [18]. At that time, the deployed junctions (weak links) consisted of point contacts, microbridges, SNS or SIS junctions (S, Superconductor; I, Insulator; N, Normal metal). For example, Taylor et al. used Josephson junction made of Sn/Sn oxide/Sn operated at 10 GHz [18].

Within the next 15 years until the early 1980s, Josephson voltage standards became more and more widespread. A few millivolts were delivered by a single underdamped Josephson junction operated on higher order constant-voltage steps at a microwave frequency around 10 GHz. To increase the output voltage, series arrays of a few individually biased Josephson junctions were used. An output voltage of 100 mV was demonstrated by Endo et al. in 1983 using 20 Josephson tunnel junctions composed of Pb(In, Au)–oxide–Pb(Au) [19].

The chapter of modern Josephson voltage standards was opened by two novel ideas around 1980: First, Levinson et al. suggested in 1977 the use of zero-bias constant-voltage steps generated in highly underdamped Josephson junctions operated at low microwave power instead of the use of normal steps observed at high microwave power used so far [20]. For the first time series arrays of Josephson junctions could be operated by a single bias source only, because the current ranges of the constant-voltage steps do overlap one another for small bias currents. Second, the series array of Josephson junctions was embedded into an adapted microwave transmission line (microstripline, cf. Fig. 2.9) as first suggested by the group at the Physikalisch-Technische Bundesanstalt (PTB) in Germany (cf. [21]). The Josephson junctions were arranged as part of a periodic superconducting microstripline with low impedance terminated by a matched load to avoid reflections and standing-wave patterns. This integrated microwave circuit allows a series connection of the Josephson junctions for the DC bias and a parallel connection of several microstripline paths with respect to the microwave drive and is still used to date for 70 GHz circuits in both conventional and programmable Josephson voltage standards. As more than 1000 Josephson junctions are already required for 1 V arrays, this progress is also intimately linked with significant improvements of the thin-film technology, to enable the fabrication of integrated circuits containing a large number of Josephson junctions similar to procedures of semiconductor industry.

First, 1 V arrays consisting of 1474 SIS junctions were realized within a cooperation between the National Bureau of Standards (NBS; now National Institute of Standards and Technology, NIST) in the USA and the PTB by Niemeyer et al. in 1984 [22]. The junctions for operation at 90 GHz were fabricated in lead/lead alloy technology with the junction barrier made by RF sputter oxidation of the base electrode. Another fabrication technology at that time was based on Josephson junctions made of $Nb/Nb_2O_5/PbBi$ (cf. [23]). First 10 V arrays containing a further increased number of junctions were realized within few years later [23, 24]. A crucial problem of lead alloy circuits was its susceptibility to damage by humidity and thermal cycling. In addition to the challenging fabrication process, the circuits worked well only for a rather limited period. The main breakthrough in a reliable process for fabrication of more robust and durable circuits was the invention of the $Nb/Al–Al_2O_3$ technology by Gurvitch et al. in 1983 [2] and its adaption to the fabrication of Josephson voltage standards.

The $Nb/Al–Al_2O_3$ technology combines the use of the durable and chemically stable metal Nb with the high critical temperature of about 9.2 K, the outstanding covering of thin Al layers on Nb, and the formation of a very homogeneous and stable oxide of Al by thermal oxidation. The thickness of the oxide barrier is about 1 nm, that one of the superconductive layers more than 150 nm and therefore roughly twice the superconducting penetration depth at least, which ensures appropriate microwave behavior for the superconducting microstriplines. For a reliable process, the trilayer defining the junctions is deposited as a sandwich structure without breaking the vacuum. The adaptation of this process made possible the fabrication of first 1 V series arrays consisting of Nb/Al–Al_2O_3/Nb Josephson junctions in 1986 [3]. Gradually, this process was used for fabrication of 1 V and 10 V arrays [24]. Nowadays, all series arrays for Josephson voltage standards are fabricated in processes fundamentally based on the invention of the Nb/Al–Al_2O_3 technology.

These conventional Josephson voltage standards for operation at 70 GHz became more and more widespread and are nowadays operated in more than 50 labs worldwide for DC measurements (cf. [6, 15]). Their widespread use has significantly been supported by the commercial availability of turn-key systems. Presently, two companies offer such DC systems: Hypres Inc., USA (www.hypres.com) and Supracon AG, Germany (www.supracon.com).

10.2.1 Design: Demands and Targets for Conventional Josephson Voltage Standards

As already discussed in Chapter 2.5, the dynamics of a Josephson junction is often described using the resistively-capacitively-shunted-junction (RCSJ) model first proposed by Stewart and McCumber in 1968 [25, 26]. Within this model, the real Josephson junction is considered as a parallel shunting of an ohmic resistance R, a capacitance C, and an ideal Josephson element. In the linear approximation, the resonance frequency is given by the plasma frequency $f_\mathrm{p} = (e \cdot j_\mathrm{c}/\pi \cdot h \cdot C_\mathrm{s})^{1/2}$ (j_c denotes the critical current density, $C_\mathrm{s} = C/A$ the specific junction capacitance, and A the junction area). Details of the behavior depend on the kind of junction, which can be characterized by the dimensionless McCumber parameter $\beta_\mathrm{c} = Q^2$ being equal to the square of the quality factor $Q = 2\beta \cdot f_\mathrm{p} \cdot R \cdot C$ of the junction. Underdamped junctions with $\beta_\mathrm{c} > 1$ show a hysteretic current–voltage characteristic, overdamped junctions with $\beta_c \leq 1$ a nonhysteretic one.

Conventional DC Josephson voltage standards are based on highly underdamped Josephson junctions typically realized by SIS junctions in $Nb/Al–Al_2O_3$ technology. The hysteretic current–voltage characteristic of these junctions caused two problems: (i) As the steps are hysteretic, a particular step cannot be selected by a simple manipulation of the DC bias. (ii) The junctions are likely to switch from one constant-voltage step to another if noise disrupts the phase-lock condition, which is the cause of the constant-voltage steps. The first problem was solved by appropriate bias electronics. The second problem was got under control by carefully optimising the design of the junctions and series arrays with respect to the materials used for the Josephson junctions. The stability requirements are in fact extreme, as was investigated by Kautz et al. in the 1980s and summarized in review papers in 1987 [27] and 1996 [9].

Zero-bias constant-voltage steps are limited to voltages below roughly half the gap voltage V_g, which typically results in output voltages of 1 mV per junction at most for Nb-based Josephson junctions. The microwave frequency for the operation of Josephson junctions should be high, as has been proved by estimations of the critical current density and further calculations of the activation energy for loss of the phase lock performed by Kautz et al. (cf. [27]). At present, most conventional DC Josephson voltage standards are therefore operated at frequencies between 70 and 100 GHz, a range where oscillators are available with an output power of about 100 mW. For example, operation of the junctions at 70 GHz corresponds to the 7th step for an output voltage of about 1 mV. For large series arrays the junctions are typically operated on the 4th or 5th step on average, resulting in an output voltage of about 580 or 725 μ V per junction, respectively. This kept the number of junctions required for a specific voltage level at an acceptable value for the fabrication technology of that time.

When the microwave frequency and the junction materials have been chosen, only three parameters must be adjusted to maximize the stability of the phase lock and to assure the generation of stable constant-voltage steps. These parameters are the critical current density j_c, the length l, and the width w of the junction, as was described by Kautz et al. (cf. [27]).

The critical current density follows from the requirement that the microwave frequency f must be higher than the plasma frequency f_p of the junctions. Simulations by Kautz et al. (cf. [27]) have shown that the ratio f/f_p should be at least $\sim$3. They regarded the average time between phase slips induced by Johnson noise as a measure of stability. The maximum value of the critical current density is then given by

$$j_c = (f/3)^2 \cdot \pi \cdot h \cdot C_s/e \tag{10.2}$$

The limitations of the length l and the width w of the junction follow from the requirement of spatial uniformity of the phase between the superconductors. The optimum length effects maximization of the phase-lock stability (cf. [27]):

$$l = \lambda_J \cdot (3j_c/\pi \cdot f \cdot C_s \cdot V_n)^{1/2} \tag{10.3}$$

where $\lambda_J = \{h/[4\pi \cdot e \cdot \mu_0 \cdot j_c \cdot (d + \lambda_{L1} + \lambda_{L2})]\}^{1/2}$ denotes the Josephson penetration depth, μ_0 the magnetic permeability of the vacuum, d the barrier thickness, and λ_{L1} and λ_{L2} the London penetration depth of the junction electrodes. The width of the junction is limited to avoid excitation of Fiske resonances. This leads to an upper limit of the width (cf. [27]):

$$w = (1/2f) \cdot [\mu_0 \cdot C_s \cdot (d + \lambda_{L1} + \lambda_{L2})]^{-1/2} \tag{10.4}$$

These equations deliver the following parameters for Josephson junctions fabricated in Nb/Al–Al_2O_3 technology and operated at 70 GHz on the 5th step: $j_c = 42$ A/cm^2, $l = 23$ μm, $w = 65$ μm. The design of Josephson series arrays was based on these parameters with typical sizes of the Josephson junctions of 20 μm × 50 μm. The critical current density was often further reduced to about 10 A/cm^2, as the constant-voltage steps of these arrays were more stable. Typical widths of the constant-voltage steps are between 20 and above 50 μA (cf. [6, 8]).

While real 1 V arrays typically consist of about 2000 Josephson junctions, 10 V arrays contain between about 14,000 and 20,000 junctions (cf. [28–32]). An essential requirement for the design of these large series arrays is the uniform microwave power distribution over all junctions, in order to enable the generation of wide and stable constant-voltage steps. The Josephson junctions of most voltage standard circuits are embedded into a microstripline with a low impedance around 5 Ω (cf. Fig. 2.9). Another microwave line was recently suggested and demonstrated for 10 V arrays by Schubert et al., who integrated the Josephson junctions into a coplanar stripline (CPS) with an impedance of 50 Ω [33].

The microwave behavior of the SIS Josephson junctions embedded into the microstripline is estimated using the RCSJ model. The ohmic resistance R_n of SIS junctions in the subgap region is about 50 Ω, while the impedance of the capacitive branch $Z_d = 1/(2\pi \cdot f \cdot C)$ is about 50 mΩ for a junction capacitance of 50 pF. High-frequency currents therefore flow mainly capacitively, resulting in a very low attenuation of the microwave power from about 1 dB/1000 junctions to 2 dB/1,000 junctions. A single branch of the microstripline can therefore contain a lot of junctions (about 3500 junctions in the real PTB design [28]) without losing the uniform microwave power distribution to all junctions. For 10 V arrays, the microstripline is split in a few stages forming parallel branches for the microwave; 10 V arrays typically consist of 4 to 16 parallel branches (cf. [28–32]).

10.2.2 Fabrication Technology and Results for Conventional Josephson Voltage Standards: A Brief Survey

Circuits for Josephson voltage standards are nowadays fabricated in processes fundamentally based on the Nb/Al–Al_2O_3 technology [2, 28–32]. Standard silicon wafers (3 inch, 100 mm, or 150 mm in diameter) are used as substrate for deposition of the different layers in thin-film technology. The SIS structure for the Josephson junctions is typically deposited in an in situ trilayer process without breaking the vacuum. Sputtered Nb is used for the superconductive layers with a thickness of typically above 150 nm, which is roughly twice the superconducting penetration depth; the Nb layers are therefore thick enough, to ensure appropriate microwave behavior. A thin sputtered Al layer with a thickness of about 10 nm is thermally oxidized to form the insulating layer of the SIS structure.

The required low critical current density of about 10 A/cm^2 is achieved with a high oxygen exposure being the product of oxygen pressure and oxidation time (e.g., 200 mbar O_2, 18 h at 30°C [8]). Most dielectric layers are realized by SiO_2 typically deposited by PECVD (plasma-enhanced chemical vapor deposition). Lithography is mostly made optically and in some cases by electron beam. The different layers are patterned by adapted fluorine-based dry etching processes using CF_4, SF_6, or C_4F_8 for Nb and CHF_3 for SiO_2. Sometimes O_2 and/or Ar are added depending on vertical or nonvertical sidewalls. The Al–Al_2O_3 layer is patterned by physical sputter etching or ion beam etching sometimes in combination with wet etching. The in situ trilayer process requires an additional wiring layer for connecting neighboring junctions by a window technology; this wiring layer is also made of Nb. A careful cleaning of the surface of the Nb top electrode is required to avoid additional weak links between both Nb layers. The dielectric of the microstripline is typically formed by a SiO_2 layer with a thickness between about 1.5 and 2 μm. The groundplane of the microstripline is made by Nb. The matched load for termination of the microstripline is realized by a resistive metal layer, e.g., Pd or AuPd.

Figure 10.1 shows the current voltage characteristics of a 10 V array fabricated at PTB without and with irradiation of 70 GHz microwaves. Arrays of this kind are nowadays operated as DC voltage standards in more than 50 labs worldwide. Typical 10 V Josephson voltage standards provide a combined uncertainty of a few parts in 10^{10} to a few parts in 10^{11} for measurements at room temperature [34, 35].

10.3 AC Josephson Voltage Standards

In spite of their successful operation, conventional DC Josephson voltage standards suffer from two main drawbacks because of the ambiguity of the constant-voltage steps: (i) It is not possible to rapidly select a particular step. (ii) The steps are only semistable so that electromagnetic interference can cause spontaneous switching between steps. Therefore these Josephson voltage standards do

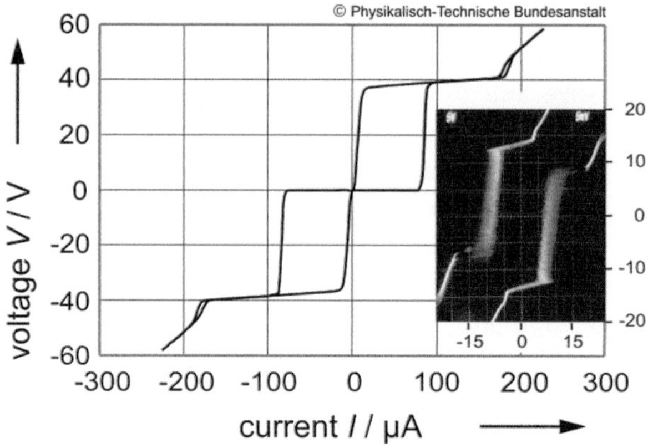

Figure 10.1 Current–voltage characteristic of a 10 V series array without and with (inset) irradiation of 70 GHz microwaves. The series array fabricated at PTB contains 13,924 SIS Josephson junctions. The inset shows the range of zero-bias constant-voltage steps between −12 and 12 V generated under microwave irradiation.

not enable systems that allow fast switching between different voltage levels or even generation of AC voltages. On the other hand, highly precise AC voltages gained increasingly in importance in the 1990s. This interest stimulated the development of corresponding measurement tools based on Josephson junctions. Two versions have been suggested and demonstrated by the NIST group in the USA, namely programmable Josephson voltage standards based on binary-divided series arrays [36] and Josephson arbitrary waveform synthesizers (JAWS) based on pulse-driven series arrays [37]. Caused by different operation principles, both versions have advantages and limitations. Promising results have been achieved using both versions that are described more detailed in this section.

The limitations of conventional DC Josephson voltage standards are mainly caused by the hysteretic current–voltage characteristic of the underdamped SIS Josephson junctions resulting in overlapping constant-voltage steps. AC Josephson voltage standards are therefore based on overdamped Josephson junctions showing a nonhysteretic current–voltage characteristic, which remains single-valued under microwave irradiation. The constant-voltage steps are consequently inherently stable and can rapidly be selected by external biasing. The overdamped Josephson junctions are typically realized using SNS or SINIS junctions (S, Superconductor; I, Insulator; N: Normal metal). As all junctions are operated on the same constant-voltage step (typically the first-order one), the number of junctions needed for 1 V or even 10 V arrays have significantly increased compared to conventional SIS arrays. Depending on the drive frequency, 10 V circuits consist of nearly 70,000 junctions (70 GHz drive) [38] and about 265,000 junctions (18 GHz drive) [39, 40], respectively. A significantly improved fabrication technology as well as sophisticated microwave designs has been required for the successful development of these giant series arrays.

Programmable Josephson voltage standards based on binary-divided arrays are operated by a sinusoidal microwave drive. In order to enable the generation of different voltage levels, the series array of junctions must be divided into segments. These segments contain numbers of junctions belonging, e.g., to a binary sequence of independently biased smaller arrays. Any integral number of constant-voltage steps permitted by the sequence of junctions can consequently be generated by these binary-divided arrays, which are therefore often called programmable Josephson voltage standards. The Josephson array is hence operated as a multi-bit digital-to-analog (D/A) converter, which results in some limitations, however. A major drawback is caused by the transients between the constant-voltage steps. As the voltage is not determined by the Josephson equation during the transition, the transients significantly contribute to the uncertainties. To reduce these uncertainties, programmable Josephson voltage standards are often operated using sampling measurement methods, which enable different applications of their very successful use as AC voltage standards (cf. [12, 14]). However, as the spectra of synthesized waveforms contain a lot of higher harmonics because of the operation as a D/A converter, binary-divided arrays are not the best choice for synthesizing AC waveforms.

A very successful method for synthesizing spectrally pure waveforms is based on pulse-driven Josephson junctions [37]. The junctions are operated by a train of short current pulses, the rate of which varies with time. A high pulse repetition rate generates high voltages; the voltage decreases with a decreasing pulse repetition rate. Arbitrary waveforms can be synthesized by modulating the pulse train using a commercial pulse pattern generator. Sometimes this version of pulse-driven Josephson arrays is therefore called Josephson arbitrary waveform synthesizer (JAWS). Spectrally pure waveforms from a few hertz up to the megahertz range have been demonstrated. For a long time, the output voltages were limited to about 275 mV because of the complex operation principle. A major breakthrough for pulse-driven series arrays was the demonstration of waveforms with RMS output voltages of 1 V in 2014 [41, 42].

10.3.1 Design: Demands and Targets for Overdamped Josephson Junctions and Series Arrays

The demands for overdamped Josephson junctions used in AC voltage standards partly differ from that for underdamped junctions used in conventional standards. The frequency range for best operation of overdamped Josephson junctions is determined by their dynamic characteristics. The most important parameter here is the characteristic voltage $V_c = I_c \cdot R_n$ (I_c denotes the critical current of the junction and R_n its normal state resistance). The characteristic voltage is related to the characteristic frequency of the junction by the Josephson equation: $f_c = (2e/h) \cdot V_c = (2e/h) \cdot I_c \cdot R_n$. Simulation by Kautz showed that the frequency of the external microwave drive should be near the characteristic frequency for best operation of the Josephson junction, i.e., wide and stable constant-voltage steps at moderate microwave power [16].

The series arrays of AC Josephson voltage standards are operated in different frequency ranges. The required microwave circuit designs are based on three different microwave lines. While low-impedance microstriplines are typically used for operation at 70 GHz, Josephson arrays operated between 10 and 20 GHz are mainly based on 50 Ω coplanar waveguide transmission lines (CPW) and 50 Ω coplanar striplines (CPS). Microstriplines require a more complex fabrication technology because of an additional ground plane and a dielectric layer. On the other hand, they offer excellent transmission characteristics around 70 GHz and a simple splitting of a single high-frequency line in two parallel ones.

The microwave behavior of overdamped Josephson junctions embedded into a microstripline completely differs from that of underdamped Josephson junctions. Now, R_n and Z_d are comparable (about 50 mΩ each), leading to a significant dissipation of the microwave current and thus to a significant attenuation of the microwave power of about 50 dB/1000 junctions (cf. [14]). The high attenuation is, however, compensated in part by an active contribution of the junctions; the junctions act as oscillators. 1 V series arrays for 70 GHz operation realized at PTB consist of 8192 junctions arranged in 64 parallel microwave branches with 128 junctions each. For 10 V arrays, up to 582 junctions each are embedded into 128 parallel microwave branches resulting in 69,632 junctions altogether (cf. [14, 43]).

The microwave behavior is completely different when the overdamped Josephson junctions are embedded into a 50 Ω CPW or CPS. Now, the large ratio of the impedance of the 50 Ω line to the low junction impedance of about 50 mΩ leads to conditions, which are similar to that of the microstripline for conventional SIS arrays: the attenuation of the microwave power is low, because the junctions are loosely linked to the CPW or CPS, respectively. The integration of many junctions in a single microwave path is therefore possible. Typical numbers for 1 V (10 V) arrays realized at NIST are 8 (32) branches containing 4096 (8400) junctions each, resulting in 32,768 (265,116) junctions altogether (cf. [39, 44]). To ensure a homogeneous microwave power distribution along 8400 junctions in each branch, the microwave design was further improved including tapered CPWs and modified power dividers at the splitter stages (cf. [39]).

While chaos can occur in underdamped junctions, it is suppressed in overdamped junctions so that the limitations for their sizes are less strict. Typical junction sizes are between a few μm and a few tens of μm. Critical current densities between some kA/cm^2 and a few 10 kA/cm^2 result in critical currents of some mA up to nearly 10 mA and step widths of several mA (cf. [10–16]).

10.3.2 Realization of Binary-Divided Josephson Voltage Standards

Different kinds of Josephson junctions have been investigated for programmable Josephson voltage standards. For the proof-of-concept experiment, Hamilton et al. at NIST used 2048 junctions of an array containing 8192 externally shunted SIS junctions operated at 75 GHz and delivering DC voltages up to about 300 mV and AC voltages in the kHz range at ± 77 mV with 150 μV resolution. As design restrictions of externally shunted SIS junctions limit the critical current and consequently the step width to a few hundred microamperes, other junction types have subsequently been investigated. The final breakthrough for programmable voltage standards was enabled by the implementation of SNS junctions by Benz in 1995 [45] on the basis of calculations by Kautz [16]. However, the resistivity of most normal metals is very low, so that the required characteristic voltage is not achieved. Only a few metals, binary alloys, or other materials show a resistivity that results in a characteristic voltage of about 30 μV, which allows an operation around 15 GHz.

The first practical 1 V arrays were realized by Benz et al. at NIST in 1997 [44]. A total of 32,768 SNS junctions containing PdAu as the normal metal were embedded into the middle of a CPW with an impedance of 50 Ω. Under microwave operation at 16 GHz, the width of the constant-voltage steps exceeded 1 mA. Different materials have been subsequently investigated and used for the normal metal of SNS junctions, for example AuPd [44], Ti [46], TiN_x [40], or $MoSi_2$ [47]. The superconducting layers of these SNS junctions are typically made by Nb.

The low drive frequency around 16 GHz requires roughly 32,000 Josephson junctions to achieve 1 V and even about 300,000 Josephson junctions for 10 V, which causes enormous challenges for the microwave design and the fabrication technology. In order to decrease the number of junctions needed for a specific voltage, the operation frequency has to be increased, which requires junctions with a higher characteristic voltage. A promising approach for operation at 70 GHz like conventional Josephson voltage standards was the use of SINIS junctions ($Nb/Al_2O_3/Al/Al_2O_3/Nb$) by Schulze et al. at PTB in 1998 [48]. The fabrication is comparable to that of conventional SIS junctions with slight oxidation processes of the Al layers. 1 V and first 10 V arrays containing 8192 and 69,120 SINIS junctions, respectively, were subsequently fabricated [49, 50], and later significantly improved for AC applications [51]. The SINIS junctions are embedded into low-impedance microstriplines as SIS junctions. Because of the microwave behavior of overdamped Josephson junctions, the number of parallel branches was increased to 64 (1 V) and 128 branches (10 V), respectively. In spite of their successful operation, a serious drawback of SINIS junctions is their sensitivity to particular steps in the fabrication process often resulting in a very few shorted junctions of a SINIS series array (typically between 0 and 2 of 10,000 junctions) probably caused by the very thin insulating oxide barriers (cf. [52]).

In search of other materials showing a high resistivity, a new type of junction has increasingly gained in importance recently: its barrier consists of a semiconductor such as Si doped with a metal and being near a metal insulator transition [53]. Although these junctions mainly behave like SNS junctions, their characteristics are more determined by the semiconductor and its interfaces to the superconductive layers. A promising version of these junctions for Josephson voltage standards was first realized by an amorphous Si barrier doped with Nb at NIST in 2006 [53]. Nb and Si are co-sputtered from two sputter targets; the Nb content is varied by adjusting the power for sputtering. If the niobium content x is tuned to a value near the metal–insulator transition around 11.5% [54], Nb_xSi_{1-x} barriers with a thickness between 10 and 30 nm combine a high resistance and a sufficient conductivity.

The characteristic voltage of these Nb_xSi_{1-x} barrier junctions can be tuned over a wide range for operation from 15 GHz to more than 70 GHz by adjusting the Nb content between about 20% and 11%. To increase the integration density of large series arrays and thus to reduce the size of the circuits, double- and triple-stacked junctions have been successfully realized. Instead of an SNS trilayer, multilayers of SNsNS or SNsNsNS are deposited as sandwich structures without breaking the vacuum (s denotes a thin superconducting layer with a thickness of about 30 nm) [38–40].

Burroughs et al. demonstrated 10 V arrays containing three-junction stacks with 268,800 Nb_xSi_{1-x} barrier junctions arranged in 32 parallel branches (CPW) with an improved microwave design for operation near 20 GHz in 2009 [55]. After further improvements, 10 V arrays containing 265,116 triple-stacked junctions showed step widths up to about 2 mA [39]. While pure metallic SNS junctions are not suitable for operation at 70 GHz, Nb_xSi_{1-x} barrier junctions also allowed the fabrication of 10 V arrays for operation at 70 GHz, as first demonstrated by Mueller et al. in 2008 within a cooperation between NIST and PTB [52]. Some 10 V arrays consisting of 69,632 junctions embedded into a microstripline were realized without any shorted junction, an array quality which had never been achieved with SINIS junctions. Step widths above 1 mA for 10 V arrays have been achieved. Figure 10.2 shows the current–voltage characteristics of a 10 V array fabricated at PTB. Using double-stacked junctions, first series arrays delivering output voltages of 20 V have also been realized [38].

The fabrication process for Nb_xSi_{1-x} barrier junctions is similar to that of conventional Josephson voltage standards except the deposition of the barrier. Nb_xSi_{1-x} barrier junctions currently enable the most reliable technology for AC Josephson voltage standards. A junction yield analysis by Fox et al. confirmed the high quality of this technology; the average number of defects per million junctions was 32.5 considering series arrays from 9 wafers with a total of 25,346,652 junctions corresponding to a total average junction yield of 99.9967% [39].

Some other kinds of junctions have been also investigated over the years. An approach for operation in a cryocooler at 10 K is based on NbN for the superconductive layers and TiN for the barrier. Although the large magnetic penetration depth of NbN limited the width of the constant-voltage steps [56], an output voltage up to 17 V was demonstrated by Yamamori et al. using an array containing about 524,000 JJs operated at 18 GHz [40]. Other versions for 70 GHz operation are based on Josephson junctions fabricated in $Nb/Al–Al_2O_3$ technology. Output voltages up to 1 V were achieved by Lacquaniti et al. using 8192 SNIS junctions based on a slightly oxidized thick Al layer (up to 100 nm) as the barrier [57]. Externally shunted SIS junctions have been already used for the first realization

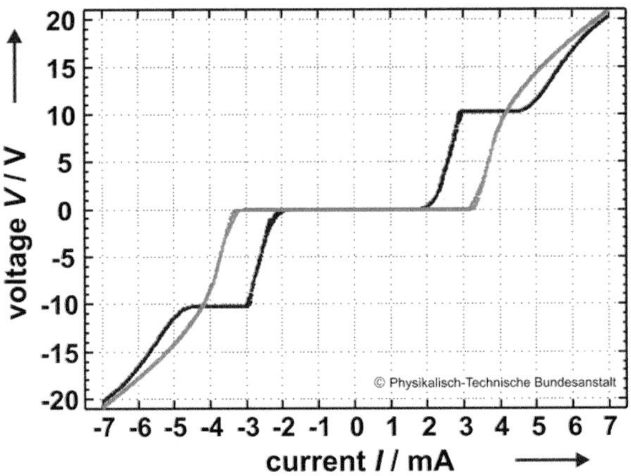

Figure 10.2 Current–voltage characteristics of a 10 V Josephson array without (grey) and with (black) microwave irradiation at a frequency of 71 GHz. The width of the constant-voltage step exceeds 1 mA. The series array consists of 69,632 Nb_xSi_{1-x} barrier Josephson junctions.

of a binary-divided array by Hamilton et al. [36]; later, Hassel et al. developed a series array of 3315 junctions that delivered an output voltage up to 1 V when operated on the third-order step [58]. As design restrictions limits the critical current of externally shunted SIS junctions and consequently the widths of the constant-voltage steps, other junction types as Nb_xSi_{1-x} barrier junctions are nowadays widely used for AC Josephson voltage standards.

10.3.3 Realization of Pulse-Driven Josephson Voltage Standards

A direct modulation of the output voltage by simply changing the frequency of an external sinusoidal microwave drive over a wide frequency range is not possible, as the constant-voltage steps nearly disappear, when the junctions are not operated near the characteristic voltage (cf. [16, 37]). The operation of Josephson junctions by a train of short current pulses, however, enables that the variation of the external drive over time (i.e., the pulse repetition frequency) directly generates voltages that vary over time. The width of the constant-voltage steps remains nearly independent from the pulse repetition frequency between zero and the characteristic frequency, if rise and fall time of the pulses are short compared to the characteristic frequency (10 GHz corresponds to 100 ps) (cf. [37]). These pulse-driven Josephson junctions allow the synthesis of spectrally pure arbitrary waveforms and are therefore of particular importance for high-precision AC measurements.

Different kinds of Josephson junctions have also been investigated for these pulse-driven Josephson voltage standards suggested and first demonstrated by Benz and Hamilton at NIST in 1996 [37]. For the proof-of-concept experiment, an array of 512 SNS junctions with a AuPd barrier was operated by unipolar pulses delivering constant-voltage steps up to 265 μV. The junctions were placed along the center conductor of a 50 Ω CPW, which assure a broadband performance required for the transmission of the broadband frequency components of the pulse train. Continuous enhancements of both the measurement setup as well as the Josephson series arrays gradually improved the spectra of the synthesized signals and increased the output voltages (cf. [11, 12, 14, 15]). The overdamped Josephson junctions now operated at clock frequencies around 15 GHz have been predominately realized by SNS junctions based on different materials for the barrier such as AuPd [37], HfTi [59], or Nb_xSi_{1-x} [41, 42, 60]. SINIS junctions have also been used for some time [61, 62].

The complex pulse operation presently requires one pulse source for each single array; a simple splitting into several parallel microwave branches is not possible. As typical commercial pulse pattern generators contain only two output channels, only two series arrays could be operated, which limited for a long time the maximum output voltage to about 275 mV by operation of two arrays containing 6400 Nb_xSi_{1-x} barrier junctions each [60]. A major breakthrough for pulse-driven series arrays was the demonstration of 1 V waveforms in 2014 by the groups at NIST and at PTB [41, 42]. Novel pulse pattern generators with four or even eight output channels allowed the simultaneous operation of four or eight arrays containing altogether about 51,000 or 63,000 Nb_xSi_{1-x} barrier junctions, respectively.

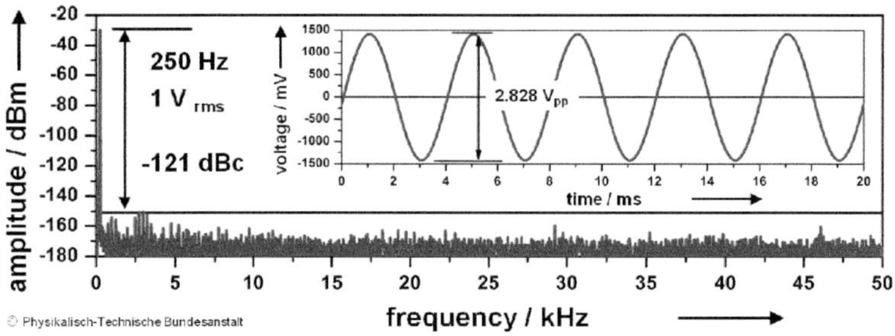

Figure 10.3 Frequency spectrum of a synthesized sine wave (inset) with an RMS output voltage of 1 V at a frequency of 250 Hz. Higher harmonics are suppressed by more than 120 dB. The sine wave is synthesized using 8 Josephson arrays consisting of 63,000 Nb_xSi_{1-x} barrier Josephson junctions in total (cf. [42]).

The pulse train for operation of the Josephson arrays is typically created by use of a first- or second-order sigma-delta modulation (cf. [11, 14]). This procedure reduces the digitization harmonics at low frequencies and shifts the noise to high frequencies; noise contributions are then removed by appropriate filtering. Josephson arbitrary waveform synthesizers enable the synthesis of spectrally pure waveforms with higher harmonics suppressed by more than 120 dB [41, 63]. Its accuracy at 1 V has been evaluated with an AC quantum voltmeter demonstrating an excellent agreement between both quantum standards of better than 1 part in 10^8 with an uncertainty of 1.2 parts in 10^8 ($k = 1$) for measurements at a frequency of 250 Hz [63]. As an example from the PTB, Fig. 10.3 shows a synthesized sine wave with an RMS output voltage of 1 V and the corresponding frequency spectrum.

10.4 Conclusions

Josephson voltage standards are nowadays well established as an important application of the Josephson effects for voltage measurements at ultimate precision. A long time of developments was needed to progress from single Josephson junctions delivering a few millivolts at most to highly integrated series arrays containing thousands of Josephson junctions for output voltages of more than

10 V. Conventional 10 V DC Josephson voltage standards with SIS junctions fabricated in Nb/Al–Al_2O_3 technology are commercially available since two decades. Binary-divided and pulse-driven series arrays based on Nb_xSi_{1-x} barrier Josephson junctions opened up the world of AC applications and, hence, have been the next step in the exciting story of the applications of the Josephson effects in metrology. Pulse-driven and binary-divided series arrays containing up to 300,000 junctions have been demonstrated for output voltages up to 1 V and 10 V or even 20 V, respectively. Their use as AC Josephson voltage standards is increasingly gained in importance currently. Advanced fabrication technologies and their permanent improvements have been essential prerequisites for this remarkable progress. The use of durable materials as Nb instead of susceptible materials has also played an important role in the fascinating story of Josephson voltage standards.

Acknowledgments

The developments described in this chapter are a result of many researchers. I would like to thank Ralf Behr, Oliver Kieler, the whole Josephson voltage standard team at PTB, and national and international colleagues in this field for many discussions, comments, and support. The work has been also supported in part within national and international research projects.

References

1. Josephson, B. D. (1962). Possible new effects in superconductive tunnelling. *Phys. Lett.* **1**, 251–253.
2. Gurvitch, M., Washington, W. A., and Huggins, H. A. (1983). High quality refractory Josephson tunnel junctions utilizing thin aluminium layers. *Appl. Phys. Lett.* **42**, 472–474.
3. Niemeyer, J., Sakamoto, Y., Vollmer, E., Hinken, J. H., Shoji, A., Nakagawa, H., Takada, S., and Kosaka, S. (1986). Nb/Al-oxide/Nb and NbN/MgO/NbN tunnel junctions in large series arrays for voltage standards. *Jpn. J. Appl. Phys.* **25**, L343–L345.

4. Semenov, V. K., and Polyakov, Yu. A. (2001). Circuit improvements for a voltage multiplier. *IEEE Trans. Appl. Supercond.* **11**, 550–553.

5. Niemeyer, J. (1998). Josephson voltage standards. In *Handbook of Applied Superconductivity*, Seeber, B., ed. (IOP, Bristol, UK), pp. 1813–1834.

6. Hamilton, C. A. (2000). Josephson voltage standards. *Rev. Sci. Instrum.* **71**, 3611–3623.

7. Yoshida, H. (2000). Application of the ac Josephson effect for precise measurements. *IEICE Trans. Electron.* **E83-C**, 20–26.

8. Kohlmann, J., Behr, R., and Funck, T. (2003). Josephson voltage standards. *Meas. Sci. Technol.* **14**, 1216–1228.

9. Kautz, R. L. (1996). Noise, chaos, and the Josephson voltage standard. *Rep. Prog. Phys.* **59**, 935–992.

10. Benz, S. P., and Hamilton, C. A. (2004). Application of the Josephson effect to voltage metrology. *Proc. IEEE* **92**, 1617–1629.

11. Jeanneret, B., and Benz, S. P. (2009). Application of the Josephson effect in electrical metrology. *Eur. Phys. J. Spec. Top.* **172**, 181–206.

12. Benz, S. (2010). Synthesizing accurate voltages with superconducting quantum-based standards. *IEEE Instrum. Meas. Mag.* **13**, 8–13.

13. Kaneko, N., Maruyama, M., Urano, C., and Kiryu, S. (2012). Review of Josephson waveform synthesis and possibility of new operation method by multibit delta-sigma modulation and thermometer code for its further advancement. *Jpn. J. Appl. Phys.* **51**, 010116 (7 pp).

14. Behr, R., Kieler, O., Kohlmann, J., Müller, F., and Palafox, L. (2012). Development and metrological applications of Josephson arrays at PTB. *Meas. Sci. Technol.* **23**, 124002 (19 pp).

15. Kohlmann, J. (2015). Quantum standards for voltage. In *Applied Superconductivity - Handbook on Devices and Applications*, Seidel, P., ed. (Wiley-VCH, Berlin, Germany), pp. 807–827.

16. Kautz, R. L. (1995). Shapiro steps in large-area metallic-barrier Josephson junctions. *J. Appl. Phys.* **78**, 5811–5819.

17. Harris, R. E., and Niemeyer, J. (2011). Quantum metrology. In *100 Years of Superconductivity*, Rogalla, H., and Kes, P., eds. (Taylor & Francis, Boca Raton, FL, USA), pp. 515–557.

18. Taylor, B. N., Parker, W. H., Langenberg, D. N., and Denenstein, A. (1967). On the use of the ac Josephson effect to maintain standards of electromotive force. *Metrologia* **3**, 89–98.

19. Endo, T., Koyanagi, M., and Nakamura, A. (1983). High-accuracy Josephson potentiometer. *IEEE Trans. Instrum. Meas.* **32**, 267–271.
20. Levinson, M. T., Chiao, R. Y., Feldman, M. J., and Tucker, B. A. (1977). An inverse ac Josephson effect voltage standard. *Appl. Phys. Lett.* **31**, 776–778.
21. Niemeyer, J. (2011). How the DC array standards were developed. In *100 Years of Superconductivity*, Rogalla, H., and Kes, P., eds. (Taylor & Francis, Boca Raton, FL, USA), pp. 528–535.
22. Niemeyer, J., Hinken, J. H., and Kautz, R. L. (1984). Microwave-induced constant-voltage steps at one volt from a series array of Josephson junctions, *Appl. Phys. Lett.* **45**, 478–480.
23. Lloyd, F., Hamilton, C. A., Beall, J., Go, D., Ono, R. H., and Harris, R. E. (1987). A Josephson array voltage standard at 10 V. *IEEE Electron Device Lett.* **8**, 449–450.
24. Pöpel, R., Niemeyer, J., Fromknecht, R., Meier, W., and Grimm, L. (1990). 1- and 10-V series array Josephson voltage standards in $Nb/Al_2O_3/Nb$ technology. *J. Appl. Phys.* **68**, 4294–4303.
25. Stewart, W. C. (1968). Current-voltage characteristics of Josephson junctions. *Appl. Phys. Lett.* **12**, 277–280.
26. McCumber, D. E. (1968). Effect of ac impedance on dc voltage-current characteristics of superconductor weak-link junctions. *J. Appl. Phys.* **39**, 3113–3118.
27. Kautz, R. L., Hamilton, C. A., and Lloyd, F. L. (1987). Series-array Josephson voltage standards. *IEEE Trans. Mag.* **MAG-23**, 883–890.
28. Müller, F., Pöpel, R., Kohlmann, J., Niemeyer, J., Meier, W., Weimann, T., Grimm, L., Dünschede, F.-W., and Gutmann, P. (1997). Optimized 1 V and 10 V Josephson series arrays. *IEEE Trans. Instrum. Meas.* **46**, 229–232.
29. Hamilton, C. A., and Burroughs, C. J. (1995). Perfomance and reliability of NIST 10-V Josephson arrays. *IEEE Trans. Instrum. Meas.* **44**, 238–240.
30. Endo, T., Sakamoto, Y., Murayama, Y., Iwasa, A., and Yoshida, H. (1995). 1-V Josephson-junction-array voltage standard and development of 10-V Josephson junction array at ETL. *IEICE Trans. Electron.* **E78-A**, 503–510.
31. Radparvar, M. (1995). Superconducting niobium and niobium nitride processes for medium-scale integration applications. *Cryogenics* **35**, 535–540.
32. Meyer, H.-G., Wende, G., Fritzsch, L., Thrum, F., Schubert, M., Müller, F., Behr, R., and Niemeyer, J. (1999). Improved primary Josephson voltage

standard with a new microwave driving source. *IEEE Trans. Appl. Supercond.* **9**, 4150–4153.

33. Schubert, M., May, T., Wende, G., Fritzsch, L., and Meyer, H.-G. (2001). Coplanar strips for Josephson voltage standard circuits. *Appl. Phys. Lett.* **79**, 1009–1011.
34. Wood, B., and Solve, S. (2009). A review of Josephson comparison results. *Metrologia* **46**, R13–R20.
35. BIPM database [Online]. Available at: http://kcdb.bipm.org.
36. Hamilton, C. A., Burroughs, C. J., and Kautz, R. L. (1995). Josephson D/A converter with fundamental accuracy. *IEEE Trans. Instrum. Meas.* **44**, 223–225.
37. Benz, S. P., and Hamilton, C. A. (1996). A pulse-driven programmable Josephson voltage standard. *Appl. Phys. Lett.* **68**, 3171–3173.
38. Müller, F., Scheller, T., Wendisch, R., Behr, R., Kieler, O., Palafox, L., and Kohlmann, J. (2013). NbSi barrier junctions tuned for metrological applications up to 70 GHz: 20 V arrays for programmable Josephson voltage standards. *IEEE Trans. Appl. Supercond.* **23**, 1101005 (5 pp).
39. Fox, A. E., Dresselhaus, P. D., Rüfenacht, A., Sanders, A., and Benz, S. P. (2015). Junction yield analysis for 10 V programmable Josephson voltage standard devices. *IEEE Trans. Appl. Supercond.* **25**, 1101505 (5 pp).
40. Yamamori, H., Yamada, T., Sasaki, H., and Shoji, A. (2008). 10 V programmable Josephson voltage standard circuit with a maximum output voltage of 20 V. *Supercond. Sci. Technol.* **21**, 105007 (6 pp).
41. Benz, S. P., Waltman, S. B., Fox, A. E., Dresselhaus, P. D., Rüfenacht, A., Howe, L., Schwall R. E., and Flowers-Jacobs, N. E. (2015). Performance improvements for the NIST 1 V Josephson arbitrary waveform synthesizer. *IEEE Trans. Appl. Supercond.* **25**, 1400105 (5 pp).
42. Kieler, O. F., Behr, R., Wendisch, R., Bauer, S., Palafox, L., and Kohlmann, J. (2015). Towards a 1 V Josephson arbitrary waveform synthesizer. *IEEE Trans. Appl. Supercond.* **25**, 1400305 (5 pp).
43. Müller, F., Scheller, T., Lee, J., Behr, R., Palafox, L., Schubert, M., and Kohlmann, J. (2014). Microwave design and performance of PTB 10 V circuits for the programmable Josephson voltage standard. *World J. Condens. Matter Phys.* **4**, 107–122.
44. Benz, S. P., Hamilton, C. A., Burroughs, C. J., Harvey, T. E., and Christian, L. A. (1997). Stable 1-Volt programmable voltage standard. *Appl. Phys. Lett.* **71**, 1866–1868.

45. Benz, S. P. (1995). Superconductor-normal-superconductor junctions for programmable voltage standards. *Appl. Phys. Lett.* **67**, 2714–2716.
46. Schubert, M., Fritzsch, L., Wende, G., and Meyer, H.-G. (2001). SNS junction on Nb-Ti base for microwave circuits. *IEEE Trans. Appl. Supercond.* **11**, 1066–1069.
47. Chong, Y., Burroughs, C. J., Dresselhaus, P. D., Hadacek, N., Yamamori, H., and Benz, S. P. (2015). Practical high-resolution programmable Josephson voltage standards using double- and triple-stacked $MoSi_2$-barrier junctions. *IEEE Trans. Appl. Supercond.* **15**, 461–464.
48. Schulze, H., Behr, R., Müller, F., and Niemeyer, J. (1998). $Nb/Al/AlO_x/Al/AlO_x/Al/Nb$ Josephson junctions for programmable voltage standards. *Appl. Phys. Lett.* **73**, 996–998.
49. Behr, R., Schulze, H., Müller, F., Kohlmann, J., and Niemeyer, J. (1999). Josephson arrays at 70 GHz for conventional and programmable voltage standards. *IEEE Trans. Instrum. Meas.* **48**, 270–273.
50. Schulze, H., Behr, R., Kohlmann, J., Müller, F., and Niemeyer, J. (2000). Design and fabrication of 10 V SINIS Josephson arrays for programmable voltage standards. *Supercond. Sci. Technol.* **13**, 1293–1295.
51. Müller, F., Behr, R., Palafox, L., Kohlmann, J., Wendisch, R., and Krasnopolin, I. (2007). Improved 10 V SINIS series arrays for applications in ac voltage metrology. *IEEE Trans. Appl. Supercond.* **17**, 649–652.
52. Mueller, F., Behr, R., Weimann, T., Palafox, L., Olaya, D., Dresselhaus, P. D., and Benz, S. P. (2009). 1 V and 10 V SNS programmable voltage standards for 70 GHz. *IEEE Trans. Appl. Supercond.* **19**, 981–986.
53. Baek, B., Dresselhaus, P. D., and Benz, S. P. (2006). Co-sputtered amorphous Nb_xSi_{1-x} barriers for Josephson-junction circuits. *IEEE Trans. Appl. Supercond.* **16**, 1966–1970.
54. Hertel, G., Bishop, D. J., Spencer, E. G. Rowell, J. M., and Dynes, R. C. (1983). Tunneling and transport measurements at the metal-insulator transition of amorphous Nb:Si. *Phys. Rev. Lett.* **50**, 743–746.
55. Burroughs, C. J., Rüfenacht, A., Dresselhaus, P. D., Benz, S. P., and Elsbury, M. M. (2009). A 10 Volt "turnkey" programmable Josephson voltage standard for DC and stepwise-approximated waveforms. *NCSLI Meas.* **4**, 70–75.
56. Yamamori, H., Urano, C., Maruyama, M., Sasaki, H., and Kohjiro, S. (2010). Large constant voltage generated with a single array including 65?536 Nb/TiN/Nb Josephson junctions. *Supercond. Sci. Technol.* **23**, 075011 (5 pp).

57. Lacquaniti, V., De Leo, N., Fretto, M., Sosso, A., Müller, F., and Kohlmann, J. (2011). 1 V programmable voltage standards based on SNIS Josephson junction series arrays. *Supercond. Sci. Technol.* **24**, 045004 (4 pp).

58. Hassel, J., Helistö, P. Grönberg, L., Seppä, H., Nissilä, J., and Kemppinen, A. A. (2005). Stimulated power generation in ES-SIS junction arrays. *IEEE Trans. Instrum. Meas.* **54**, 632–635.

59. Hagedorn, D., Kieler, O., Dolata, R., Behr, R., Müller, F., Kohlmann, J., and Niemeyer, J. (2006). Modified fabrication of planar sub-μm superconductor-normal metal-superconductor Josephson junctions for use in a Josephson arbitrary waveform synthesizer. *Supercond. Sci. Technol.* **19**, 294–298.

60. Benz, S. P., Dresselhaus, P. D., Rüfenacht, A., Bergren, N. F., Kinard, J. R., and Landim, R. P. (2009). Progress toward a 1 V pulse-driven ac Josephson voltage standard. *IEEE Trans. Instrum. Meas.* **58**, 838–843.

61. Kohlmann, J., Müller, F., Behr, R., Hagedorn, D., Kieler, O., Palafox, L., and Niemeyer, J. (2006). Development of Josephson junction series arrays for synthesis of ac voltages and arbitrary waveforms. *J. Phys. Conf. Ser.* **43**, 1385–1388.

62. Urano, C., Maruyama, M., Kaneko, N., Yamamori, Y., Shoji, A., Maezawa, M., Hashimoto, Y., Suzuki, H., Nagasawa, S., Satoh, T., Hidaka, M., and Kiryu, S. (2009). Operation of a Josephson arbitrary waveform synthesizer with optical data input. *Supercond. Sci. Technol.* **22**, 114012 (4 pp).

63. Behr, R., Kieler, O., Lee, J., Bauer, S., Palafox, L., and Kohlmann, J. (2015). Direct comparison of a 1 V Josephson arbitrary waveform synthesizer and an ac quantum voltmeter. *Metrologia* **52**, 528–537.

Index

For Product Safety Concerns and Information please contact our EU representative GPSR@taylorandfrancis.com Taylor & Francis Verlag GmbH, Kaufingerstraße 24, 80331 München, Germany